Wearable Solar Cells

Wearable Solar Cells

Mechanisms, Materials, and Devices

Hao Sun
Zhibin Yang
Longbin Qiu
Huisheng Peng

WILEY-VCH

Authors

Dr. Hao Sun
Frontiers Science Center for Transformative Molecules
School of Chemistry and Chemical Engineering
Zhangjiang Institute for Advanced Study
Key Laboratory of Green and High-End Utilization of Salt Lake Resources (Chinese Academy of Sciences)
Shanghai Jiao Tong University
Shanghai 200240
China

Dr. Zhibin Yang
School of Chemistry and Chemical Engineering
Shanghai Jiao Tong University
Shanghai 200240
China

Dr. Longbin Qiu
SUSTech Energy Institute for Carbon Neutrality
Department of Mechanical and Energy Engineering
Southern University of Science and Technology
Shenzhen 518055
China

Prof. Huisheng Peng
State Key Laboratory of Molecular Engineering of Polymers
Department of Macromolecular Science and Laboratory of Advanced Materials
Fudan University
Shanghai 200438
China

Cover Image: © Andrey Suslov/Getty Images

Library of Congress Card No.: applied for

British Library Cataloguing-in-Publication Data
A catalogue record for this book is available from the British Library.

Bibliographic information published by the Deutsche Nationalbibliothek
The Deutsche Nationalbibliothek lists this publication in the Deutsche Nationalbibliografie; detailed bibliographic data are available on the Internet at <http://dnb.d-nb.de>.

Print ISBN: 978-3-527-35055-1
ePDF ISBN: 978-3-527-83822-6
ePub ISBN: 978-3-527-83823-3
oBook ISBN: 978-3-527-83824-0

Typesetting: Straive, Chennai, India
Printing and Binding: CPI Group (UK) Ltd, Croydon, CR0 4YY

C9783527350551_051223

Contents

About the Authors

Dr. Hao Sun is currently a tenure-track associate professor at Shanghai Jiao Tong University. He received his BE in Polymer Materials and Engineering from Sun Yat-Sen University in 2012 and his PhD degree in Polymer Chemistry and Physics from Fudan University in 2017 and then worked at Stanford University as a Postdoc Fellow. His research interests include fiber electronics and energy storage devices. He has published over 60 papers and authorized ten invention patents. He also contributes to five book chapters. He received the Materials Research Society (MRS) Graduate Student Gold Award and the Geneva International Exhibition of Inventions Silver Medal.

Dr. Zhibin Yang is currently a tenure-track associate professor at Shanghai Jiao Tong University. He received his bachelor's degree from East China University of Science and Technology in 2009 and his PhD from Fudan University in 2014. And then he successively worked as a postdoc at the University of Washington and the University of North Carolina at Chapel Hill between 2014 and 2019. His research interests focus on photovoltaics. Till now, he has published over 80 papers and authorized 11 invention patents. He was also awarded the IUPAC-SOLVAY International Award for Young Chemists, the MRS Graduate Student Silver Award, etc.

Dr. Longbin Qiu is currently a tenure-track assistant professor at the Southern University of Science and Technology. He received his BE in Polymer Materials and Engineering from Sun Yat-Sen University in 2011 and his PhD in Polymer Chemistry and Physics from Fudan University in 2016 and then worked at Okinawa Institute of Science and Technology as a Postdoc Fellow. His current interests focus on flexible and wearable electronics and scalable and stable photovoltaic devices. He has published over 80 papers and authorized three invention patents.

Dr. Huisheng Peng is currently a university professor at Fudan University. He received his BE in Polymer Materials from Donghua University in China in 1999, his MS in Macromolecular Chemistry and Physics from Fudan University in China in 2003, and his PhD in Chemical and Biomolecular Engineering from Tulane University in the USA in 2006. He then worked at Los Alamos National Laboratory in the USA between 2006 and 2008 and joined the Laboratory of Advanced Materials and Department of Macromolecular Science at Fudan University in 2008. He focuses on fiber electronics.

Preface

The ever-growing demands for wearable devices have facilitated the development of the corresponding power sources in both flexible energy harvesting and storage devices. In particular, flexible solar cells in a wearable format can be ideal candidates to ease the frequent charging process, thus promoting the operation duration of wearable devices. However, conventional solar cells that are bulky and rigid cannot fully meet wearable requirements such as portability, flexibility, and breathability. To this end, a variety of wearable solar cells in a novel fiber format have attracted extensive attention attributed to their unique and promising features. For instance, their high flexibility ensures stable performances under complex deformations such as twisting and stretching, so they can adapt to irregular substrates effectively in various applications. More importantly, fiber solar cells can be woven into textiles that breathe freely or integrated into multifarious structures to fit the human body.

This book provides an overview of the basic mechanisms and state-of-the-art development of different types of wearable solar cells, including but not limited to dye-sensitized solar cells, polymer solar cells, and perovskite solar cells. In particular, wearable solar cells in a novel fiber format are highlighted as they represent an emerging and attractive field in the past decade. In addition, the integration and functionalization of energy harvesting and storage devices in a wearable format will be covered with a variety of promising applications. We also summarize the current challenges and future opportunities for wearable solar cells, particularly for the continuous production of long-fiber solar cells and the weaving techniques for wearable applications. It covers both fundamental and application aspects, which are helpful for people with various backgrounds. For example, what are the unique features of fiber solar cells compared to conventional planar ones? At the highly curved interfaces of fiber electrodes, how can charge transport and separation occur? The readers may find the answers in this book.

This book can be a useful resource for scientists and engineers in industry, which provides insights for emerging fields. It may bridge the academic and industrial communities, inspiring professors and students in material science, energy, electrochemistry, applied physics, nanoscience and nanotechnology, and polymer science and engineering.

We would like to acknowledge our graduate students who participated in writing this book: Chapter 1 was drafted by Zhaofeng Ouyang, Chapter 3 by Jinglin Sun, Chapter 4 by Shuo Wang, Chapter 5 by Qiuchen Xu, Chapter 8 by Jie Zhou, Tianyu Wen, Yang Shi, and Yue Wu, and Chapter 9 by Bin Yuan.

November 12, 2023
Shanghai

Hao Sun
Zhibin Yang
Longbin Qiu
Huisheng Peng

1 Introduction

1.1 Development of Wearable Solar Cells

Wearable devices can be worn close to the human body, have a good fit on irregular body surfaces, and perform ideal functions when worn. They are considered to be the next-generation electronic devices with broad applications. They thus enable a variety of desirable and unfulfilled functions, including but not limited to energy harvesting, storage, sensing, display, and actuation in a wearable format, which can reshape our traditional lifestyle [1]. Wearable solar cells, an important branch of wearable devices, can convert light into electric energy, namely the photovoltaic conversion process, in a wearable format and can be readily integrated with other wearable devices such as supercapacitors, batteries, sensors, displays, and actuators [2, 3]. These appealing features and potential applications have greatly facilitated the development of wearable solar cells in the past decades.

Although thin-film solar cells have been investigated as potential candidates in wearable applications, their limited flexibility generally results in inferior interfacial stability that severely degrades their photovoltaic conversion performance under large deformation in wearable applications [4]. More importantly, the two-dimensional configuration sacrifices breathability and cannot meet the requirements for breathability and comfortability. Therefore, the emergence of fiber-shaped solar cells has offered a promising approach to realizing fully wearable solar cells [5]. In 2002, a pioneering work reported a fiber-shaped dye-sensitized solar cell (DSSC), in which the photoactive materials were layer-by-layer coated on a metal wire, shedding light on the feasibility of fiber-shaped solar cells with promising wearability [6]. This also inspired the production of a fiber-shaped polymer solar cell five years later, which consisted of photoactive materials coaxially incorporated into an optical fiber [7]. In 2014, the first perovskite solar cell in a fiber format was developed with all-solid-state characteristics, which were highly appealing for the requirement for practical applications [8]. Over the past twenty years, a variety of exciting progresses has been achieved in device structure, fiber electrode, photoactive material, and electrolyte, as well as the functionalization, integration, weaving, and application of the obtained fiber-shaped and textile solar cells. These will be mainly introduced in the other chapters of this monograph with more details.

Wearable Solar Cells: Mechanisms, Materials, and Devices, First Edition.
Hao Sun, Zhibin Yang, Longbin Qiu, and Huisheng Peng.

1.2 Characteristics of Fiber-Shaped Solar Cells

As mentioned above, wearable solar cells in the fiber format can afford a variety of attractive characteristics that benefit wearable applications. For instance, fiber-shaped solar cells generally demonstrate excellent flexibility owing to their unique one-dimensional configuration, which can tolerate multiple deformations such as bending, twisting, and even stretching with specific device design [9]. This overcomes the long-standing bottleneck for traditional silicon-based solar cells, which are highly rigid and cannot fully meet the requirement of flexibility for wearable applications [10].

For instance, fiber-shaped solar cells can absorb light from all directions, recognized as the three-dimensional light harvesting capability originating from the unique one-dimensional configuration of the device. For traditional solar cells, such as Si-based solar cells, the power conversion efficiencies (PCE) are highly dependent on the incident light angle. In contrast, fiber-shaped solar cells can harvest light from all directions without significant performance fluctuation. This is an attractive feature for practical indoor applications in which scattered light is highly abundant. This unique feature allows rational device design to further improve the photovoltaic conversion performance toward practical applications.

The weavability of fiber-shaped solar cells represents another key advantage relevant to their production and application. Ideally, fiber-shaped solar cells can be woven into various forms, such as clothing, curtains, and tents, either alone or with other common fibers through proven textile manufacturing techniques such as knitting and weaving [11]. The weavability is highly dependent on the strength and flexibility of the fiber-shaped solar cells as the basic building blocks, which play a key role in the continuous weaving process. The resulting fabrics and textiles can not only inherit the flexibility of the fiber devices but, more importantly, show breathability different from all their bulky and planar counterparts, which provides the best wearability for the resulting devices. This allows for producing wearable solar cells with the best flexibility and breathability that cannot be easily achieved for state-of-the-art solar cells.

Despite the above advantages, fiber-shaped solar cells still suffer from several disadvantages that hinder their practical production and application [12]. For instance, the electrical resistance of fiber-shaped solar cells generally rises when the fiber length increases, which requires the use of highly conductive fiber electrodes. In addition, the light shielding issue based on either twisting or coaxial structure remains a general challenge for fiber-shaped solar cells. These should be taken into serious consideration and will be introduced in more detail in the corresponding chapters.

1.3 Functionalization and Integration

Functionalization and integration have been widely recognized as an important branch for fiber-shaped solar cells, which accommodate complex and harsh applications. For instance, fiber-shaped devices are required to be remotely controlled when a physical detachment is not available in aerospace, which gives birth to a novel magnetic response fiber-shaped solar cell [13]. The Fe_3O_4 nanoparticles were incorporated into multi-walled carbon nanotube (MWCNT) fiber via a dry-spinning method, offering a high saturation magnetization of 17.9 emu/g. With the above Fe_3O_4/MWCNT hybrid fiber as the counter electrode and a modified Ti wire as the working electrode, the obtained fiber-shaped solar cells could be reversibly attached to and detached from a substrate controlled by a magnet. This work also sheds light on smart wearable solar cells through rational material design for practical applications.

The safety issue of fiber-shaped solar cells is very important because they are closely attached on the human body. This inspires the development of safe materials to promote device safety. For instance, a polymer-ionic liquid (IL) gel electrolyte was developed to replace the flammable carbonate and ether solvents, which significantly improved the safety and long-term stability of the obtained fiber-shaped solar cells [14]. Hydrophobic polyvinylidene fluoride-hexafluoropropylene copolymer and 1-butyl-3-methylimidazolium bis(trifluoromethanesulfonyl) imide (BMImTFSI) IL served as the gelata and solvent, respectively, which could maintain the quasi-solid state from room temperature to 300 °C. The resulting fiber-shaped solar cells thus exhibited improved safety and long-term stability and could retain 90% of the original PCE after 30 days.

Efficient integration of fiber-shaped solar cells for enhancement of wearable applications also represents an important direction for real-world applications of fiber-shaped solar cells. To simultaneously realize energy harvesting and energy storage, Bae et al. integrated nanogenerator, supercapacitor, and DSSCs into a single fiber, which could convert the mechanical and optical energies into electric energy and store it in the supercapacitor [15]. Specifically, ZnO nanowires decorated with Kevlar fiber served as a common electrode, and the transparent graphene on Cu mesh was wound around the common electrode as another electrode. Although the total PCE of the integrated device was only 0.02%, this pioneering work showed the feasibility of rational integration of fiber-shaped electronics, which opens a new avenue for further improvement.

1.4 Applications of Wearable Solar Cells

The unique characteristics of fiber-shaped solar cells have made them promising candidates in a broad range of wearable applications, including but not limited to consumer electronics, the Internet of Things (IoT), and smart healthcare, which promise to innovate the conventional function paradigm. Their unrivaled flexibility enables the incorporation of irregular devices and substrates and provides a reliable power supply to diverse electronic devices with high performance.

For instance, wearable solar cells can power consumer electronics in a novel and fascinating way. Owing to their high flexibility and wearability, they can be easily integrated onto the surface of electronic devices without compromising their original flexibility and functionality. In addition, they can significantly extend the life span of consumer electronics as a secondary power source. They can form self-sustaining electronics, for example, a self-powered sports sensor that can generate electric power from outdoor or indoor light, which could overcome the bottleneck of wearable power supplies.

The IoT has been rapidly growing in the past few years, which can realize the connection of numerous electronic devices (e.g. wireless sensors, actuators, transmitters, and wearable devices) for efficient information interaction. All these devices require a stable power supply, which demands self-sustainable power sources such as wearable solar cells [16]. Therefore, fiber-shaped solar cells can be ideal candidates to support the IoT in a highly flexible and portable way. To achieve this goal, the power conversion efficiency, lifetime, and integration of fiber-shaped solar cells should be taken into serious consideration. Overall, fiber-shaped solar cells can significantly contribute to the development of IoT and open a new avenue to achieve the goal of "Internet of Everything" for sufficient information interaction and usage.

Another promising application of fiber-shaped solar cells lies in smart healthcare, such as wearable medical devices. For example, a wearable health monitoring device equipped with a fiber-shaped solar cell panel can continuously track the heart rate, blood pressure, and oxygen levels. It can potentially avoid replacing energy storage devices such as batteries to reduce the infection risk during surgery. Therefore, they can be potentially used to power implantable devices such as pacemakers, defibrillators, or neural stimulators [17]. Of course, there are still some concerns, such as toxicity, biocompatibility, and device stability for implementable applications. This interdisciplinary and innovative direction has attracted broad attention in recent years, which may create a brand new "wearing" way to extend the application of fiber-shaped solar cells.

In summary, we introduce the development of wearable solar cells, particularly in a flexible fiber format, which demonstrates a variety of appealing advantages such as high flexibility, wearability, and weavability. The exploration of functionalization and integration has made them promising candidates in a variety of applications, including consumer electronics, IoT, and smart healthcare devices. We will introduce more details on the background and progress of wearable solar cells in the other chapters of this monograph, providing a comprehensive view of this emerging and highly promising field.

References

1 Shi, X., Zuo, Y., Zhai, P. et al. (2021). Large-area display textiles integrated with functional systems. *Nature* 591 (7849): 240–245.

2 Yang, Z., Sun, H., Chen, T. et al. (2013). Photovoltaic wire derived from a graphene composite fiber achieving an 8.45% energy conversion efficiency. *Angew. Chem. Int. Ed.* 52 (29): 7545–7548.

3 Kim, J., Campbell, A.S., de Ávila, B.E.-F. et al. (2019). Wearable biosensors for healthcare monitoring. *Nat. Biotechnol.* 37 (4): 389–406.

4 O'Regan, B. and Grätzel, M. (1991). A low-cost, high-efficiency solar cell based on dye-sensitized colloidal TiO_2 films. *Nature* 353 (6346): 737–740.

5 Sun, H., Yang, Z., Chen, X. et al. (2013). Photovoltaic wire with high efficiency attached onto and detached from a substrate using a magnetic field. *Angew. Chem. Int. Ed.* 52 (32): 8276–8280.

6 Fan, X., Chu, Z.Z., Wang, F.Z. et al. (2008). Wire-shaped flexible dye-sensitized solar cells. *Adv. Mater.* 20 (3): 592–595.

7 Liu, J., Namboothiry, M.A.G., and Carroll, D.L. (2007). Optical geometries for fiber-based organic photovoltaics. *Appl. Phys. Lett.* 90 (13): 133515.

8 Qiu, L., Deng, J., Lu, X. et al. (2014). Integrating perovskite solar cells into a flexible fiber. *Angew. Chem. Int. Ed.* 53 (39): 10425–10428.

9 Ren, J., Zhang, Y., Bai, W. et al. (2014). Elastic and wearable wire-shaped lithium-ion battery with high electrochemical performance. *Angew. Chem. Int. Ed.* 53 (30): 7864–7869.

10 Zhang, Z., Chen, X., Chen, P. et al. (2014). Integrated polymer solar cell and electrochemical supercapacitor in a flexible and stable fiber format. *Adv. Mater.* 26 (3): 466–470.

11 Wu, C., Kim, T.W., Guo, T. et al. (2017). Wearable ultra-lightweight solar textiles based on transparent electronic fabrics. *Nano Energy* 32: 367–373.

12 Hashemi, S.A., Ramakrishna, S., and Aberle, A.G. (2020). Recent progress in flexible–wearable solar cells for self-powered electronic devices. *Energy Environ. Sci.* 13 (3): 685–743.

13 Sun, H., Yang, Z., Chen, X. et al. (2013). Photovoltaic wire with high efficiency attached onto and detached from a substrate using a magnetic field. *Angew. Chem.* 125 (32): 8434–8438.

14 Li, H.P., Guo, J.J., Sun, H. et al. (2015). Stable hydrophobic ionic liquid gel electrolyte for stretchable fiber-shaped dye-sensitized solar cell. *ChemNanoMat* 1 (6): 399–402.

15 Bae, J., Park, Y.J., Lee, M. et al. (2011). Single-fiber-based hybridization of energy converters and storage units using graphene as electrodes. *Adv. Mater.* 23 (30): 3446–3449.

16 Shi, Q., Dong, B., He, T. et al. (2020). Progress in wearable electronics/photonics—moving toward the era of artificial intelligence and internet of things. *InfoMat* 2 (6): 1131–1162.

17 Kim, D.H., Shin, H.J., Lee, H. et al. (2017). In vivo self-powered wireless transmission using biocompatible flexible energy harvesters. *Adv. Funct. Mater.* 27 (25): 1700341.

2

Working Mechanisms of Solar Cells

2.1 Introduction

Solar energy is the most abundant energy around the earth. All the energy on earth could be sourced back to the solar energy. The power that reaches the surface of the earth is in the magnitude of 10^{17} W [1, 2]. On the other hand, for the whole human society, the energy requirement is only around tens of 10^{12} W [3]. However, the solar energy cannot be used directly, and most of the energy is reflected and transferred to the heat in the environment [2]. Solar cells are a kind of device that can convert light energy directly to electricity, which makes great use of abundant solar energy and is especially applicable for portal and wearable electronics. The solar cells work based on the photovoltaic phenomenon, which was developed a hundred years ago. It was until the 1950s that the first practical silicon-based solar cell was fabricated and showed potential practical applications for energy conversion, especially in a clean way and also providing electricity in outer space [4]. How does the solar cell work, and how much energy could be converted by the solar cell? It is essential to understand the basic principles and mechanisms of the photovoltaic process. In this chapter, we will briefly introduce the working mechanism of typical solar cells and the efficiency limitations of these cells. The readers could find more detailed information regarding the basic device physics in the textbook. The working mechanism for traditional silicon-based solar cells is first summarized to elucidate the physical principle in photovoltaics [5, 6]. Aiming at the wearable and flexible applications, especially, the main efforts are then made to discuss the different mechanisms for different types of solar cells, i.e. dye-sensitized solar cells [3], polymer solar cells [7], and perovskite solar cells [8]. The resulting advantages and disadvantages of the above solar cells are highlighted, particularly from a viewpoint of making wearable photovoltaic devices, to indicate their advancements and challenges in the past decade.

Wearable Solar Cells: Mechanisms, Materials, and Devices, First Edition.
Hao Sun, Zhibin Yang, Longbin Qiu, and Huisheng Peng.

2.2 Working Mechanism and Efficiency Limitation for Solar Cells

Solar cells are fabricated based on semiconductors and, in the early days, were mainly based on Germanium (Ge), Silicon (Si), and GaAs from the development of semiconductor electronic devices [9]. Nowadays, the main product for commercialized solar cells is Si cells, which occupy over 90% of the market [10]. There are second-generation thin-film solar cells like cadmium telluride (CdTe) solar cells and copper indium gallium selenide (CIGS) solar cells, which are being developed and increasing market sharing [10]. The recently developed thin film solar cells, like dye-sensitized solar cells, polymer solar cells, and perovskite solar cells, are promising for flexible and wearable applications. However, they are still in the lab proof-of-concept stages.

To understand how much energy is to be converted, the input power from the solar should be elucidated first (Figure 2.1) [11]. The sunlight incident from outer space to the surface of the ground, where the solar cells are mounted. The solar power is daytime- and season-sensitive due to the varying incidence angle [12]. Due to the different incidence angles, the light paths are different (Figure 2.1a). The light spectrum that reaches the outer space of the earth is defined as air mass 0 (AM0). While with a direct angle incidence reaching the ground surface, the spectrum is defined as AM1.0. The other light incidence spectra are defined as the inverse of the cosine of the incidence angle [11]. The solar spectra measured with different air masses have been obtained, and it has been set that the standard solar spectrum for the solar simulator is AM1.5. Interestingly, it is based on the measured spectrum, the surface temperature of the Sun is supposed to be 6000 K when comparing a theoretical black body emission spectrum (Figure 2.1b). Due to the reflection and scattering of light in the cloud and also in the space, the real spectra on the ground surface are different from the ones in outer space. For a normal solar cell, the light spectrum that is to be considered is the AM 1.5G (global), which includes direct light and also reflective light. While for the application of concentrated solar cells, the spectra to be used are the AM 1.5D (direct), which only includes the direct incidence light [11]. Standard light intensity for AM 1.5G is 100 mW/cm^2, while for most of the wearable and portable sensor systems, the energy requirement is only microwatt to several mW. The input energy from the Sun is beyond enough for application [13].

The most applied photovoltaic devices are silicon solar cells, with several typical high-performance structures shown in Figure 2.2 [5]. The silicon solar cells have a theoretical efficiency limit of 29.4% [5], and currently, the highest certified record efficiency is 26.8% [14], close to its limitation. The basic model for Si solar cells is a p–n junction, with a hole majority carriers (p-type) and electron majority carriers (n-type) silicon materials forming a junction, as shown in Figure 2.3a. By forming a p–n junction, the majority of electrons in n-type semiconductor and majority of holes in p-type semiconductor would diffuse through the junction due to the density gradient. And a voltage potential would be formed due to the movement of electrons and holes. With a balance of the drift and diffusion movement of electrons and holes, a built-in potential (V_{bi}) would be obtained, with a higher electric potential in the

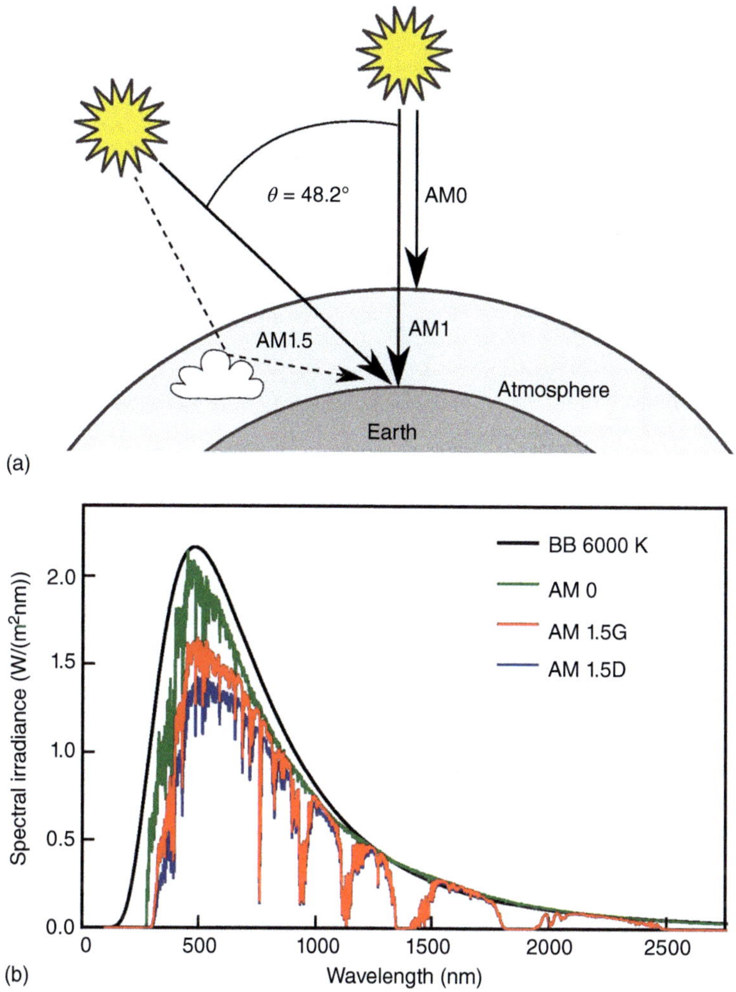

Figure 2.1 (a) The incident light from the Sun and the definition of air mass (AM). (b) The spectral irradiance of the sunlight passing through different paths. Source: Rühle [11], © 2016/Elsevier.

n-type side and a lower potential in the p-type side. One should note that for electrons, the direction of the driven force is from the lower electric potential pointing to the higher potential [6].

Under light illumination, the photons will be absorbed, and electron–hole pairs formed. With photons that have energy larger than the band gap, the semiconductors could absorb the photons, and the electrons in the valence band could be excited into the conduction band, forming an electron–hole pair [16]. Due to the driven force of the built-in potential (Figure 2.3b), the electrons in the junction will move toward to the n-type side and then transfer to the external circuit. Connecting to the variant load in the external circuit, there will be a potential drop which is referred to as V. And the effective driven force for the movement of electrons will be

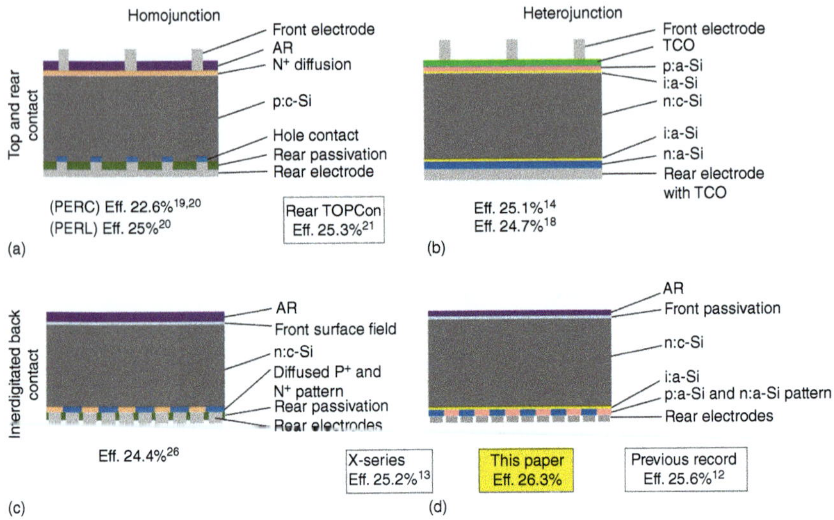

Figure 2.2 Schematic drawing showing the typical structure of silicon solar cells. (a) A top/rear contacted homojunction solar cell. (b) A top/rear contacted heterojunction solar cell. (c) A homojunction interdigitated back contact solar cell. (d) A heterojunction cell interdigitated back contact. Source: Yoshikawa et al. [5], © 2017/Springer Nature.

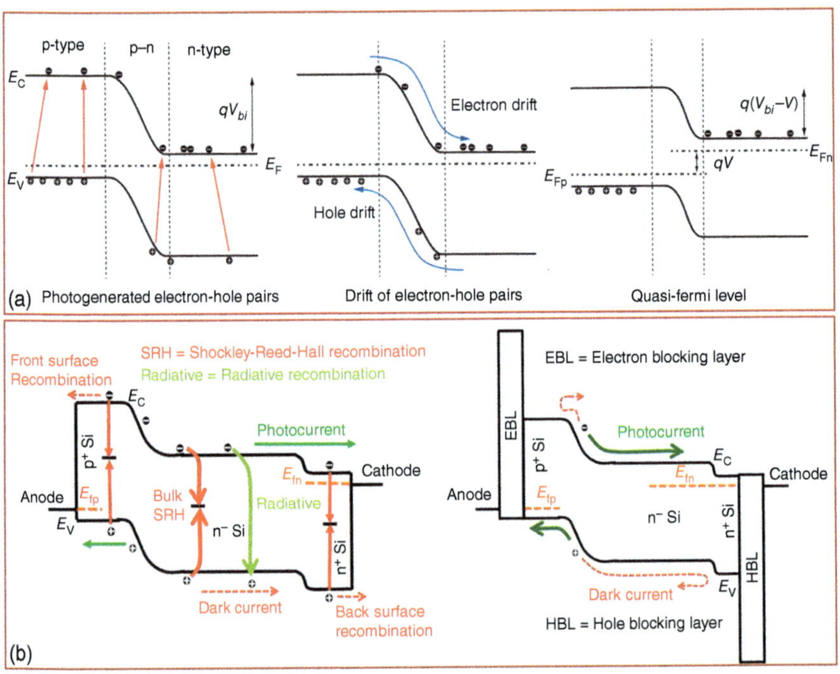

Figure 2.3 (a) Schematic drawing showing the photogeneration of electron-hole pairs in a semiconductor. (b) The typical energy level diagram shows the formation of p–n junction and built-in potential under illumination. Source: Islam and Saraswat [15], © 2018/Springer Nature/CC BY 4.0.

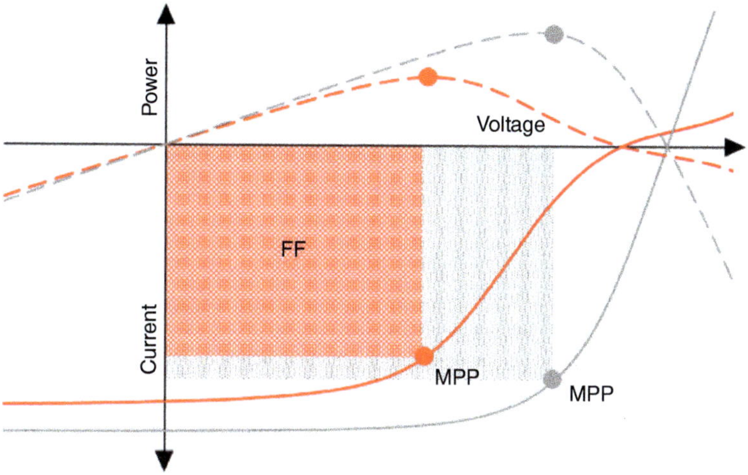

Figure 2.4 A typical current density–voltage (J–V) curves (solid line) and power as a function of external circuit voltage (dash line) for solar cells. Source: Saive [17], © 2019/IEEE/CC BY 4.0.

altered as $q(V_{bi} - V)$ as shown in Figure 2.3b. The external circuit potential V is also equal to the difference of quasi-fermi level for electrons and holes $[(E_{Fn} - E_{Fp})/q]$. By varying the load from zero to infinite, short-circuit and open-circuit conditions are obtained. With a specific load, the maximum energy output could be achieved, and the typical I–V curve and power output-voltage curve are shown in Figure 2.4. The output power could be calculated by multiplying the external circuit current I with voltage drop V:

$$P_{out} = I \times V \tag{2.1}$$

While the power input could be represented as:

$$P_{in} = A \times \Phi \tag{2.2}$$

where Φ is the incidence of light intensity.

Thus, the power conversion efficiency (PCE) could be calculated as:

$$\eta = \frac{P_{out}}{P_{in}} = \frac{I \times V}{A \times \Phi} = \frac{J \times V}{\Phi} \tag{2.3}$$

where J is the representation of current density.

The most important efficiency and the solar cell efficiency are defined at the maximum power point (J_{max} and V_{max}). To correlate the solar cell efficiency with the intrinsic properties of photoactive materials, the current density at short circuit condition (J_{SC}) and the output voltage at open circuit condition (V_{OC}) are incorporated into the calculation of efficiency.

$$\eta = \frac{J_{max} \times V_{max}}{\Phi} = \frac{J_{SC} \times V_{OC} \times \mathrm{FF}}{\Phi} \tag{2.4}$$

The fill factor (FF) is defined as:

$$FF = \frac{J_{max} \times V_{max}}{J_{SC} \times V_{OC}} \tag{2.5}$$

For a solar cell, the external circuit current is a function of the output voltage, which means the three parameters (J_{SC}, V_{OC}, and FF) are not completely independent. Obtaining the function between the J and V could be the key to the estimation of the theoretical efficiency limit based on photoactive material properties. There are three main models that describe the function describing the relation between J and V and will be discussed in detail, namely (i) detailed balance limitation, (ii) drift-diffusion model, and (iii) equivalent circuit model. The realistic model considering recombination process, as shown in Figure 2.3b, will also be included. The mechanism and theoretical model are all applicable to the other solar cells, including dye sensitized solar cells, polymer solar cells, and perovskite solar cells, which will be discussed in the following sections.

2.2.1 Detailed Balance Limitation

The theoretical efficiency limit of solar cells is of interest to optimize the design of photoactive materials and device structures. The most popular model was the so-called S–Q limit that based on thermodynamic law (detailed balance limit) proposed in 1961 [18]. The basic two assumptions for the detailed balance limit are (i) the photons with energy larger than the band gap will be absorbed, and one photon converts to one electron–hole pair and (ii) the excited electrons will be relaxed to the conduction band edge for all the electrons that has energy much higher than the band gap. The number of photons that could be absorbed depends on the incident light spectra. For comparable results from around the community, a standard incident light spectrum was introduced, which is AM 1.5G (Figure 2.1b). It could be easily understood that with a smaller band gap, a larger number of photons will be absorbed. However, in that case, the energy potential for each excited electron will also be smaller. The total energy output depends on the product of a number of photons (related to J) and energy potential (related to V).

To be specific, the photogenerated carriers that are excited by incident photons could be calculated by integrating the number of photons with energy higher than the band gap, as shown in Eq. (2.6).

$$J_{SC} = q \times \int_0^{\infty} \alpha(E)\Theta(E)dE = q \times \int_{E_g}^{\infty} \Theta(E)dE \tag{2.6}$$

Here the $\alpha(E)$ is the photon absorption coefficient as a step function of energy. When the photon energy is larger than bandgap E_g, the coefficient is 1. When the photon energy is smaller than the band gap, the coefficient is 0. And the $\Theta(E)$ is the photon number density in the energy range of E to $E+dE$.

Based on Kirchhoff's law, there will be light emission from the photoactive materials, and the emission spectrum is related to the temperature of the cell (black body). This light emission is assigned to the radiative recombination process, which is unavoidable. The number of photons emitted is called the reverse saturation current in a p–n junction [19]. Other than radiative recombination that would reduce the photogenerated electron–hole pairs, there is also non-radiative recombination process. For a conserve of the particles (photons, electrons), the external circuit current density J at a steady state of the p–n junction could be described as:

$$J_{ext} = J_{SC} - J_r(V) + J_{nr}(0) - J_{nr}(V) = J_{SC} - J_r(0) + J_r(0) - J_r(V) + J_{nr}(0) - J_{nr}(V) \tag{2.7}$$

where J_{SC} is the maximum current density calculated based on the energy band gap. $J_r(V)$ is the reverse current density is related to radiative recombination. $J_r(0)$ is the carrier generation from environmental thermal energy in the dark condition. $J_{nr}(0)$ and $J_{nr}(V)$ are the electron–hole pair generation process and recombination process that are related to non-radiative process, respectively. Based on the conventional semiconductor p–n junction current density–voltage relation:

$$J_r(V) = J_r(0) \times \exp\left(\frac{qV}{kT}\right) \tag{2.8}$$

$$J_{nr}(V) = J_{nr}(0) \times \exp\left(\frac{qV}{kT}\right) \tag{2.9}$$

where q is the elementary electron; k is the Boltzmann constant; T is temperature.

We can rewrite the above equation by defining a new factor f_c, which is the faction of radiative recombination as: [18]

$$f_c = \frac{J_r(0) - J_r(V)}{J_r(0) - J_r(V) + J_{nr}(0) - J_{nr}(V)} = \frac{J_r(0)}{J_r(0) + J_{nr}(0)} \tag{2.10}$$

and

$$J_{ext} = J_{SC} - J_r(0) + J_r(0) \times \frac{1 - \exp\left(\frac{qV}{kT}\right)}{f_c} \tag{2.11}$$

When f_c equals to unity, there is only a radiative recombination process, and it shows the highest V_{OC}. And $J_r(0)$ is relatively small compared with J_{SC} and could be neglected.

$$V_{OC} = \frac{kT}{q} \times \ln\left[\frac{J_{SC}}{J_r(0)} + 1\right] \tag{2.12}$$

The output energy could be calculated by multiplying the external current density J_{ext} and voltage potential across the load V.

$$P_{out} = V \times J_{ext} = V \times \left[J_{SC} - J_r(0) + J_r(0) \times \frac{1 - \exp\left(\frac{qV}{kT}\right)}{f_c}\right] \tag{2.13}$$

By plotting the above equation, the maximum power output voltage V_{max} and current density J_{max} could be obtained. Thus, the PCE could be obtained from the above-mentioned Eqs. (2.3) and (2.4). And the PCE could be described as $V_{OC}*J_{SC}*FF/P_{in}$. For AM 1.5G, the P_{in} is defined as 100 mW/cm^2. The theoretical

PCE is only a function of the energy band gap of a photoactive material based on a detailed balance limit [11].

However, the calculation is based on the assumption that the light absorbance coefficient is infinite and could absorb all the light. On the other hand, the mobility is also assumed to be infinite, that all the generated electron–hole pairs could be separated into corresponding electrodes. The above calculation is also only based on the spectrum and thermodynamic theory. Regarding the faction of radiative recombination f_c, the calculation tells us no information about the origin of the non-radiative recombination process. Based on the electric magnetic theory and the electronic properties of the functional materials, the device performance might be more realistic.

2.2.2 Drift-Diffusion Model

For the drift-diffusion model, the basic equations are the one-dimensional continuity equation and the Poisson's equation [20, 21].

$$\frac{\partial}{\partial t}N_{n(p)}(x,t) \mp \frac{1}{q}\frac{\partial}{\partial x}J_{n(p)}(x,t) = G(x,t) - R(x,t) \tag{2.14}$$

Here, $N_{n(p)}(x, t)$ is the number density of electrons (holes) in the position of x and at time t. And $J_{n(p)}(x, t)$ is the current vector that flows through position x at time t. $G(x, t)$ and $R(x, t)$ are the generation and recombination rate of electrons or holes in position x at time t. The continuity equation is the conservation of particles. Due to the planar structure of solar cells and the direction of the voltage potential, the electrons (holes) only move in the direction that is perpendicular to the solar cell plane. And one-dimensional continuity equation is considered. The change of the electrons (holes) in an infinite thin space is equal to the change of the carrier density plus the change from the flow. When in a steady state, the change of current density is zero. The net change equals to the generation of carriers ($G(x, t) - R(x, t)$, the difference between generation and recombination). While for the Poisson's equation, it is the correct electrical response of a device in a steady state following Maxwell's equation.

Other than the continuity equation and the Poisson's equation, the movement of carriers follows the drift-diffusion equation. The drift of carriers describes the moving under the electric field, and the diffusion describes the moving force of the density gradient. Based on Einstein's relation, the diffusion coefficient (D) and carrier mobility (μ) are related as $D_{n(p)} = \mu_{n(p)}{*}k{*}T/q$. The current flow from the movement of electrons and holes could be described as:

$$J_{n(p)}(x,t) = \mu_{n(p)}\left(qEN_{n(p)} \pm kT\frac{\partial}{\partial x}N_{n(p)}\right) \tag{2.15}$$

Solving the above equations gives the J–V curves, which could deduce the energy output and the PCE of the resulting devices. The generation rate $G(x, t)$ should be calculated based on the absorbance coefficient for specific active and functional materials [22]. And the specific recombination process, including surface recombination and Shockley–Reed–Hall recombination (Figure 2.3b) [15, 23], could be incorporated into the continues equation[21, 24]. Compared with the detailed balance limitation model, here, the recombination process could be more specific and included into the charge generation/recombination part. The readers could refer to a more detailed information regarding the model and numerical simulation of solar cell performance based on the drift-diffusion model [20, 21].

2.2.3 Equivalent Circuit Model

The third model that is used to analyze a solar cell is the equivalent circuit model, which considers the solar cell as a constant current generator that is in parallel connection to a diode (Figure 2.5) [25]. In an idealized model with no extra energy loss in the circuit (no series connected resistance and the infinite value for the shunting path resistance), the generated current only flows through the external circuit, and the modelized diode is constant, of which, depending on band gap of the photoactive materials. Connecting to different external load, the output voltage and current $J(V)$ varies, and the voltage across the load and the diode is the same. In this case, the current flowing through the diode obeys the current-voltage characteristic, similar to the detail balance limit [18].

In the realistic equivalent circuit model, the recombination process and the other energy loss processes should be considered as series resistance and parallel resistance that are connected to the circuit. Although the specific mechanism for the cause of series and parallel resistance is not clear, this model could be used to fit or simulate the current density-voltage curves by solving the equation as follows:

$$J(V) = J_{SC} - J_r(V) - J_{bulk} - J_{surf}(V) - J_{sh}(V) \tag{2.16}$$

All above three above-mentioned models are useful for understanding the working mechanism and cause of energy loss during the fabrication and working of solar cells. To sum up, for efficient solar cells, there are several steps for generating electricity under light illumination, including light absorption and excitation, generation of electron–hole pairs (or excitons for some kinds of solar cells), separation of electron and holes, and injection into selective carrier transport layers, and the working of flowing of electrons in the external circuit [1]. These models

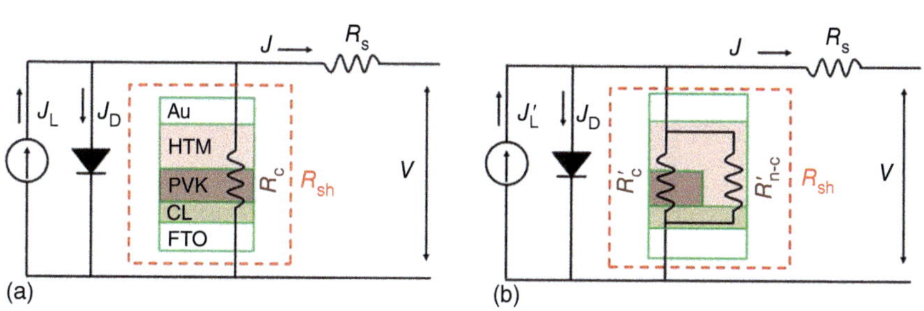

Figure 2.5 Equivalent circuit model for solar cells by taking perovskite solar cell as a demonstration. (a) A full coverage of the active layer with less recombination through the shunting path. (b) A non-full coverage of the active layer with a higher shunting current. Source: Li et al. [25], © 2017/Springer Nature/CC BY 4.0.

were built based on Si solar cells in the early ages; however, they could be applied to newly developed solar cells that are compatible with flexible and wearable structures. While the specific mechanism for each step might be different, overall, the models above are applicable. Among these new-generation solar cells promising for flexible and wearable applications are dye-sensitized solar cells, polymer solar cells, and perovskite solar cells. The working mechanism and respective advantages and disadvantages will be discussed in the following sections.

2.3 Working Mechanism of Dye-Sensitized Solar Cells

The typical structure of dye-sensitized solar cells is shown in Figure 2.6 [3]. Unlike silicon solar cells with a homojunction or heterojunction, dye-sensitized solar cells are composed of a photoanode, counter electrode, and electrolytc (Figure 2.6a). The absorption of light ascribes to the dye molecular that chemically bonds to the n-type semiconductor metal oxide. With the more dye molecules, the more incident light could be absorbed. However, the stacking of the dye molecular is not conductive for the photogenerated carrier transported. The development of mesoporous TiO_2 realized an extremely high surface area in a small volume. In this case, a monolayer dye molecular absorption was first achieved, and practical dye-sensitized solar cells with an efficiency of around 7% were obtained [26].

The charge carrier generation and transportation flow could be described as follows (Figure 2.6b) [3]. The light incidence from the photoanode side is absorbed by the dye molecular with suitable band gap. The electrons in the highest occupied molecular orbital (HOMO) level will be excited by photons with energy higher than the band gap to the lowest unoccupied molecular orbital LUMO level (labeled process 0 in Figure 2.6b). One specific energy level alignment is required for the efficient injection of excited carriers in the LUMO to the conduction band minimum of TiO_2 (labeled process 2 in Figure 2.6b). There are holes remaining in the HOMO level of the dye, or one can consider that the dye has been oxidized

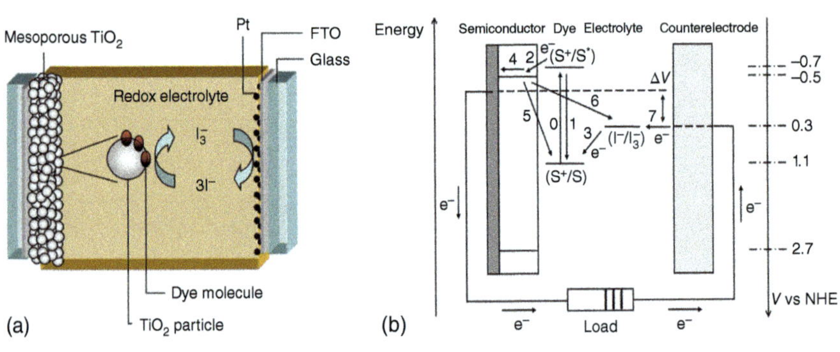

Figure 2.6 (a) The typical structure of dye-sensitized solar cells. (b) The flow of carrier generation, transportation, and recombination. Source: Hagfeldt et al. [3], © 2010/American Chemical Society.

during the photogeneration process. Thus, a reduction of the dye will be required. Typical oxidation–reduction pairs like I^-/I_3^- are applied in the electrolyte [27]. In the photoanode side, then I^- anion and the oxidized dye reacted to reduce the dye to its original condition, and the I^- oxidized to I_3^- (labeled process 3 in Figure 2.6b). The reduction of I_3^- to I^- happens in the counter electrode to balance the concentration of these pairs. The electrons that come from the excited carriers that are injected into the TiO_2 are collected by the electrode to the external circuit (labeled process 4 in Figure 2.6b). After working in the external circuit, the electrons are collected by the counter electrode, and the reduction of I_3^- to I^- is catalytic by the counter electrode (labeled process 7 in Figure 2.6b) [28]. In this case, there is no net change in the system, but only solar energy from photons to excited carriers and then to external load as electrical energy. The difference between the redox potential of the I^-/I_3^- pair and the conduction band minimum of TiO_2 is the maximum electrical potential to drive the circuit [29]. Under a light illumination in a stabilized power output condition, the initial aligned Fermi level will split into the quasi-fermi level for electrons and holes. Depending on the excited carrier concentration in the TiO_2 conduction band, the quasi-fermi level for the electrons will shift close to the conduction band minimum [30, 31]. On the other hand, the quasi-fermi level for holes is close to the redox potential of I^-/I_3^- under the counter electrode catalyst [32]. The difference between the quasi-fermi levels for electrons and holes in open circuit conditions is the output open-circuit voltage. And the band gap of the dye determines the absorption edge of incidence light, which determines the short circuit current density [33]. To further enhance light absorption, co-sensitized solar cell structure was designed with more than one kind of dye were absorbed with different band gaps [34].

Considering the carriers' generation and transportation flow and the interface structure of the solar cells, there are several recombination processes for the excited carriers (Figure 2.6b) [35]. The first recombination process is that the excited electrons in the LUMO level recover to the HOMO level (labeled process 1 in Figure 2.6b). There is competition between the recombination process 1 and the injection process 2. And the electrons in the conduction band of TiO_2 might be recombined by the oxidized dye (labeled process 5 in Figure 2.6b) or the electrolyte with I_3^- to I^- (labeled process 6 in Figure 2.6b). There is competition between the carrier collection process 4 and the recombination processes 5 and 6. How to enhance the carrier generation, extraction, and transportation efficiency while, in the meantime, reducing the recombination process are the keys for high-performance of dye-sensitized solar cells. Recently, with the development of high absorbance coefficient dye, high performance counter electrodes, efficient electrolyte, and high-quality photoanode materials, the efficiency of dye-sensitized solar cells has been over 15% [36]. By replacing the liquid electrolyte with solid materials like spiro-MeOTAD [37] or quasi-solid ionic liquid [38], solid state dye-sensitized solar cells have been further developed [39]. For more detailed information regarding the materials and structures, the readers could refer to the following chapters.

2.4 Working Mechanism of Polymer Solar Cells

Polymer solar cells consist of polymer semiconductors as photoactive materials sandwiched between hole-selective and electron-selective layers (Figure 2.7) [40]. The design of polymer solar cells could be traced back to the development of conducting polymers. In the early stage, planar-junction structures formed by donor (p-type) and acceptor (n-type) materials were applied[41]. The photoactive materials absorb photons and are expected to excite to the LUMO level and then transport to the external circuit (Figure 2.7a), however, out of expectation that the early designed device exhibited low performance. It is found that the carrier lifetime is low in the photoactive polymers, and the carrier diffuses an ultra-short path before recombination [42]. To absorb enough incidence light, thick enough photoactive polymers are required. The balance of the carrier extraction and photons absorption could not be matched with the planar–junction structure. The design of bulk heterojunction solved the balance issue of carrier extraction and photon absorption by forming a continues interfused phase of donor and acceptor (Figure 2.7b) [43].

Nowadays, polymer solar cells are mainly based on bulk heterojunction structure by mixing donor and acceptor to form a continues interfused phase with fined domain size (Figure 2.8) [7, 44]. The photogeneration of carriers is different from other thin film solar cells, with the initial generation of excitons [45]. Excitons are binding electron-hole pairs with a binding energy higher than the environmental thermal energy [46]. In this case, the excitons could not dissociate into free carriers and migrate along to the donor/acceptor interface. The excitons will move toward to the direction of the hole selective layer and separate into free carriers of electrons and holes. After dissociation, the free carriers of holes and electrons migrate along the donor and acceptor domains, respectively. The domain size dimension is around 20 nm, similar to the diffusion length of excitons, which helps dissociate the excitons into free carriers before recombination [47].

To determine the performance of a polymer solar cell, the energy levels of donor and acceptor materials are required. It is investigated that mainly the donor polymer materials absorb the incidence light, and the optical band gap of the donor polymer is related to the accessible maximum J_{SC} [40, 48]. The most used acceptor materials

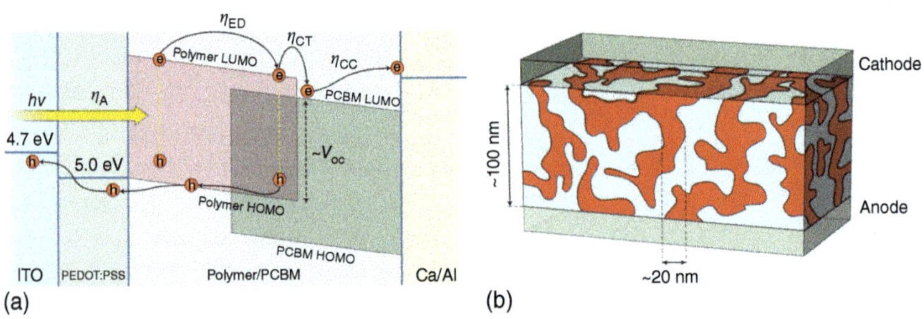

Figure 2.7 Typical structure of bulk heterojunction polymer solar cells and the carrier generation and transportation process. (a) Energy level alignment of a conventional polymer solar cell. (b) The bulk heterojunction structure of the active layer. Source: Li et al. [40], © 2012/Springer Nature.

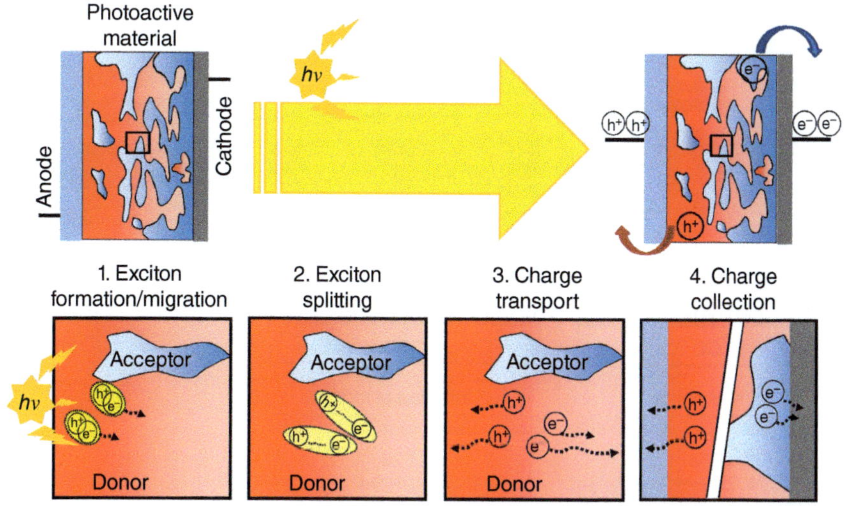

Figure 2.8 Polymer solar cells with a gradient bulk-heterojunction structure and the generation of excitons and free carriers under light illumination. Source: Facchetti [7],

in the early stage were the derivative materials of C60, like (6,6)-phenyl-C61 butyrie acid methyl ester (PCBM) and [6,6]-Phenyl C71 butyric acid methyl ester (PC70BM) [49]. The excitons were separated in the heterojunction interface, with the electrons in the LUMO of the acceptor and the holes in the HOMO in the donor (Figure 2.7a). The difference in the quasi-fermi level is equal to the energy difference between the acceptor LUMO level and the donor HOMO level that determines the maximum V_{OC} to be obtained [50]. With minimized non-radiative recombination and an active layer thickness close to the free carrier diffusion length, a high-efficiency polymer solar cell with the bulk heterojunction was reported.

To further enhance the device performance, one should consider the recombination processes and energy level alignment structure [51]. The optical band gap of the acceptor is normally larger than donor materials. And the LUMO level of acceptor should be much close to but slightly lower than the LUMO level of donor, for efficient exciton separation into free carriers and a maximum quasi-fermi level of free electrons. New polymer materials and processing conditions have been developed. And full polymer fullerene-type polymer solar cells have been developed recently with more and more knowledge and understanding regarding the composition-structure-properties of the semi-conducting polymers [7]. The performance has been much improved for the bulk heterojunction to recently over 19% [52, 53].

2.5 Working Mechanism of Perovskite Solar Cells

Perovskite solar cells were initially developed from dye-sensitized solar cells by using perovskite nanocrystals working as a dye (Figure 2.9a) [8]. And the first report

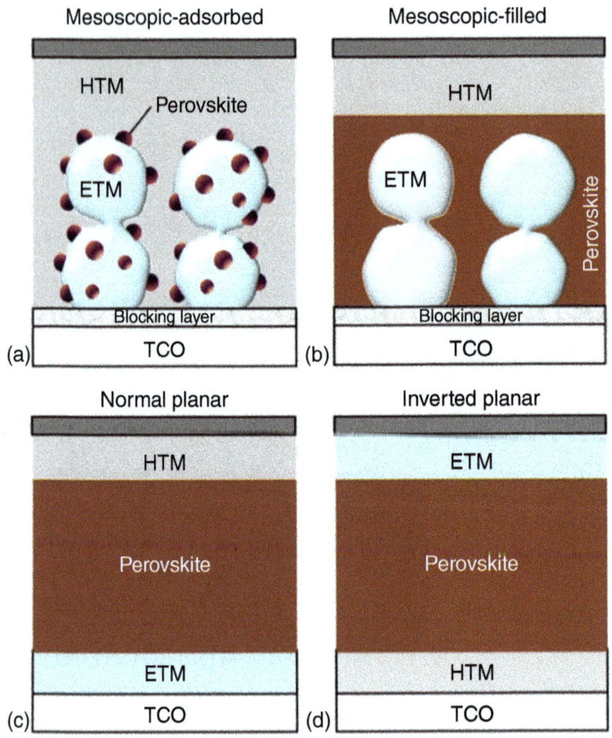

Figure 2.9 Typical structures of perovskite solar cells. (a) The perovskite-sensitized type of perovskite solar cells. (b) The mesoscopic-filled perovskite solar cells. (c) The normal structure-based planar perovskite solar cells. (d) The inverted structure-based planar perovskite solar cells. Source: Kim et al. [8], © 2020/American Chemical Society.

showed an efficiency of 3.8% with liquid electrolyte [54]. However, the perovskite material is easily dissolved in the liquid electrolyte, and the stability is only tens of minutes. Further optimization of the liquid electrolyte component, the efficiency of early-stage perovskite nanocrystal-sensitized solar cells has reached 6.5%[55]. Surprisingly, by replacing the liquid electrolyte with solid electrolyte learning from conventional dye-sensitized solar cells, the efficiency of solid-state perovskite solar cells has reached 9.7% and a further 10.9% [56, 57]. It is found that unlike the organic dye, which could only absorb light and inject the charge carriers to the photoanodes like TiO_2, the perovskite materials were discovered with a high carrier mobility and long diffusion length [58, 59]. Perovskite materials are intrinsically high-performance semiconductors. Thus, a thicker perovskite layer and thinner TiO_2 mesoporous layer structure were developed (Figure 2.9b). The thickness of the TiO_2 was reduced step by step from tens of micrometers to hundreds or tens of nanometers [60]. And the understanding of the working mechanism is getting more comprehensive and different from the dye-sensitized solar cells. Furthermore, planar junction structures like p–i–n or n–i–p structures applied in polymer solar cells were also introduced into perovskite solar cells and achieved an ultrahigh record performance (Figure 2.9c,d,).

The working mechanism is more like other inorganic thin film solar cells like CdTe and CIGS, with perovskite being the photoactive layer. With light illumination, the electrons in the valance band of perovskite are excited into the conduction band. Unlike the organic solar cells and polymer solar cells, where the binding energy of electrons and holes is higher than thermal energy from the environment, the binding energy of excited carriers for perovskite materials is much lower [61]. Free charge carriers instead of excitons are generated under light illumination. Under the built-in potential, the carriers transport to the charge selection layers for electrons and holes, respectively. The electron transport layer (ETL) layer extracts electrons and block holes, while the HTL extracts holes and block electrons [62]. The electrons transport from the cathode to the external circuit and to the anode to recombine the holes from HTL. The net change is the energy conversion from light to electricity.

The energy level diagram of perovskite solar cells is depicted in Figure 2.10 [63]. Under dark conditions, the Fermi level is aligned across the whole device and induces built-in potential due to the transfer of electrons and holes in the interface [64]. Under illumination, the Fermi level splits into quasi-fermi levels

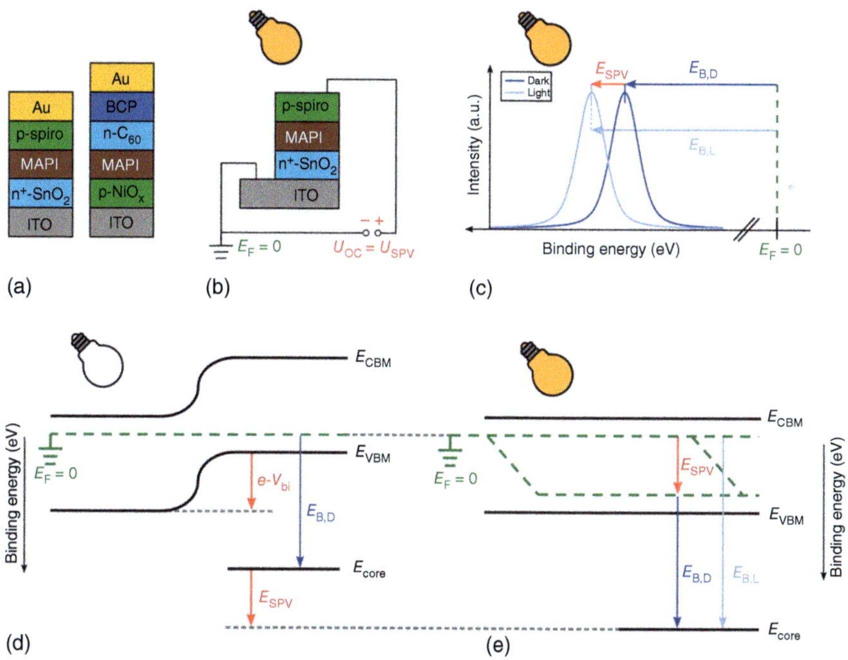

Figure 2.10 Normal structure and inverted structure perovskite solar cells and the energy level diagram under dark and illumination conditions. (a) normal (n-i-p) structure and inverted structure (p-i-n) perovskite solar cells. (b) Schematic showing the working condition of a n-i-p structure solar cell. (c) The light induced surface potential (SPV). (d) The energy level diagram of a solar cell under dark condition. (e) The energy level diagram of a solar cell under light illumination connected to a certain external circuit. Hellmann et al. [63], © 2020/John Wiley & Sons.

of electrons and holes. The photogenerated electrons and holes transport to the ETL and HTL layers, respectively. The difference between the quasi-fermi level for electrons and holes in the open circuit condition is the V_{OC} of the solar cells [65]. On the other hand, the light absorption ability depends on the band gap of perovskite materials, which could be tuned from 1.2 to 2.8 eV, for various potential applications [66]. There are also recombination processes for perovskite solar cells. Due to the polycrystalline features, there are plenty of defects in the grain boundaries and the bulk. Interestingly,the perovskite is highly defect tolerant, with the defect states mainly in the shallow position [67]. Although the defect density is several orders of magnitude higher than silicon, the perovskite solar cell efficiency is still approaching that of silicon solar cells.

One important issue to be noted is that the defect state is related to the stability of perovskite solar cells [68]. Perovskite materials are ionic compounds with a formula of ABX_3. More information about the materials' structure and composition is provided in detailed in Chapter 7. The A-site cation and the X-site halogen are volatile, leaving vacancies in the materials, especially in the surface and grain boundaries. There is movement of electrons and holes, and also the movement of the ion vacancy (Figure 2.11) [20]. The movement of ion vacancy deteriorates the device's performance and induces the hysteresis effect when scanning the solar cells from different directions [69]. The hysteresis phenomenon makes it difficult to measure the true performance of perovskite solar cells. Regarding the mechanism and reduction of ion migration and hysteresis, the readers could refer to several recent review articles [70–73] ,which will not be discussed in detail in this chapter.

2.6 Comparison of Solar Cells for Flexible and Wearable Application

The above-mentioned solar cells that are fabricated under a low-temperature process have the potential for flexible and wearable applications. Among them, the dye-sensitized solar cells include a high-temperature sintering of photoanode before dye absorption, which limits the usage of flexible polymer substrate [74]. While the coating of TiO_2 on a PET substrate with a low temperature limits the cell performance [75]. On the other hand, the usage of liquid electrolyte might be easily leakage and requires robust encapsulation. The development of solid electrolytes in dye-sensitized solar cells was applied in the early-stage perovskite-sensitized solar cells and achieved great success. Most of the works of planar-type flexible solar cells are now focusing on perovskite solar cells and polymer solar cells rather than dye-sensitized solar cells. However, when considering flexible and wearable applications, the ability to fabricate on arbitrary and rough surfaces is required. Due to the various applicable deposition process for photoanode and the chemical absorption of dye in a solution, the design of high-quality photoanode in various shape and structure are accessible, for example, in a fiber structure.

The polymer solar cells are based on polymer materials are attractive, with intrinsic flexibility, lightweight, and solution-compatible coating process for large-scale

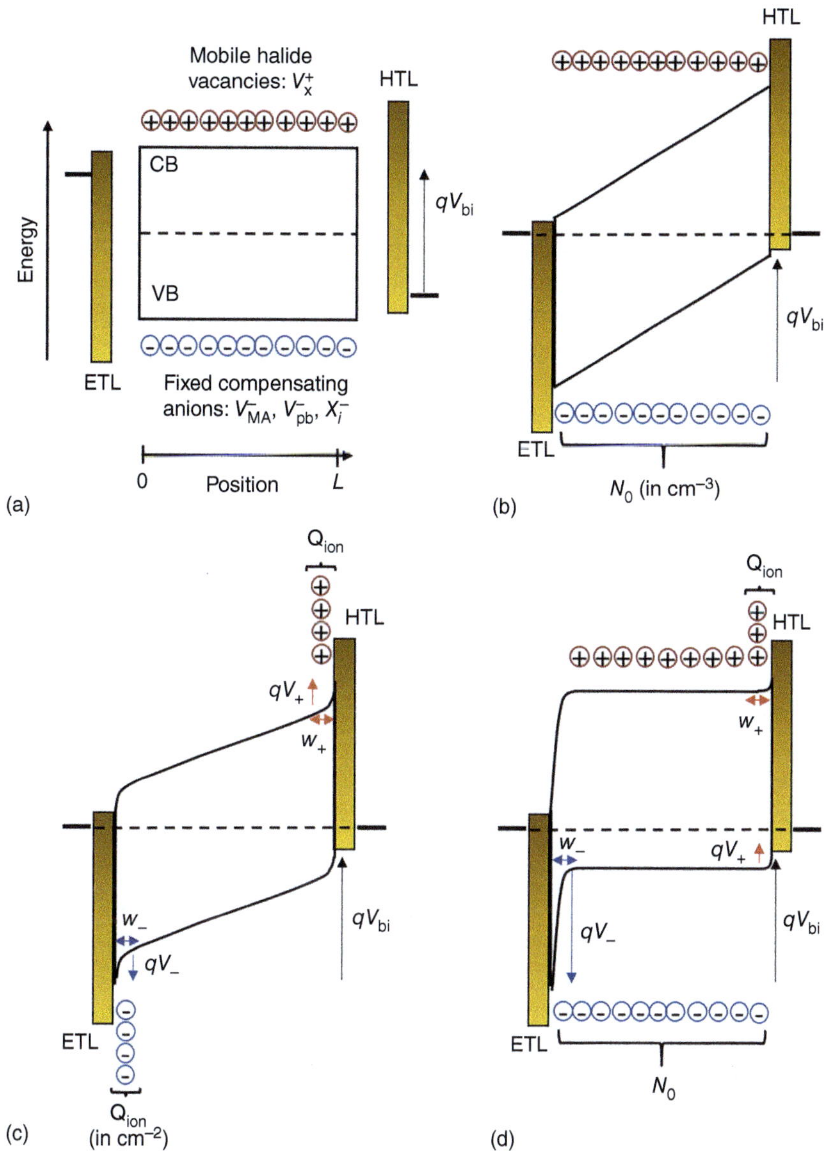

Figure 2.11 The model and energy level diagram for the simulation of perovskite solar cells with ion vacancy migration. (a) The perovskite film before contact to the functional layer. (b) the deposition of HTL and ETL on each side of the perovskite layer that induces a built-in potential. (c) The case of both cations and anions are mobile under built-in potential. (d) The case of cations is mobile, while anions are immobile. Source: Bertoluzzi et al. [20], © 2020/Elsevier.

roll-to-roll processes [76]. Furthermore, in recent decades, the performance of small-area cells has been rapidly improved from around 10% to approaching 20% [77] and large-area modules with an area of $54\,cm^2$ showed efficiency over 13% [76]. The ultra-high absorbance and the short carrier diffusion length determine that a thin layer of polymer bulk heterojunction should be applied. This also further enhances the flexibility of solar cells under various deformations. However, regarding arbitrary surfaces, it is difficult to deposit polymer layers uniformly and conformably [78]. When considering a wearing application based on textiles and fibers, the effective coating of high-quality polymer materials is still a grand challenge. The strategies for improving the polymer solar cells in rigid and planar-type substates are waiting to be incorporated into wearable polymer solar cells with arbitrary shaped [79]. When considering in another way by attaching ultrathin flexible polymer solar cells to arbitrary surfaces might be a potential alternative [80]. In this case, rollable polymer solar cells could be used to power portable or even implanted electronics [81].

The perovskite solar cells, it is compatible with a low-temperature solution coating process [82] under 150 °C and a vapor-based deposition process [83, 84], which has potential for conformal coating on an arbitrary surface. Due to the small binding energy of excitons, a high carrier mobility, and long diffusion length, the thickness of the perovskite could be much thicker than requirement while maintaining or enhancing the cell performance [85]. Ultrathin perovskite solar cell with a high power-to-weight ratio was developed [86]. By further modulating the covering layer thickness, the functional layer and the photoactive layer could be placed in the neutral plane, which induces no stress or strain during bending deformation [87]. However, there is solvable metal lead in perovskite solar cells, which is toxic and forbidden in most of the countries [88]. And the operational stability and moisture and oxygen sensitivity of perovskite solar cells need be further improved [89, 90]. Even though the performance of flexible perovskite solar cells has reached over 24%, its application for wearing might be doubted. The development of non-lead-based perovskite solar cells are promising alternative for wearable applications [91]. On the other hand, it is important to mention that perovskite solar cells could work efficiently under a weak indoor light illumination [92], with an efficiency over 40% [93]. The generated power will be high enough when fabricated in a large area for various wearable and portable sensors and electronics. Although dye-sensitized solar cells and polymer solar cells could also both function under a weak light, their performance is much lower than perovskite solar cells [94]. It is the large range band gap tunability and also the high absorbance coefficient that ensures the high energy conversion under indoor light. One should know that indoor light is mainly in the visible range rather than the standard solar light spectra [95].

2.7 Conclusions

In conclusion, in this chapter, we describe the current basic understanding of the operation mechanism of photovoltaic cells. The generation of excited carriers under light illumination and the transportation and extraction of carriers by respective selection layers are the key processes. Specifically for various solar cells, the charge carrier generation, separation/extraction, collection, and non-radiative recombination processes are different. The potential that drives the movement of carriers should be maximized by functional and active materials with suitable energy level diagrams [68]. The recombination process should be minimized by passivating the front and rear surfaces with blocking layers. The three basic models for describing maximum efficiency guide the design of new materials and structures for photovoltaic cells. Except for the perovskite solar cells showing performance approaching the theoretical efficiency, both dye-sensitized and polymer solar cells are still far beyond the theoretical limit due to the unexpected non-recombination process and unbalanced energy level alignment. However, for the application of flexible and wearable solar cells, the fundamental issue is not focusing on the operation mechanism but on the realization of flexible and wearable applications by structure design. The wearable application requires the solar cells to be compatible and conformal to arbitrary surfaces. The remaining issues are still focusing on the efficient deposition of high-quality functional selective layers and photoactive layers. The proved effective strategies for improving solar cell performance have not been able to be applied to the reported wearable structured solar cells, especially in a fiber or textile format. In the following chapter, the research regarding the development of solar cells in fiber and textile format will be discussed in more detail.

References

1 Breeze, P. (2019). Chapter 13 – Solar power. In: *Power Generation Technologies*, 3e (ed. P. Breeze), 293–321. Newnes.

2 Kannan, N. and Vakeesan, D. (2016). Solar energy for future world: a review. *Renewable Sustainable Energy Rev.* 62: 1092–1105.

3 Hagfeldt, A., Boschloo, G., Sun, L. et al. (2010). Dye-sensitized solar cells. *Chem. Rev.* 110 (11): 6595–6663.

4 Green, M.A. (2009). The path to 25% silicon solar cell efficiency: history of silicon cell evolution. *Prog. Photovoltaics Res. Appl.* 17 (3): 183–189.

5 Yoshikawa, K., Kawasaki, H., Yoshida, W. et al. (2017). Silicon heterojunction solar cell with interdigitated back contacts for a photoconversion efficiency over 26%. *Nat. Energy* 2 (5): 17032.

6 Bilal, B. and Najeeb-ud-Din, H. (2021). Fundamentals of and recent advances in carrier selective passivating contacts for silicon solar cells. *J. Electron. Mater.* 50 (7): 3761–3772.

7 Facchetti, A. (2013). Polymer donor–polymer acceptor (all-polymer) solar cells. *Mater. Today* 16 (4): 123–132.

8 Kim, J.Y., Lee, J.-W., Jung, H.S. et al. (2020). High-efficiency perovskite solar cells. *Chem. Rev.* 120 (15): 7867–7918.

9 Schubert, M.B. and Werner, J.H. (2006). Flexible solar cells for clothing. *Mater. Today* 9 (6): 42–50.

10 Chu, S., Cui, Y., and Liu, N. (2016). The path towards sustainable energy. *Nat. Mater.* 16 (1): 16–22.

11 Rühle, S. (2016). Tabulated values of the Shockley–Queisser limit for single junction solar cells. *Sol. Energy* 130: 139–147.

12 Tress, W., Domanski, K., Carlsen, B. et al. (2019). Performance of perovskite solar cells under simulated temperature-illumination real-world operating conditions. *Nat. Energy* 4 (7): 568–574.

13 Wen, Z., Yeh, M.H., Guo, H. et al. (2016). Self-powered textile for wearable electronics by hybridizing fiber-shaped nanogenerators, solar cells, and supercapacitors. *Sci. Adv.* 2 (10): e1600097.

14 Green, M.A., Dunlop, E.D., Siefer, G. et al. (2022). Solar cell efficiency tables (Version 61). *Prog. Photovoltaics Res. Appl.* 31 (1): 3–16.

15 Islam, R. and Saraswat, K. (2018). Limitation of optical enhancement in ultra-thin solar cells imposed by contact selectivity. *Sci. Rep.* 8 (1): 8863.

16 Savin, H., Repo, P., von Gastrow, G. et al. (2015). Black silicon solar cells with interdigitated back-contacts achieve 22.1% efficiency. *Nat. Nanotechnol.* 10 (7): 624–628.

17 Saive, R. (2019). S-shaped current–voltage characteristics in solar cells: a review. *IEEE J. Photovolt.* 9 (6): 1477–1484.

18 Shockley, W. and Queisser, H.J. (1961). Detailed balance limit of efficiency of p-n junction solar cells. *J. Appl. Phys.* 32 (3): 510–519.

19 Tress, W. (2017). Perovskite solar cells on the way to their radiative efficiency limit – insights into a success story of high open-circuit voltage and low recombination. *Adv. Energy Mater.* 7 (14): 1602358.

20 Bertoluzzi, L., Boyd, C.C., Rolston, N. et al. (2020). Mobile ion concentration measurement and open-access band diagram simulation platform for halide perovskite solar cells. *Joule* 4 (1): 109–127.

21 Li, D., Song, L., Chen, Y. et al. (2020). Modeling thin film solar cells: from organic to perovskite. *Adv. Sci.* 7 (1): 1901397.

22 Cui, P., Wei, D., Ji, J. et al. (2019). Planar p–n homojunction perovskite solar cells with efficiency exceeding 21.3%. *Nat. Energy* 4 (2): 150–159.

23 Allen, T.G., Bullock, J., Yang, X. et al. (2019). Passivating contacts for crystalline silicon solar cells. *Nat. Energy* 4 (11): 914–928.

24 Riquelme, A.J., Valadez-Villalobos, K., Boix, P.P. et al. (2022). Understanding equivalent circuits in perovskite solar cells. Insights from drift-diffusion simulation. *Phys. Chem. Chem. Phys.* 24 (26): 15657–15671.

25 Li, Y., Ding, B., Chu, Q.Q. et al. (2017). Ultra-high open-circuit voltage of perovskite solar cells induced by nucleation thermodynamics on rough substrates. *Sci. Rep.* 7: 46141.

26 O'Regan, B. and Grätzel, M. (1991). A low-cost, high-efficiency solar cell based on dye-sensitized colloidal TiO_2 films. *Nature* 353 (6346): 737–740.

27 Chen, T., Qiu, L., Cai, Z. et al. (2012). Intertwined aligned carbon nanotube fiber based dye-sensitized solar cells. *Nano Lett.* 12 (5): 2568–2572.

28 Wu, J., Lan, Z., Lin, J. et al. (2015). Electrolytes in dye-sensitized solar cells. *Chem. Rev.* 115 (5): 2136–2173.

29 Sharma, K., Sharma, V., and Sharma, S.S. (2018). Dye-sensitized solar cells: fundamentals and current status. *Nanoscale Res. Lett.* 13 (1): 381.

30 Lobato, K., Peter, L.M., and Würfel, U. (2006). Direct measurement of the internal electron quasi-fermi level in dye sensitized solar cells using a titanium secondary electrode. *J. Phys. Chem. B* 110 (33): 16201–16204.

31 Zhang, S., Yanagida, M., Yang, X. et al. (2012). Directly determine an additive-induced shift in quasi-fermi level of TiO_2 films in dye-sensitized solar cells. *Jpn. J. Appl. Phys.* 51 (10S): 10NE15.

32 Pinto, A.L., Parola, A.J., Leal, J.P. et al. (2019). Dye-sensitized solar cells using fluorone-based ionic liquids with improved cell efficiency. *Sustainable Energy Fuels* 3 (12): 3510–3517.

33 Richhariya, G., Kumar, A., Tekasakul, P. et al. (2017). Natural dyes for dye sensitized solar cell: a review. *Renewable Sustainable Energy Rev.* 69: 705–718.

34 Cole, J.M., Pepe, G., Al Bahri, O.K. et al. (2019). Cosensitization in dye-sensitized solar cells. *Chem. Rev.* 119 (12): 7279–7327.

35 Ondersma, J.W. and Hamann, T.W. (2013). Recombination and redox couples in dye-sensitized solar cells. *Coord. Chem. Rev.* 257 (9): 1533–1543.

36 Ren, Y., Zhang, D., Suo, J. et al. (2022). Hydroxamic acid preadsorption raises efficiency of cosensitized solar cells. *Nature* 613 (7942): 60–65.

37 Hsu, C.-Y., Chen, Y.-C., Lin, R.Y.-Y. et al. (2012). Solid-state dye-sensitized solar cells based on spirofluorene (spiro-OMeTAD) and arylamines as hole transporting materials. *Phys. Chem. Chem. Phys.* 14 (41): 14099–14109.

38 Chen, J., Peng, T., Fan, K. et al. (2011). Iodine-free quasi solid-state dye-sensitized solar cells based on ionic liquid and alkali salt. *J. Mater. Chem.* 21 (41): 16448–16452.

39 Mahmood, A. (2015). Recent research progress on quasi-solid-state electrolytes for dye-sensitized solar cells. *J. Energy Chem.* 24 (6): 686–692.

40 Li, G., Zhu, R., and Yang, Y. (2012). Polymer solar cells. *Nat. Photonics* 6 (3): 153–161.

41 Hiramoto, M., Fujiwara, H., and Yokoyama, M. (1992). p-i-n like behavior in three-layered organic solar cells having a co-deposited interlayer of pigments. *J. Appl. Phys.* 72 (8): 3781–3787.

42 Sajjad, M.T., Ruseckas, A., and Samuel, I.D.W. (2020). Enhancing exciton diffusion length provides new opportunities for organic photovoltaics. *Matter* 3 (2): 341–354.

43 Yu, G., Gao, J., Hummelen, J.C. et al. (1995). Polymer photovoltaic cells: enhanced efficiencies via a network of internal donor-acceptor heterojunctions. *Science* 270 (5243): 1789–1791.

44 Zhang, Q., Chen, Z., Ma, W. et al. (2019). Optimizing domain size and phase purity in all-polymer solar cells by solution ordered aggregation and confinement effect of the acceptor. *J. Mater. Chem. C* 7 (40): 12560–12571.

45 Grancini, G., Maiuri, M., Fazzi, D. et al. (2013). Hot exciton dissociation in polymer solar cells. *Nat. Mater.* 12 (1): 29–33.

46 Zhu, L., Zhang, J., Guo, Y. et al. (2021). Small exciton binding energies enabling direct charge photogeneration towards low-driving-force organic solar cells. *Angew. Chem. Int. Ed.* 60 (28): 15348–15353.

47 Liu, J., Yin, Y., Wang, K. et al. (2022). Domain size control in all-polymer solar cells. *iScience* 25 (4): 104090.

48 Scharber, M.C., Mühlbacher, D., Koppe, M. et al. (2006). Design rules for donors in bulk-heterojunction solar cells—towards 10% energy-conversion efficiency. *Adv. Mater.* 18 (6): 789–794.

49 Doumon, N.Y., Dryzhov, M.V., Houard, F.V. et al. (2019). Photostability of fullerene and non-fullerene polymer solar cells: the role of the acceptor. *ACS Appl. Mater. Interfaces* 11 (8): 8310–8318.

50 Brabec, C.J., Cravino, A., Meissner, D. et al. (2001). Origin of the open circuit voltage of plastic solar cells. *Adv. Funct. Mater.* 11 (5): 374–380.

51 Proctor, C.M., Kuik, M., and Nguyen, T.-Q. (2013). Charge carrier recombination in organic solar cells. *Prog. Polym. Sci.* 38 (12): 1941–1960.

52 Zhang, M., Zhu, L., Zhou, G. et al. (2021). Single-layered organic photovoltaics with double cascading charge transport pathways: 18% efficiencies. *Nat. Commun.* 12 (1): 309.

53 Zhu, L., Zhang, M., Xu, J. et al. (2022). Single-junction organic solar cells with over 19% efficiency enabled by a refined double-fibril network morphology. *Nat. Mater.* 21 (6): 656–663.

54 Kojima, A., Teshima, K., Shirai, Y. et al. (2009). Organometal halide perovskites as visible-light sensitizers for photovoltaic cells. *JACS* 131 (17): 6050–6051.

55 Im, J.-H., Lee, C.-R., Lee, J.-W. et al. (2011). 6.5% efficient perovskite quantum-dot-sensitized solar cell. *Nanoscale* 3 (10): 4088–4093.

56 Kim, H.-S., Lee, C.-R., Im, J.-H. et al. (2012). Lead iodide perovskite sensitized all-solid-state submicron thin film mesoscopic solar cell with efficiency exceeding 9%. *Sci. Rep.* 2 (1): 591.

57 Lee, M.M., Teuscher, J., Miyasaka, T. et al. (2012). Efficient hybrid solar cells based on meso-superstructured organometal halide perovskites. *Science* 338 (6107): 643–647.

58 Stranks, S.D., Eperon, G.E., Grancini, G. et al. (2013). Electron-hole diffusion lengths exceeding 1 micrometer in an organometal trihalide perovskite absorber. *Science* 342 (6156): 341–344.

59 Xing, G., Mathews, N., Sun, S. et al. (2013). Long-range balanced electron- and hole-transport lengths in organic-inorganic $CH_3NH_3PbI_3$. *Science* 342 (6156): 344–347.

60 Zhao, Y. and Zhu, K. (2013). Charge transport and recombination in perovskite $(CH_3NH_3)PbI_3$ sensitized TiO_2 solar cells. *J. Phys. Chem. Lett.* 4 (17): 2880–2884.

61 Baranowski, M. and Plochocka, P. (2020). Excitons in metal-halide perovskites. *Adv. Energy Mater.* 10 (26): 1903659.

62 Qiu, L., Ono, L.K., and Qi, Y.B. (2018). Advances and challenges to the commercialization of organic–inorganic halide perovskite solar cell technology. *Mater. Today Energy* 7: 169–189.

63 Hellmann, T., Das, C., Abzieher, T. et al. (2020). The electronic structure of MAPI-based perovskite solar cells: detailed band diagram determination by photoemission spectroscopy comparing classical and inverted device stacks. *Adv. Energy Mater.* 10 (42): 2002129.

64 Jiang, Y., Leyden, M.R., Qiu, L. et al. (2018). Combination of hybrid CVD and cation exchange for upscaling Cs-substituted mixed cation perovskite solar cells with high efficiency and stability. *Adv. Funct. Mater.* 28 (1): 1703835.

65 Caprioglio, P., Stolterfoht, M., Wolff, C.M. et al. (2019). On the relation between the open-circuit voltage and quasi-fermi level splitting in efficient perovskite solar cells. *Adv. Energy Mater.* 9 (33): 1901631.

66 McMeekin, D.P., Sadoughi, G., Rehman, W. et al. (2016). A mixed-cation lead mixed-halide perovskite absorber for tandem solar cells. *Science* 351 (6269): 151–155.

67 Zhang, X., Turiansky, M.E., Shen, J.-X. et al. (2022). Defect tolerance in halide perovskites: a first-principles perspective. *J. Appl. Phys.* 131 (9): 090901.

68 Qiu, L., He, S., Ono, L.K. et al. (2020). Progress of surface science studies on ABX_3-based metal halide perovskite solar cells. *Adv. Energy Mater.* 10 (13): 1902726.

69 Richardson, G., O'Kane, S.E.J., Niemann, R.G. et al. (2016). Can slow-moving ions explain hysteresis in the current–voltage curves of perovskite solar cells? *Energy Environ. Sci.* 9 (4): 1476–1485.

70 Zai, H., Ma, Y., Chen, Q. et al. (2021). Ion migration in halide perovskite solar cells: mechanism, characterization, impact and suppression. *J. Energy Chem.* 63: 528–549.

71 Ren, X., Zhang, L., Yuan, Y. et al. (2021). Ion migration in perovskite solar cells. *J. Semicond.* 42 (1): 010201.

72 Zhu, W., Wang, S., Zhang, X. et al. (2022). Ion migration in organic-inorganic hybrid perovskite solar cells: current understanding and perspectives. *Small* 18 (15): e2105783.

73 Bi, E., Song, Z., Li, C. et al. (2021). Mitigating ion migration in perovskite solar cells. *Trends Chem.* 3 (7): 575–588.

74 Qiu, L., Jiang, Y., Sun, X. et al. (2014). Surface-nanostructured cactus-like carbon microspheres for efficient photovoltaic devices. *J. Mater. Chem. A* 2 (36): 15132.

75 Qiu, L., Wu, Q., Yang, Z. et al. (2015). Freestanding aligned carbon nanotube array grown on a large-area single-layered graphene sheet for efficient dye-sensitized solar cell. *Small* 11 (9-10): 1150–1155.

76 Qin, F., Sun, L., Chen, H. et al. (2021). 54 cm(2) Large-area flexible organic solar modules with efficiency above 13. *Adv. Mater.* 33 (39): e2103017.

77 Green, M.A., Dunlop, E.D., Hohl-Ebinger, J. et al. (2022). Solar cell efficiency tables (Version 60). *Prog. Photovoltaics* 30 (7): 687–701.

78 Zhang, Z., Yang, Z., Wu, Z. et al. (2014). Weaving efficient polymer solar cell wires into flexible power textiles. *Adv. Energy Mater.* 4 (11): 1301750.

79 Liu, P., Gao, Z., Xu, L. et al. (2018). Polymer solar cell textiles with interlaced cathode and anode fibers. *J. Mater. Chem. A* 6 (41): 19947–19953.

80 Kaltenbrunner, M., White, M.S., Głowacki, E.D. et al. (2012). Ultrathin and lightweight organic solar cells with high flexibility. *Nat. Commun.* 3 (1): 770.

81 Park, S., Heo, S.W., Lee, W. et al. (2018). Self-powered ultra-flexible electronics via nano-grating-patterned organic photovoltaics. *Nature* 561 (7724): 516–521.

82 Li, Z., Klein, T.R., Kim, D.H. et al. (2018). Scalable fabrication of perovskite solar cells. *Nat. Rev. Mater.* 3 (4): 18017.

83 Jiang, Y., He, S., Qiu, L. et al. (2022). Perovskite solar cells by vapor deposition based and assisted methods. *Appl. Phys. Rev.* 9 (2): 021305.

84 Qiu, L., He, S., Jiang, Y. et al. (2021). Metal halide perovskite solar cells by modified chemical vapor deposition. *J. Mater. Chem. A* 9 (40): 22759–22780.

85 Liu, Z., Qiu, L., Juarez-Perez, E.J. et al. (2018). Gas-solid reaction based over one-micrometer thick stable perovskite films for efficient solar cells and modules. *Nat. Commun.* 9 (1): 3880.

86 Kaltenbrunner, M., Adam, G., Glowacki, E.D. et al. (2015). Flexible high power-per-weight perovskite solar cells with chromium oxide-metal contacts for improved stability in air. *Nat. Mater.* 14 (10): 1032–1039.

87 Lee, G., Kim, M.-C., Choi, Y.W. et al. (2019). Ultra-flexible perovskite solar cells with crumpling durability: toward a wearable power source. *Energy Environ. Sci.* 12 (10): 3182–3191.

88 Schileo, G. and Grancini, G. (2021). Lead or no lead? Availability, toxicity, sustainability and environmental impact of lead-free perovskite solar cells. *J. Mater. Chem. C* 9 (1): 67–76.

89 Di Giacomo, F., Fakharuddin, A., Jose, R. et al. (2016). Progress, challenges and perspectives in flexible perovskite solar cells. *Energy Environ. Sci.* 9 (10): 3007–3035.

90 Khenkin, M.V., Anoop, K.M., Katz, E.A. et al. (2019). Bias-dependent degradation of various solar cells: lessons for stability of perovskite photovoltaics. *Energy Environ. Sci.* 12 (2): 550–558.

91 Abate, A. (2017). Perovskite solar cells go lead free. *Joule* 1 (4): 659–664.

92 Wojciechowski, K. and Forgács, D. (2022). Commercial applications of indoor photovoltaics based on flexible perovskite solar cells. *ACS Energy Lett.* 7 (10): 3729–3733.

93 Muhammad, B.T., Kar, S., Stephen, M. et al. (2022). Halide perovskite-based indoor photovoltaics: recent development and challenges. *Mater. Today Energy* 23: 100907.

94 Wang, K.L., Zhou, Y.H., Lou, Y.H. et al. (2021). Perovskite indoor photovoltaics: opportunity and challenges. *Chem. Sci.* 12 (36): 11936–11954.

95 Bulloch, A., Wang, S., Ghosh, P. et al. (2022). Hysteresis in hybrid perovskite indoor photovoltaics. *Philos. Trans. A Math Phys. Eng. Sci.* 380 (2221): 20210144.

3

Flexible Film Solar Cells

3.1 Overview of Flexible Film Solar Cells

Crystal silicon solar cells have dominated the photovoltaic market in the last three decades, which are produced based on silicon wafers. However, the thick silicon wafer possesses 50% of the cost of the entire module due to its low absorption coefficient. One feasible way to reduce the cost is replacing the Si wafter with other semiconductors with high absorption coefficients that are suitable for thin film solar cells. Based on this strategy, lots of materials have been studied for thin film solar cells, such as doping amorphous hydrogenated Si (a-Si: H), CdTe and copper indium diselenide ($CuInSe_2$), polymers, metal halide perovskite, etc. Among these thin film solar cells, the dye-sensitized solar cells (DSSCs), organic solar cells (OSC), and perovskite solar cells (PSCs) can be fabricated in solution processing methods at low cost. In addition, these devices can be fabricated on flexible and conductive substrates for lightweight and bendable flexible film solar cells, which show great potential in outdoor applications [1–4]. For instance, they are widely used as the power sources of various portable and wearable electronics, such as solar-powered umbrellas [5], tents, and [6] bags [7]. Therefore, we present a review of flexible thin film solar cells, including the following aspects: (i) Materials for flexible transparent conductive electrodes, such as metal oxides, polymers, carbon nanomaterials, and other materials. (ii) Fabrication techniques for solution processable thin-film solar cells. (iii) Recent progress of flexible OSCs, DSSCs, and PSCs. Finally, critical challenges and perspectives are discussed for flexible film solar cells.

3.2 Flexible Transparent Conductive Electrodes

Flexible transparent conducting electrodes (TCEs) are the foundation for constructing flexible solar cells, which should possess several vital strengths: (i) high electrical conductivity along with low sheet resistance to reduce device series resistance; (ii) high optical transparency for photons; (iii) appropriate work function for better charge extraction; (iv) high flexibility; and (v) excellent endurance for water, oxygen, and mechanical stress. Herein, we mainly discuss the key parameters (conductivity

Wearable Solar Cells: Mechanisms, Materials, and Devices, First Edition.
Hao Sun, Zhibin Yang, Longbin Qiu, and Huisheng Peng.

and transparency) of transparent conductive electrodes. The high carrier density is required for high electrical conductivity rather than high optical transparency, whereas this physical contradiction usually restricts their compatibility [8]. The reported transparent conductive electrodes include metal oxides, conductive polymers, and carbon nanomaterials.

3.2.1 Metal Oxides

Transparent conducting oxides, including tin-doped indium oxide (ITO), fluorine-doped tin oxide (FTO), aluminum-doped zinc oxide (AZO), and Ga-doped ZnO (GZO), are usually prepared via electromagnetic sputtering techniques [9, 10]. In particular, ITO is predominant as TCE in flexible solar cells [11]. Zardetto et al. analyzed the electrical conductivity, optical transmittance, and thermal and mechanical endurance of ITO/polyethylene terephthalate (PET) and ITO/polyethylene naphthalate (PEN) [12]. With an annealing temperature below 235 °C, the sheet resistances of ITO/PET and ITO/PEN are maintained at ~15 Ω/□, which is close to 10 Ω/□ for ITO/glass. When the annealing temperature reached 250 °C, the resistance of the deformed PEN substrate could increase by almost 20-fold. Meanwhile, the critical temperature was only 150 °C for ITO/PET (Figure 3.1a). In addition, the transmittances of the ITO/PET and ITO/PEN were slightly lower than the glass-based counterparts in the visible light range. Still, they were much lower in the ultraviolet light range, which was attributed to the strong ultraviolet absorption of polymer substrates. For the mechanical property, when they are bent

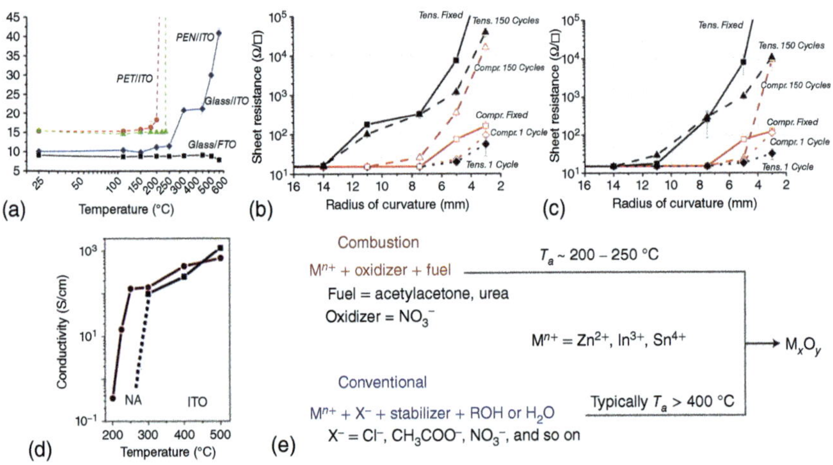

Figure 3.1 (a) Room temperature sheet resistances of conductive PET/ITO, PEN/ITO foils, and glass/ITO, glass/FTO substrates after undergoing a thermal treatment at different temperatures for 30 minutes. The sheet resistance of PET/ITO (b) and PEN/ITO (c) as a function of radius of curvature under tensile (in black) and compressive (in red) stress. (d) The electrical conductivity of ITO films versus annealing temperature. The red and blue lines refer to ITO films synthesized via combustion and conventional processes, respectively. (e) Schematics of the combustion process compared to conventional sol–gel processes. Source: Zardetto [12]/John Wiley & Sons and Kim et al. [13]/Springer Nature.

at a curvature radius lower than 7 mm, both ITO/PET and ITO/PEN showed sharply increased sheet resistances (Figure 3.1b,c).

The large bandgaps of the most transparent conducting oxides result in poor electrical conductivities, which negatively impacts the power conversion efficiencies (PCEs) of flexible solar cells. In principle, n-type or p-type doping via vacuum fabrication can improve the carrier mobility of transparent conducting oxides [14]. After decades of development, sol–gel approach becomes an ideal substitution. With the adjustment of the dopant concentration in precursor solutions, the doping concentration of transparent conducting oxides can be readily modulated [15]. Unfortunately, owing to the charge scattering at the grain boundaries of small nanoparticles, the deposited films via sol–gel method require a post-annealing process for better crystallinity and conductivity, which hinders using most flexible substrates with poor thermal stability. To address this issue, a laser annealing technique has been developed to reduce the sheet resistances under low-temperature conditions [16, 17]. For instance, Kim et al. developed a combustion process by mixing fuels and oxidizers, which not only reduced the annealing temperature of metal oxides to <250 °C, but also improved the conductivity of ITO films to 130 S/cm (Figure 3.1d) [13]. The contrastive schematics between the combustion process and the conventional sol-gel process is illustrated in Figure 3.1e.

Despite the high efficiencies achieved by transparent conducting oxides for flexible solar cells, the intrinsic brittleness of crystal is prone to cause a cracking problem that sacrifices the mechanical stability of the resulting flexible devices. To this end, conductive polymers and carbon materials with better flexibility have drawn broad attention in recent years.

3.2.2 Conductive Polymers

Conducting polymers possess the advantages of high electrical conductivity, solution processibility, low cost, and good environmental stability. A variety of conducting polymers have been synthesized and applied in different fields, such as poly (3,4-ethylene dioxythiophene) (PEDOT), polyaniline (PANI), polyacetylene (PA), poly(*p*-phenylene-vinylene) (PPV), polypyrrole (PPY), polyfuran (PF), and other polythiophene (PTh) derivatives.

Among various types of conductive polymers, PEDOT is currently the most possible material to replace the conventional ITO electrode for TCE application. PEDOT and its derivatives possess excellent properties of high electrical conductivity and stability. Normally, polyelectrolyte PSS is added to the PEDOT solution for better solubility. PEDOT: PSS has been widely used in flexible solar cells for its high conductivity, optical transmittance, and roll-to-roll compatibility [18, 19].

However, the pristine PEDOT: PSS films usually exhibit poor electrical conductivities, which are determined by many factors, for instance, the phase separation between PEDOT and PSS, the crystallization of PEDOT, and the ratio of PSS. Considering the insulating nature of PSS, reducing the ratio of PSS could increase the conductivity. However, the optimization of the ratio of PSS in PEDOT: PSS only resulted in an improved conductivity of 1 S/cm, which was not qualified for TCE applications. Thus, great efforts have been made to improve the conductivity of PEDOT: PSS. For instance, post-surface treatment with acids or doping additives such as tetrahydrofuran (THF), ethylene glycol (EG), *N*,*N*-dimethylformamide (DMF), dimethylsulfoxide (DMSO), sorbitol, glycerol, surfactants, and ionic liquids have been explored. In 2008, Na et al. improved the PEDOT: PSS film conductivity to >500 S/cm through the optimization of the doping ratio of DMSO [20]. Upon adjusting the doping parameter of such dopants within PEDOT: PSS films, the conductivity can be further improved. For instance, Hu et al. raised the conductivity of PEDOT: PSS to >4000 S/cm by applying 5 wt% glycol and 0.5 wt% Zonyl FS-300 as the doping agents [21, 22]. Posttreatment had also been explored for improving the conductivity of PEDOT: PSS films. Ouyang et al. utilized dilute sulfuric acid (H_2SO_4) to enhance the conductivity of PEDOT: PSS from 0.3 to 3065 S/cm [23]. Lee et al. further enhanced the conductivity to 4380 S/cm via exposure to H_2SO_4 fume [24]. Song et al. investigated the effect on PEDOT: PSS conductivity with multiple gentle acid treatment at room temperature (Figure 3.2a) [25]. Undoubtedly, the gentle acid treatment is more desirable for printed flexible photovoltaic devices. Posttreatment with solvents or acids can further boost the conductivities of PEDOT: PSS to >6000 S/cm [27–29]. The representative PEDOT films with high conductivities have been summarized in Figure 3.2b [26].

The optical transmittance of PEDOT: PSS electrodes is studied based on parameters of film thickness, chain configuration, doping ratio, etc., which should be above 80% in the visible region for photovoltaic application [30–32]. For example, the acid treatment will effectively improve the transmittance as well as the electrical conductivity of PEDOT: PSS electrodes. As shown in Figure 3.2c, higher transparency of PEDOT: PSS films is observed in the region of $\lambda \geq 400$ nm after methanesulfonic acid treatment because removing the residual PSS moieties results in a better stack of PEDOTs [25].

3.2.3 Carbon Nanomaterials

Due to the excellent properties of high optical transmittance, electrical conductivity, and mechanical strength, carbon nanomaterials, including graphene and carbon nanotube (CNT), have been widely studied as next-generation TCEs for flexible solar cells in recent years.

3.2.3.1 Graphene

Graphene was discovered by Novoselov and Geim via a micromechanical exfoliation method in 2004. The major synthesis approaches of graphene are summarized in Figure 3.3 [33]. Chemical exfoliation is the most efficient and mature approach for graphene preparation. However, the residue oxygenated groups and defective sites lead to a lower conductivity. Therefore, liquid-phase exfoliation of graphite without the oxidation process becomes an alternative route to obtain graphene sheets. By employing high surface tension solvents or water solutions with surfactant/polymers as stabilizers, monolayer graphene flakes can be prepared through sonication [34, 35]. Moreover, some vapor phase methods are usually used to

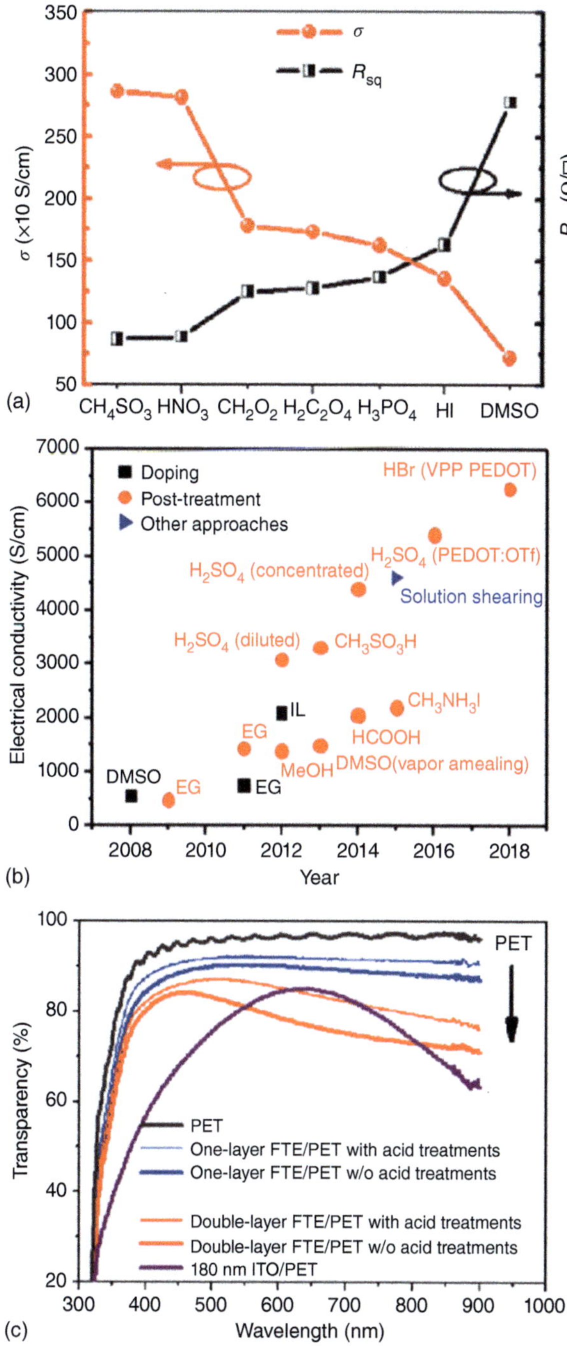

Figure 3.2 (a) Comparison of square resistance and electrical conductivity of the PEDOT: PSS with diverse acid treatments at room temperature and DMSO doping. (b) Summary of highly conductive PEDOT films obtained via doping or posttreatment in recent years. (c) Transparency spectra of PET substrates (≈200 μm), PEDOT: PSS/PET and ITO/PET substrates. Source: Song et al. [25]/John Wiley & Sons and Zhang et al. [26]/American Chemical Society.

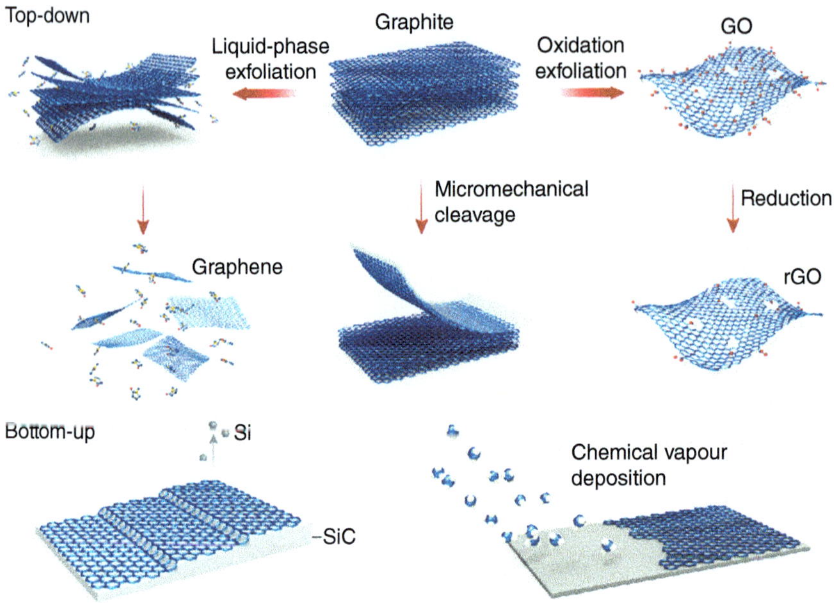

Figure 3.3 Major fabricated approaches of graphene synthesis. Source: Wang et al. [33]/Springer Nature.

fabricate large-area and uniform graphene films to replace the solution-based processed approach. Those methods mainly involve chemical vapor deposition (CVD) and plasma-enhanced chemical vapor deposition (PECVD), Where CVD is the most readily accessible and typical one [36, 37].

Due to the high conductivity and transparency of graphene, Wang et al. first adopted rGO films as a TCE for DSSCs. The ~10 nm graphene electrode exhibits a large sheet resistance of 1.8 kΩ/□ with a transparency of >70% under the wavelength of 1000–3000 nm [38]. The graphene electrode prepared with a CVD method shows an improved performance with sheet resistance of 770 Ω/□ and transmittance of 90% [39]. As shown in Figure 3.4a, with the thickness of graphene TCE increased from 6 to 30 nm, the sheet resistance reduced from 1350 to 210 Ω/□, and the transmittance reduced from 91% to 72% in the visible range [40].

In order to enhance the conductivities of graphene TCE, Sung et al. modify the graphene with MoO_3. As a result, the sheet resistance of graphene films reduced from ~2000 to 780 Ω/□, and the work function increased from 4.23 to 4.71 eV due to the hole doping effect (Figure 3.4b) [41]. Bae et al. reported the large-area (~30 in.) CVD monolayer graphene sheets with roll-to-roll production and wet-chemical doping, which exhibited a low sheet resistance of ~125 Ω/□ with ~97.4% optical transmittance. They further prepared a doped four-layer graphene film with a low sheet resistance of ~30 Ω/□ and high transmittance of ~90% through the layer-by-layer stacking method (Figure 3.4cd) [42].

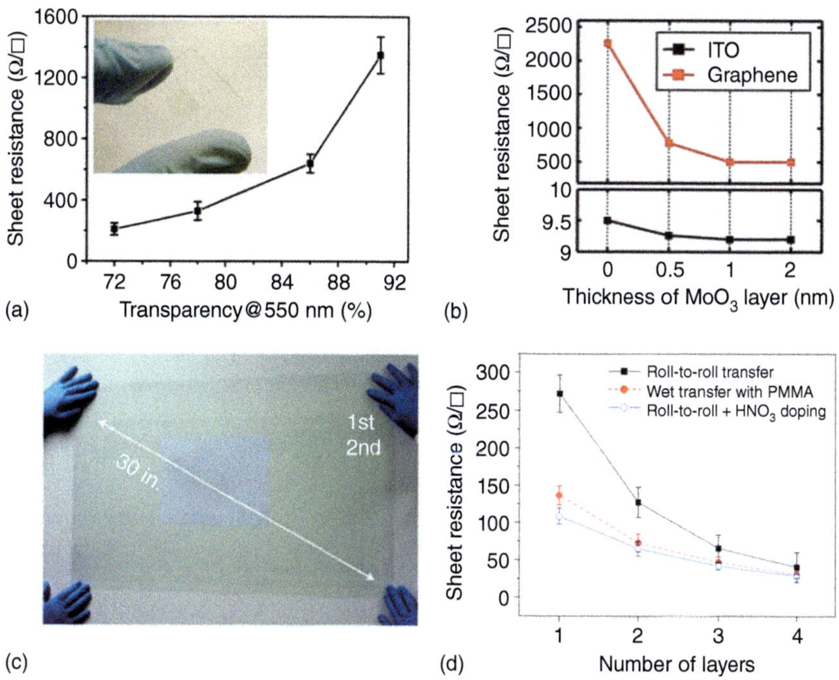

Figure 3.4 (a) Sheet resistance versus transparency curves of graphene films. (Inset: Optical image of a graphene film on glass). (b) Relationship between sheet resistance and MoO_3 thickness for graphene and ITO. (c) A transparent ultra-large area graphene film transferred on a 35-in. PET sheet and (d) Sheet resistances of transferred graphene films using different methods. Source: Wang et al. [40]/AIP Publishing, Sung et al. [41]/John Wiley & Sons, and Bae et al. [42]/Springer Nature.

3.2.3.2 Carbon Nanotube

Single- or multi-layered graphene scroll to a hollow cylinder of carbon, which can be considered as the structure of CNT [43]. Arc-discharge deposition, laser ablation, and CVD are the three main approaches to synthesizing CNTs. The CVD process, which is different from the arc discharge and laser ablation with relatively high temperatures, can prepare CNTs at relatively lower temperatures, from 500 to 1200 °C; therefore, it is more appealing for the current semiconductor industry. Moreover, upon varying the structure and composition of catalysts, the wall number, diameter, and length of CNTs can be precisely controlled in a CVD process. For instance, Yamada et al. exploited the different thicknesses of catalyst to selectively synthesize double-walled CNT (DWCNT) arrays [44].

CNTs are also studied for their applications in TCEs due to their high conductivity. Lots of methods, including vacuum-assisted filtration, spray coating, dip coating, and spin coating, are used to fabricate transparent CNT films. Pure single-walled CNT (SWCNT) films prepared via filtration or coating processes have excellent transparency and conductivity. Pasquier et al. first used the SWCNT films as transparent electrodes in polymer solar cells. The films (thickness <100 nm) fabricated by vacuum-assisted filtration exhibit higher transparency than ITO/PET at wavelengths from 250 to 850 nm [45]. Tenent et al. reported a modified ultrasonic spraying method to acquire CNTs with high uniformity. The electrical conductivity and optical transmittance of CNT films are 150 Ω/□ and 78%, respectively [46]. The electrical conductivity of CNT films can be further improved by doping, acidic treatment, adding metal nanomaterials, etc. Through $SOCl_2$ doping and PEDOT passivation, the SWCNT films showed a lower sheet resistance of 160 Ω/□ with an optical transparency of 87% [47]. Acid treatment has the combined effect of doping and residual surfactant removal, which represents another effective method to decrease the sheet resistance. For example, treated with 12 M HNO_3 for 60 minutes, the sheet resistance of CNT films was reduced to 40–70 Ω/□ with transmittance of 70–80% in the visible region (Figure 3.5a,b) [48].

On the other hand, tuning the work function of CNT films by doping process has been widely reported. Lee et al. showed the free-standing CNT films doped with trifluoromethane-sulfonic acid (TFMS) vapor for PSCs (Figure 3.5c). As shown in Figure 3.5d, after 40 seconds doping by TFMS, the sheet resistance of the CNT sheet was significantly depressed from 153 to 121 Ω/□, as well as the work function ascended from 4.75 to 4.96 eV [49]. Similarly, polymer modification could tune the conductivity and work function of CNT sheets. Zhou et al. fabricated 2 μm thick CNT: PEI film by modification with PEI/IPA solution. Its sheet resistance and work

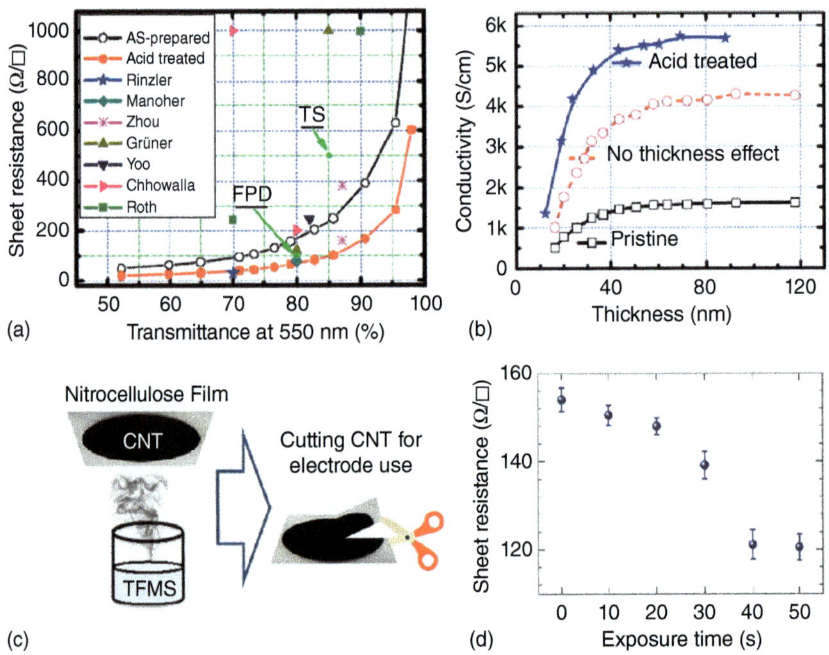

Figure 3.5 (a) Sheet resistance versus transmittance at 550 nm before and after nitric acid treatment with previously reported values for comparison, and (b) the corresponding conductivity as a function of film thickness before and after acid treatment. (c) Schematic illustration showing ex-situ doping of CNT via vaporized TFMS, and (d) measured sheet resistance with different exposure times to TFMS vapor.

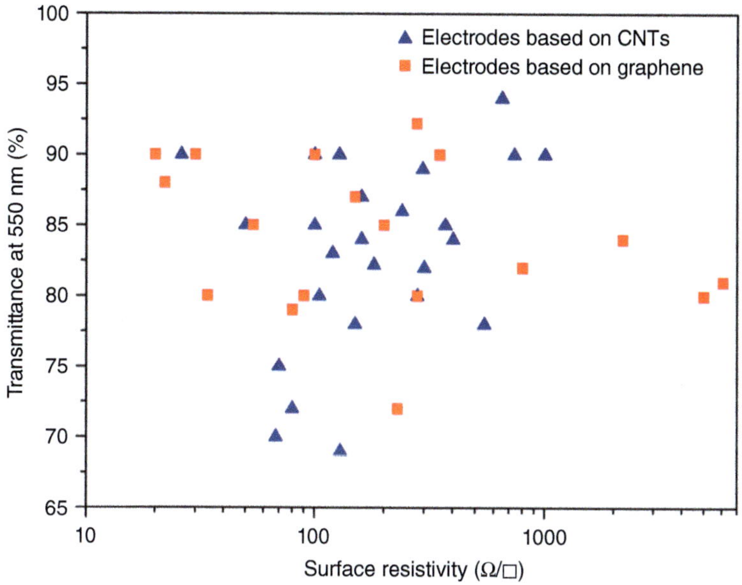

Figure 3.6 Relationship between optical transmittance and surface resistivity. Source: Yang et al. [51]/American Chemical Society.

functions were reduced to ~40.5 Ω/□ and 4.45 eV (The initial work function was 4.68 eV), respectively. As a result, the photovoltaic performance of the device is enhanced due to a smaller charge transfer barrier between the electron transporting layer and CNT films [50].

As summarized in Figure 3.6, many high-performance TCEs based on CNTs and graphene have been developed in recent years [51]. These electrodes generally possess medium sheet resistances of 60–200 Ω/□ and high transmittances of 80–90%. However, High optical transmittance and low sheet resistance are mutually contradictory, i.e. the increase in conductivity generally leads to a decrease in transmittance, and vice versa. These electrodes can be used in devices with low conductivity requirements (such as touch screens), but still cannot satisfy the demand for photovoltaic devices.

3.3 Fabrication Techniques of Flexible Film Solar Cells

The main manufacturing methods of large-area film solar cells are usually divided into addition-, subtraction-, and pattern-based techniques. Addition-based technique, such as printing, coating, and vacuum evaporation, involves depositing additional components onto the substrates without damnification. Subtraction-based technique technology, such as laser ablation and photolithography, are generally used in large-area module fabrication via removed materials from the substrates. The pattern-based technique includes wetting/de-wetting, imprinting, or bonding methods [52]. Among them, solution-processable techniques are the most

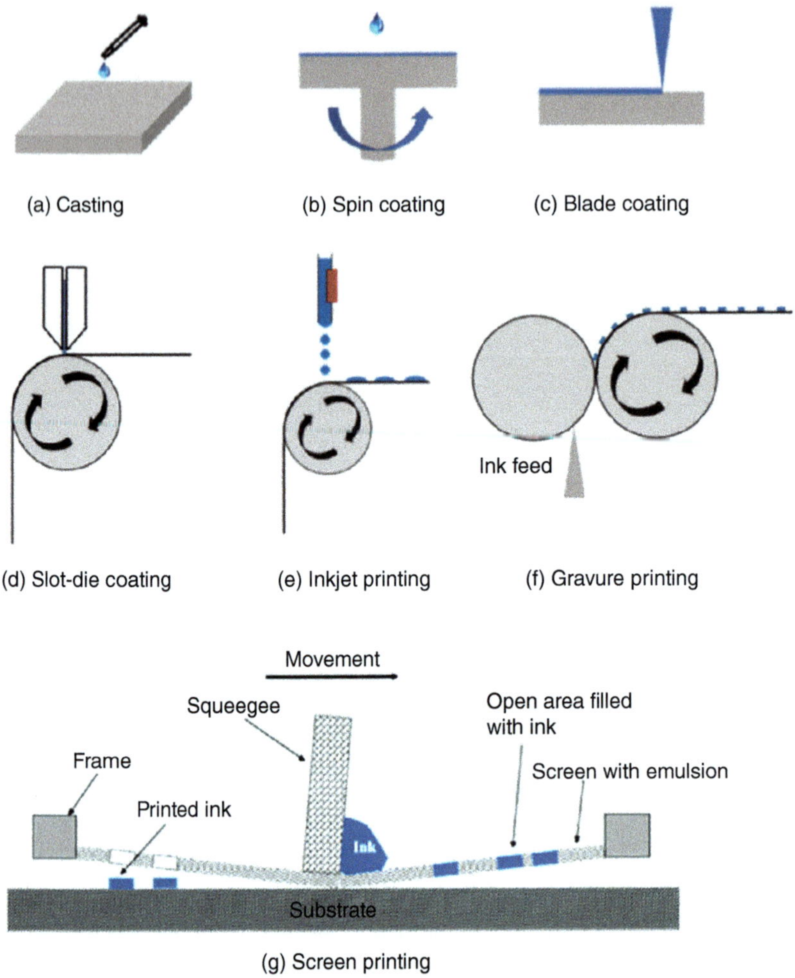

Figure 3.7 Single substrate-based fabrication technologies. (a) Casting. (b) Spin coating. (c) Blade coating. (d) Slot-die coating. (e) Inkjet printing. (f) Gravure printing. (g) Screen printing.

generally and widely used in practical applications and, therefore, summarized in this section (Figure 3.7).

3.3.1 Spin Coating

To prepare a thin film by the spin-coating method, dropping ink onto the substrate and spinning the substrate to volatilize most of the solvent can form a thin film. The film thickness and surface/bulk morphology are highly related to the spinning speed, annealing temperatures, and the physical characteristics of the ink, e.g. concentration, viscosity, volatility, etc. Zabihi et al. reviewed the morphology of PEDOT:PSS thin films deposited through the spin-coating method. The significant change in surface profile and topography can be distinguished, whereas the films dried at high temperature resemble shrunken lamellae, and the films dried at room temperature exhibit needle-like (spike) or blade-like crystals. The shrunken lamella and crystals are interpreted as PSS-rich layer and PEDOT oligomers phase area. In the meantime, the film structure and morphology can be smoothed by spinning speed [53]. Normally, the film thickness can be elaborated by the empirical relationship:

$$d = k\omega^{\alpha} \tag{3.1}$$

where ω is the angular velocity, k and α are empirical constants related to solvent properties, and d is the film thickness [54]. Generally, the spin-coated film morphology is high quality and reproducible; thus, the small-area rigid and flexible solar cell-based spin-coating can achieve high PCE.

Although spin coating is widely used in labs for thin film research, it is unsuitable for large-scale production due to its disadvantages, including nonhomogeneity, solution waste, and incompatibility with continuous production.

3.3.2 Blade Coating

Being regarded as an alternative method to spin coating, blade coating is the first attempt toward the large-area fabrication of thin films. It has a high utilization ratio of material (>95%) and a roll-to-roll compatible feature. Based on a blade dragging process on the substrate, homogeneous thin films can form with adjustable thickness by varying the distance (usually 10–500 μm) between blade and substrate. Besides, the film thickness is also related to surface energy of substrate, physical property (tension and viscosity) of ink, geometry of the blade, as well as the blade coating speed. The thin film thickness is elaborated by the following equation:

$$d = 0.5\left(g\frac{c}{\rho}\right) \tag{3.2}$$

where d, g, c, and ρ are the film thickness, the distance between the blade and substrate, the concentration of the ink, and the density of the material, respectively. The blade speed is usually controlled in the range of 1–100 mm/s. As a result of the very fast blading speed and low deposition temperature, the ink cannot be dried immediately. Thus, a nitrogen knife with a fixed distance is usually applied to accelerate the drying process at room temperature.

The blade-coating method has been widely used in PSCs in recent years. Initially, it is not easy to prepare a high-quality perovskite film via the blade coating method at room temperature. Therefore, comprehensive optimizations (such as post-annealing, solvent engineering, gas quenching, and vacuum-dry) have been attempted to control the film quality through tuning the drying and crystallization processes. In most cases, substrate temperatures are commonly set between 40 and 165 °C for blade coating. Wu et al. reported that a high-quality $MAPbI_3$ perovskite film could be achieved by a high-temperature annealing process at $MAPbI_3$ at 150 °C [55]. To accelerate the solvent evaporation for rapid crystallization of the perovskite, N_2 or dry air flow is employed for blade coating, which is compatible with a roll-to-roll process for fast production [56]. For instance, Gao et al. utilized a multi-flow air knife to rapidly dry the perovskite film in ambient air and obtained uniform perovskite films with a low surface roughness of 4.98 nm and size of $10 \times 10\,mm^2$ [57]. Alternatively, ink engineering plays a key role in the blade-coating process of perovskite films. It involves composition management of ink, solvent selection, additive modification, etc. For ink engineering, Deng et al. used a combination of volatile and noncoordinating solvents (acetonitrile and 2-methoxy ethanol) with nonvolatile and coordinating solvents (DMSO) by a certain ratio to realize the uniform perovskite films under a fast blading speed of 99 mm/s at room temperature (Figure 3.8a). Through forming strong bonding with Pb^{2+} in perovskite precursor, the coordinating solvent results in the formation of an intermediate phase to promote the crystallinity of perovskite, while N_2 blowing could accelerate the evaporation of the volatile solvent to achieve smooth perovskite films. Figure 3.8b shows a photograph image of an as-coated perovskite film on a 15 cm by 15 cm flexible substrate. Finally, the produced perovskite modules with an area of over $60\,cm^2$ acquired over 16% efficiency [58]. In addition, additive engineering is another way to control the crystallization of the perovskite film. Deng et al. introduced a surfactant of L-α-Phosphatidylcholine into the precursor and prepared a homogeneous film with root-mean-square roughness of 14.5 nm over 1 cm (see Figure 3.8c). The derived perovskite PV module achieved stabilized efficiencies of 15.3% and 14.6% with aperture areas of 33.0 and $57.2\,cm^2$, respectively [59].

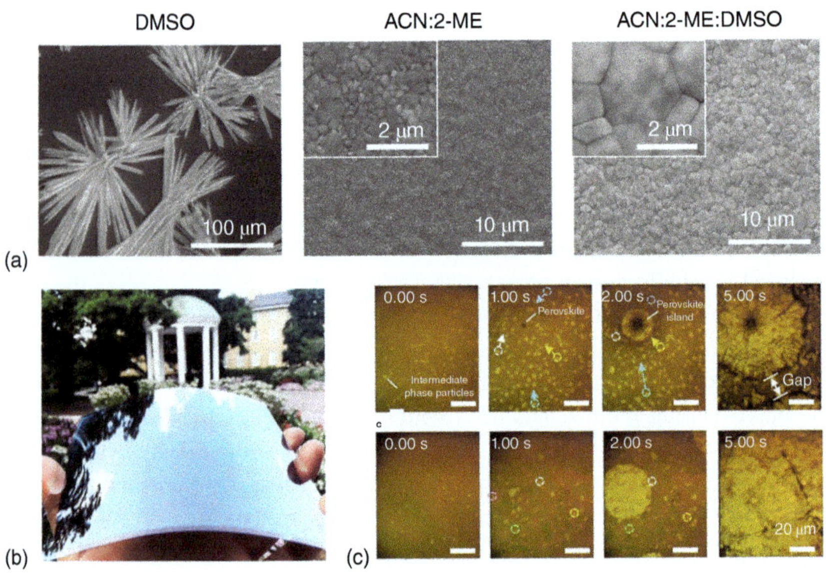

Figure 3.8 (a) SEM images of perovskite films prepared with different solvent or solvent mixtures. (b) Photograph image of an as-coated perovskite film on a 15 cm by 15 cm flexible substrate. (c) *In situ* microscopy observations of drying process for perovskite ink without (top row) and with (bottom row) surfactants. Source: Deng et al. [58]/American Association for the Advancement of Science – AAAS and Deng et al. [59]/Springer Nature.

3.3.3 Slot-Die Coating

Slot-die coating is a kind of film deposition technique similar to blade coating but with continuous ink supply. As shown in Figure 3.11a, the slot-die head is consisted of two moveable metal blades and possesses a slit on the bottom side. The slot-die head and ink reservoir are interconnected through a solution pumping system and supply ink at a certain rate to the slit. The film quality deposited via slot-die-coated method is determined by the capillary number (C_a) of the system. When C_a is below 0.02, the parameters of gap height (h_0) and actual line velocity of web/roller (V) will emerge with a significant effect on film thickness. For a system with high-viscosity solutions, the flow rate (f_r) of ink and V become the key factors of coating process [47]. The film thickness can be calculated by the formula as follows:

$$d = \frac{f_r}{nwV} \times \frac{\rho_l}{\rho_s} \tag{3.3}$$

where d, f_r, n, w, V, ρ_l, and ρ_s are the film thickness, the flow rate, the number of coating stripes, width of one coating stripe, actual line velocity of web/roller, density of the solid in liquid, and density of the solid in a dried film, respectively. Compared to the other methods, slot-die coating has several advantages: (i) It could print striped patterns with a resolution of several hundreds of micrometers through attaching a mask onto a slit. (ii) It can prevent changes in the solution concentration because all process is generated in a closed system. (iii) It is possible to improve the fluid uniformity by giving a high temperature to the solution. (iv) No precursor solution is wasted.

Similar to blade-coating, numerous approaches, including substrate temperature control, solvent engineering, gas quenching, and additive engineering, have been used in the slot-die coating process to control nucleation and crystallization for achieving high-quality perovskite films. Bu et al. fabricated a flexible perovskite solar cell *via* slot-die coating method (Figure 3.9). The interfacial layer SnO_2 in the device was passivated with KOH to reduce the trap density, which yielded a high PCE of 15.22% [60].

Besides the active layer, the charge transport layers can be coated via the slot-die method. For example, several hole transport materials (HTM) including the polymers such as PEDOT: PSS [61, 62], P3HT [63], and PTAA [59] or small molecules like Spiro-OMeTAD and BifluoOMeTAD [62, 64], and several electron transport materials (ETM) consisting of the C_{60} [61, 65], PCBM [61, 65] and inorganic metal oxide nanoparticles like NiO_x, [65] ZnO [62–64], SnO [66], or TiO_2 have been successfully deposited by slot-die coating.

3.3.4 Screen Printing

As a versatile printing technique, screen printing is suitable for 2D pattern printing. A moved squeegee can press the ink onto the empty areas through a stencil mesh to form patterns. Therefore, the layer thickness is highly related to the screen volume, which can be calculated by following the empirical formula:

$$d = Vk_p\frac{c}{\rho} \tag{3.4}$$

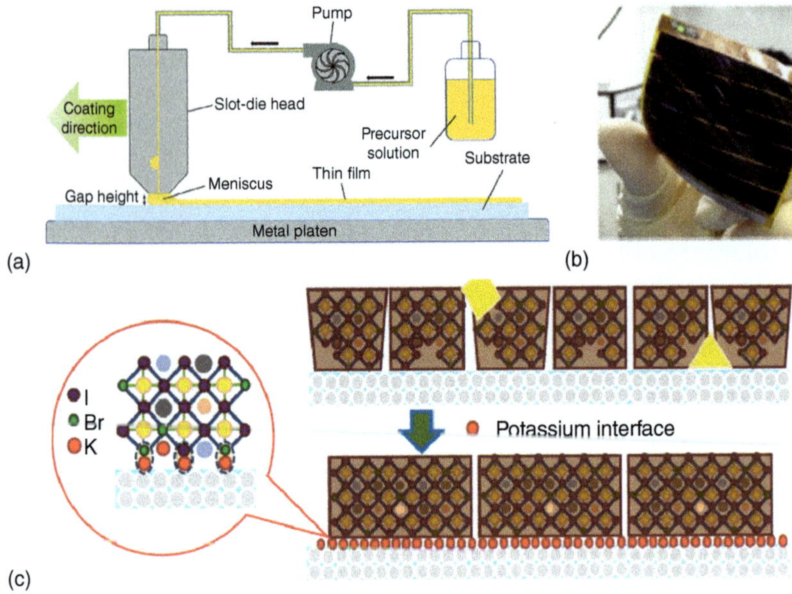

Figure 3.9 (a) Schematic illustration of the key steps involved in slot-die coating method. (b) Photograph of the flexible PSC. (c) Schematic showing the effect of KOH on the perovskite growth process. Source: (b) Bu et al. [60]/Springer Nature.

where d, V, k_p, c, and ρ are the film thickness, the screen volume, the material pick-out ratio, the solution concentration, and the material density of the film, respectively. Screen printing requires a relatively high viscosity, ranging from 50 to 5000 cP, to avoid ink flow between the blank areas on the screen. Meanwhile, the ink should be low volatility to avoid drying during the ink flowing and film formation process [67]. Screen printing is usually used to fabricate metallic electrodes because it can achieve relatively thicker film with 10–500 μm than other printing technologies. In addition, it combines the advantages of low production cost and high throughput, all of which make it a promising candidate for all-printed flexible PV modules for industrial-scale fabrication.

As mentioned above, solution-processed screen printing has been widely used to prepare thicker films, such as screen-printed PEDOT: PSS and rGO transparent electrodes, as well as many carrier transport layers. As shown in Figure 3.10a, the highly efficient rotary screen printing can provide very fast printing speeds of >10 m/min for PEDOT: PSS films of flexible organic solar modules. It is worth mentioning that the mesh-structured transparent electrode not only shows higher optical transmittance but also obtains a lower sheet resistance of 60 Ω/□ on PET substrates than ITO/PET [68]. The schematic and microscopic images of mesh-structured PEDOT: PSS transparent electrodes fabricated by screen printing procedure were respectively displayed in Figure 3.10b,c [69]. Upon adjusting the mesh count, stainless steel mask thickness, as well as wire diameter, the PEDOT: PSS meshes with different width/period ratios can be acquired. The screen-printed GO patterns on flexible PET substrates are depicted in Figure 3.10d, further

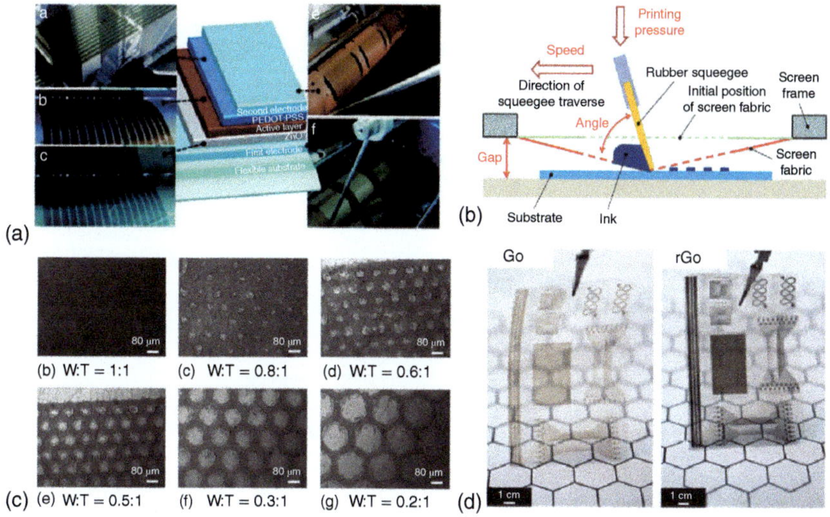

Figure 3.10 (a) Roll-to-roll-processed organic solar modules with screen-printed PEDOT: PSS TEs. (b) Schematics of the screen-printing process of PEDOT: PSS meshes, and (c) microscopic images of screen-printed PEDOT: PSS meshes with different width/period ratios. Source: Heo et al. [62]/John Wiley & Sons, Inc. (d) Optical images of screen-printed GO (left) and rGO (right) films on PET substrate. Source: Overgaard et al. [70]/John Wiley & Sons, Inc.

reduced to rGO (right) with HI under gentle conditions. Through this approach, the hundreds of nanometer thickness of graphene transparent electrodes can be simply produced with large-scale patterns [70]. In addition, the screen printing approach can achieve stripe-type transparent electrodes, which acquire the higher efficiency of large-area solar cells than pain-type devices [71].

3.3.5 Roll-to-Roll System

In contrast with vacuum deposition methods, the solution-processed roll-to-roll manufacturing system depends on simple technique and low cost for fabricating a variety of industrial products. As shown in Figure 3.11, the roll-to-roll system comprises an unwinder and rewinder, an edge-position controller, dancer roll, in-feeder, preheater, dryers, and out-feeder [72]. Besides optimizing the precursor solution, it is also necessary to tune the web handling of different roll-to-roll system parts for realizing high-quality flexible PSCs. For instance, the eccentricity and velocity of the backup roller, as well as the solution flow rate, are vital for controlling the thickness of active layers. The effective drying duration can be altered by the speed of the roller or the length of the drying section. Furthermore, the surface tension between the substrate and solution also affects the quality and thickness of the obtained film [73].

Many deposition approaches, including hot casting, antisolvent deposition, and air knife, have been widely applied in the Roll-to-Roll system for fabrication of flexible PSCs. Kim et al. recently reported all-printed flexible PSCs with PCEs of 9.7%

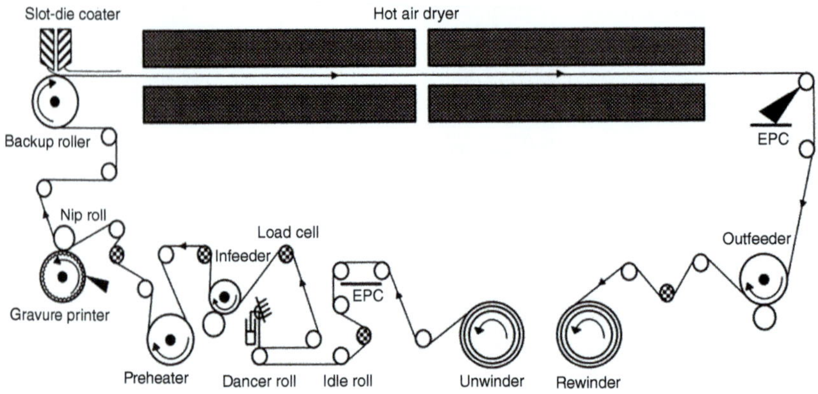

Figure 3.11 General configuration of a roll-to-roll printing and coating system.

based on the structure of ITO/SnO_2/$MAPbI_3$/spiro-OMeTAD/Au layers. First, SnO_2 and PbI_2/DMSO solution were successively printed on the PEN substrate, and then DMSO was extracted by IPA bath to generate a porous PbI_2 film. Then, the PbI_2 film was immersed in a bath of MAI/IPA to form $MAPbI_3$ film [74]. In their other work, flexible solar cells with a multi-layer of SnO_2/perovskite/P3HT were fabricated by the roll-to-roll process and achieved a high PCE of the PSC of 13.8% [75].

3.4 Flexible Organic Solar Cells

Organic materials are highly suitable for flexible photovoltaics due to their special merits of flexibility. Normally, a flexible organic solar cell (OSC) mainly consists of a flexible substrate, TCE, active layer (donor-acceptor blend), electron/hole transporting layers (ETL/HTL), top electrode, and passivation layers (Figure 3.12a) [76]. The mechanism of OSCs is depicted in Figure 3.12b, which can be divided into six key steps [77]. (i) The photoactive layer absorbs photons to produce bound electron–hole pairs (excitons). (ii) Excitons diffuse toward the donor/acceptor interface. (iii) Energy offset drives exciton to dissociate into free holes and electrons at the interfaces. (iv) Free holes and electrons are transmitted along the opposed transport orbit under an internal electric field. (v) Electrodes harvest holes and electrons. (vi) Exciton recombination leads to the loss of excitons.

Generally speaking, flexible OSCs exhibit lower PCEs than rigid ones due to the different transparency of the substrates. On the other hand, the flexible OSCs require higher mechanical stability than rigid ones due to their application scenarios for wearing. Herein, photovoltaic performance and mechanical robustness are the main concerns in studying flexible OSCs.

Generally, upon employing a p-type polymer as the electron donor and a n-type small molecule as the electron acceptor, OSCs with bulk heterojunction active layer achieve great grades over a long period. However, from a material point of view, the dominated fullerene and its derivatives as electron acceptors have the

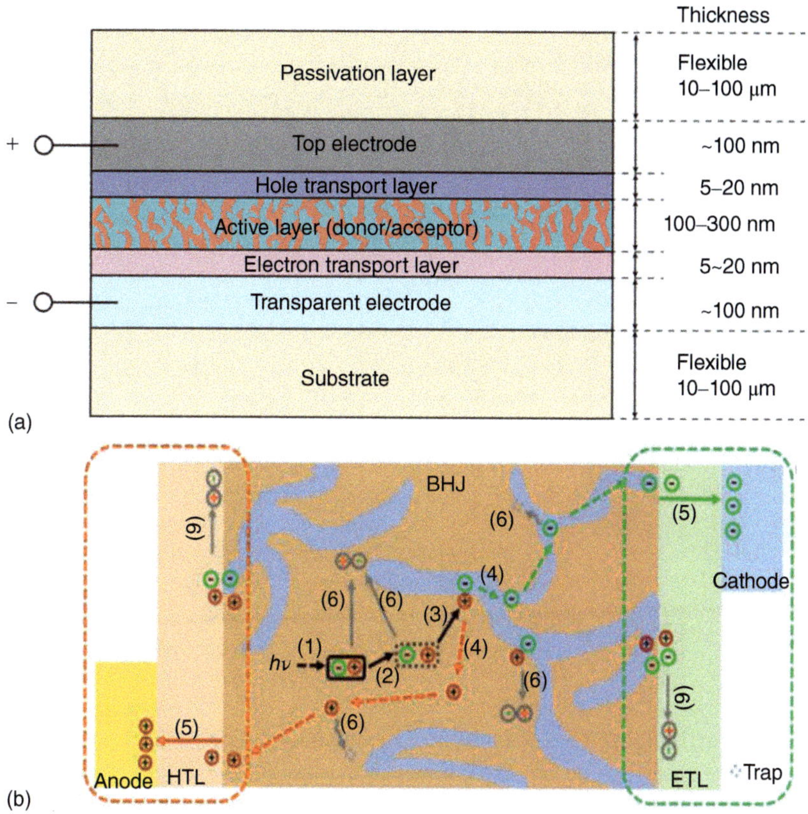

Figure 3.12 (a) A schematic illustration presented thickness for flexible OSCs with an inverted structure. (b) Work principles of OSCs based on bulk heterojunctions.

inherent restriction of narrow light-absorbing range, low absorption coefficient, and nonadjustable energy levels. Therefore, the maximum PCE of fullerene-based OSCs is limited to ~12% [78]. Fortunately, due to optoelectronic properties that can be modulated *via* altering the D and A units, non-fullerene acceptors (NFAs) became the ideal replacement. As a result, the PCEs of OSCs have been rapidly improved in recent years. Currently, Y6 or its derivatives-based OSCs have acquired exceeded 19% of PCEs in single-junction devices [79, 80]. In the case of flexible organic solar cells, the PCE has vastly boosted from 4% to more than 10% since 2014, and a high PCE value of 14.06% was reported in 2019 [81]. Yan et al. utilized the PM6 polymer donor, Y6 non-fullerene acceptor, and third component of $PC_{71}BM$ to fabricate ternary OSCs (Figure 3.13). By tuning the blend ratio of PM6:Y6:$PC_{71}BM$, the optimized morphologies of the blend films can be obtained with an optimal blend ratio of 1 : 1 : 0.2. Meanwhile, a record PCE of 14.06% on flexible substrate could be realized in organic solar cells. In 2021, the same research group reported finely tuning film morphology by incorporating crystalline small molecules into the host binary systems [82]. In this work, the authors designed and synthesized two small molecular donors with similar chemical structures, G17 and G19. Different

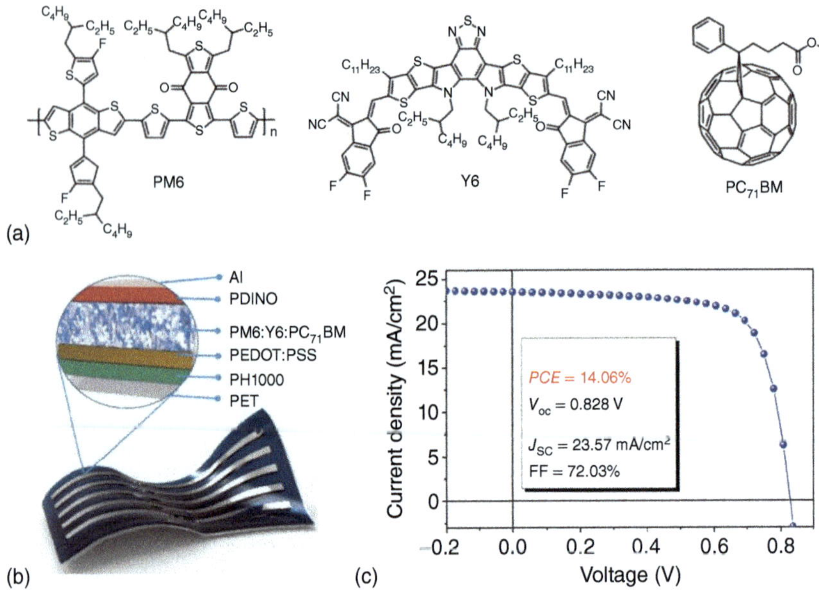

Figure 3.13 (a) Chemical structures of PM6, Y6, and PC71BM. (b) The schematic structure of flexible device, and (c) *J*–*V* curves. Source: (b) Yan, T., et al. [81], © 2019/WILEY-VCH Verlag GmbH & Co. KGaA, Weinheim.

from the G17 with amorphous orientation, the silicon-substituted G19 possessed an extremely ordered edge-on orientation, which brings G19-based ternary flexible OSCs device with a large improvement of PCE (15.9%) by doping G19 into D18-Cl: Y6 host system. These results show that an effective method *via* doping highly ordered molecular donors is offered to construct highly efficient flexible OSCs. Furthermore, the design and interface modification of functional charge transport layers is also critical for highly efficient and stable flexible OSC devices. In particular, the hole interface layer, the typical representation of PEDOT: PSS, serves as an essential constituent in OSCs. Recently, Lin et al. introduced the hole interface layer of PEDOT: MoO_3 to improve the carrier extraction and the film hydrophobicity while employing the lamination technique to produce flexible organic solar cells [83]. This attractive and effective method led to interface-modified flexible ITO-free OSCs with PCE improvement by 30% compared with the conventional laminated device, indicating that the anode interface engineering has a significant effect on improving the performance of roll-to-roll laminated and self-encapsulated OSCs.

The cycling stability of a fixed bending radius or angle is often evaluated for the mechanical tolerance of a flexible solar cell. However, a lack of proper scientific mechanical analysis causes inaccurate calculations about strain in critical device layers corresponding to bending radii or angles. Actually, the layer with weak tolerance to its strain response can benefit the mechanical robustness of the device. Therefore, applying minimal bending strain for the weakest layer (usually the electrode layer) is an important strategy to enhance bending stability. Previous works indicated that brittle ITO film occurs in cracking phenomenon with 2% strain, which was considered one of the weakest layers inside organic solar cells [84].

Thus, other materials-based TCEs, such as graphene [85], PEDOT: PSS [86], silver nanowires [87], and silver nanonetworks [88] can be promising candidates to replace ITO film to effectively increase flexibilities of mechanical robustness for flexible OSCs. A novel example has been reported recently [89]. Zeng et al. used an ionic liquid (IL)-type reducing agent containing Cl^- and a dihydroxyl group to precisely regulate the reduction process of silver (Ag) in Ag NW-based flexible transparent electrodes. The Ag^+ concentration can be controlled via the dissolution of AgCl, while it can be reduced to Ag metal by the dihydroxyl group. As a result, the Ag growth in situ at the junction site of the AgNWs provides an atomic-level contact between the Ag NWs and the reduced Ag to enhance the mechanical stability (Figure 3.17). The flexible OSCs based on this chemically welded electrode achieved a record PCE of 15.82% (active area: 1.0 cm^2), showing robust bending durability even under extreme conditions (Figure 3.14).

Moreover, the durability of bending for organic semiconductor films has been widely investigated [90]. For instance, organic materials are used for carrier injection layers, which can replace some oxide materials for improvement of mechanical robustness [91]. Xu et al. improved the interface adhesion force by inserting a thermoplastic elastomer (polystyrene-block poly(ethylene-ran-butylene)-block-polystyrene (SEBS)) between the organic photoactive layer and MoO_3 hole transporting layer, which improved the poor bending durability with decreased the modulus [92]. This interface bonding layer could remain over 90% of the initial PCE after 10,000 cyclic bending. In addition, decreasing the crystallinity of active

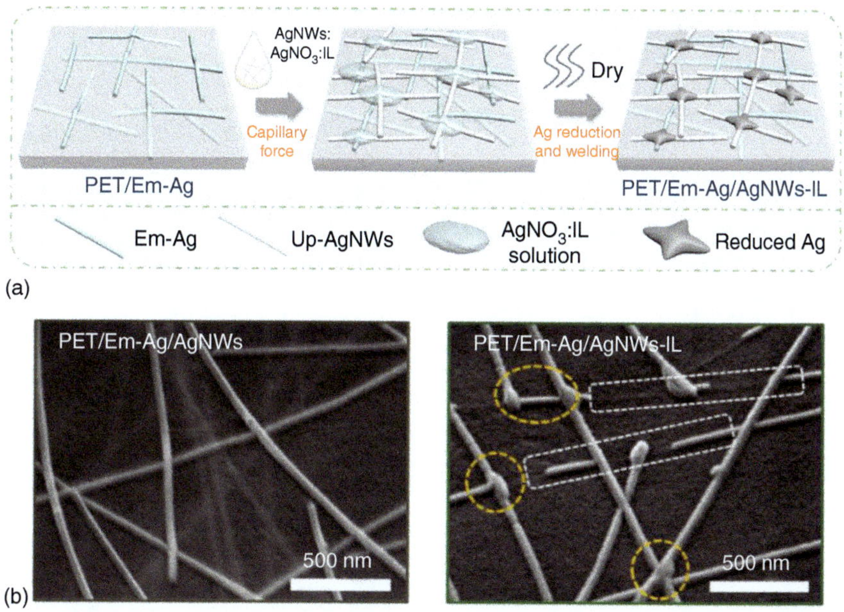

Figure 3.14 (a) Schematic diagram displaying the Ag reduction and welding processes involved in the fabrication of AgNW FTEs. (b) SEM images of the PET/Em-Ag/AgNWs and the PET/Em-Ag/AgNWs-IL FTEs. Source: (b) Zeng et al. [89], © 2022/American Chemical Society.

layers can also help to improve the flexibility. Song et al. introduced polymer PY-IT into PM6:BTP-eC9 blend film to form the ternary system [93]. This ternary system contained entangled structure morphology to dissipate the local energy, which also limited the diffusion and crystallization of BTP-eC9 acceptor and successfully improved the ductility and morphological stability compared with the binary blend. The flexible OSCs based on PET substrate thus exhibited a PCE of 16.52%.

3.5 Flexible Dye-Sensitized Solar Cells

A typical structure of flexible DSSCs is shown in Figure 3.15a. The photoelectrical conversion mechanism of DSSCs was briefly illustrated in Figure 3.15b, which has been demonstrated in Chapter 2. When the photons from the sunlight are injected into the photoanode, dye molecules adsorbed on the surface of the semiconductor film will turn from the ground state to the excited state by assimilating photons' energy, and then the photo-excited electrons can be injected into the lowest conduction band of the semiconductor, and eventually diffuse to the conductive electrode under the effect on electron concentration difference. Then, the photo-excited electrons transfer to the counter electrode through the external circuit. Meanwhile, the dye molecule loses an electron to form an oxidized state (D^+). Its oxidized state is further reduced by iodides (I^-), while iodides are oxidized to I_3^-, which breaks the stable balance of the I^-/I_3^- redox couple in the electrolyte. In the process, the I_3^- is transferred to the conductive electrode and reduced to I^- via accepting the electron from the outside circuit. Thus, a circle of the whole working procedure can be shown in this Eqs. (3.5)–(3.8): [94].

$$2\text{Dye} + \text{hv} \rightarrow 2\text{Dye}^+ + 2\text{e}^- \tag{3.5}$$

$$3\text{I}^- \rightarrow \text{I}_3^- + 2\text{e}^- \tag{3.6}$$

$$2\text{Dye}^+ + 2\text{e}^- \rightarrow 2\text{Dye} \tag{3.7}$$

$$\text{I}_3^- + 2\text{e}^- \rightarrow 3\text{I}^- \tag{3.8}$$

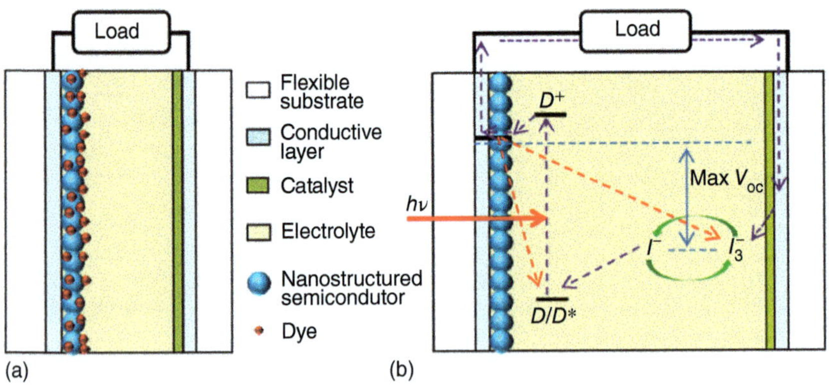

Figure 3.15 (a) Typical structure and (b) operational principle of FDSSCs.

The PCEs of flexible DSSCs are mainly limited by the TCE substrate, dye, and electrolyte. Semiconductor thin films on the photoanodes are the necessary pathways for electron transport in DSSCs; therefore, their structure and physical properties can impact the performance of flexible DSSCs. Owing to the excellent light-scattering capability, high surface area, and easy fabrication, mesoporous TiO_2 has been widely used in flexible DSSCs. As displayed in Figure 3.16a, Liu et al. reported that Ti meshes supported TiO_2 nanorods/Er^{3+}-Yb^{3+} co-doped TiO_2 nanoparticles offer a relative efficiency promotion of 86% than TiO_2 nanorods supported bare TiO_2 nanoparticles [95]. Song et al. used electrospun and ultrasonic treatment to obtain bilayer photoanode composed of the smaller diameter CNTs/TiO_2 nanorods and the larger diameter Mg^{2+} doped nanorods for flexible dye-sensitized solar cells, which achieved a PCE of 3.9% (Figure 3.16b) [96]. Miclau et al. utilized α-CrOOH as passivation for 3R-$CuCrO_2$ delafossite-based DSSCs (Figure 3.16c). The p-type 3R-$CuCrO_2$ film was passivated by n-type α-CrOOH nanomaterial, which could form a p–n junction to facilitate the charge separation. On the other hand, the α-CrOOH layer acted as a barrier layer for charge recombination between the holes in the valence band and the electrolyte or dye, therefore making a remarkable enhancement of short-circuit current density (J_{SC}) and open-circuit voltage (V_{OC}) of flexible DSSCs [97]. In addition, UV-O_3 Plasma treatment can be an effective way to improve external quantum efficiency (EQE) through adding functional groups on semiconductor surfaces and promoting dye absorption.

Mechanical stability is vital for flexible DSSCs, especially for devices fabricated on plastic substrates. The plastic substrate-based photoanodes are usually fabricated

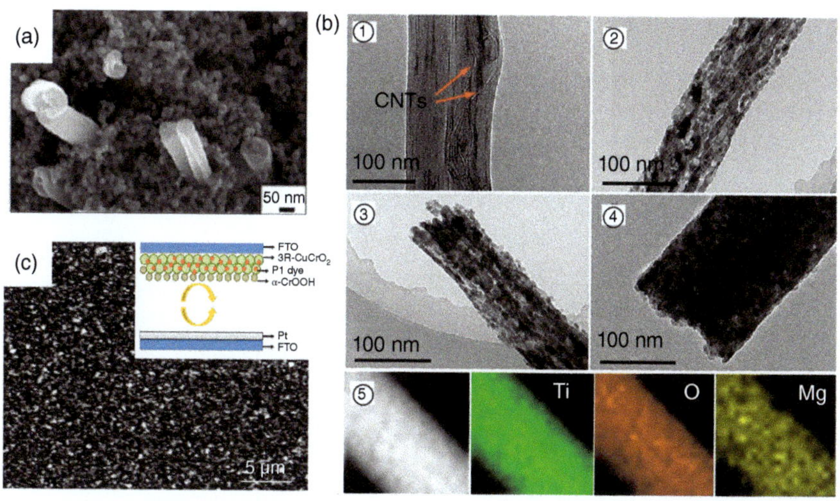

Figure 3.16 (a) TiO2 NRs/Er^{3+} and Yb^{3+} co-doped TiO_2 NPs structure (b) TEM images of the PVAc/TiP/MWCNTs NFs ①, CNTs/TiO_2 NFs ②, CNTs/TiO_2 NRs ③, Mg^{2+}/TiO_2 NRs with larger diameters ④, and TEM-EDS mapping images of the Mg^{2+}/TiO_2 NRs ⑤ with the elements mappings of Ti, O, and Mg. (c) SEM micrographs of a-CrOOH. (Inset: Schematic diagram of DSSC.) Source: (a) Liu et al. [95]/Elsevier. (b) Song et al. [96]/Elsevier. (c) Miclau et al. [97]/Elsevier.

without high-temperature annealing treatment or through binder-free methods, which results in the formation of cracks between the electrolyte and conductive layer in the drying process. In addition, the interconnection between the semiconductor film and conductive layer is easy to broken when flexible DSSCs suffer from internal stretch or outside force compression, which accelerates the degeneration of the photoanode. Figure 3.17 illustrates the schematic and microscopic morphology of photoanodes before and after outside force bending [98].

Profit from reliable high-temperature resistance, lower resistance, and great flexibility, metallic foils (Ti foil), meshes (Ti, stainless steel mesh), and wires (Ti wire) were considered ideal substrates for flexible DSSCs. For example, Liu et al. fabricated Ti mesh-supported photoanode, and the corresponding flexible Er^{3+}/Yb^{3+} co-doped DSSC achieves 8.1% IPCE [95].

The counter electrode works as a highly effective catalysis to facilitate reduction of I^-/I_3^-, and its robustness is also an indispensable part of the performance of flexible

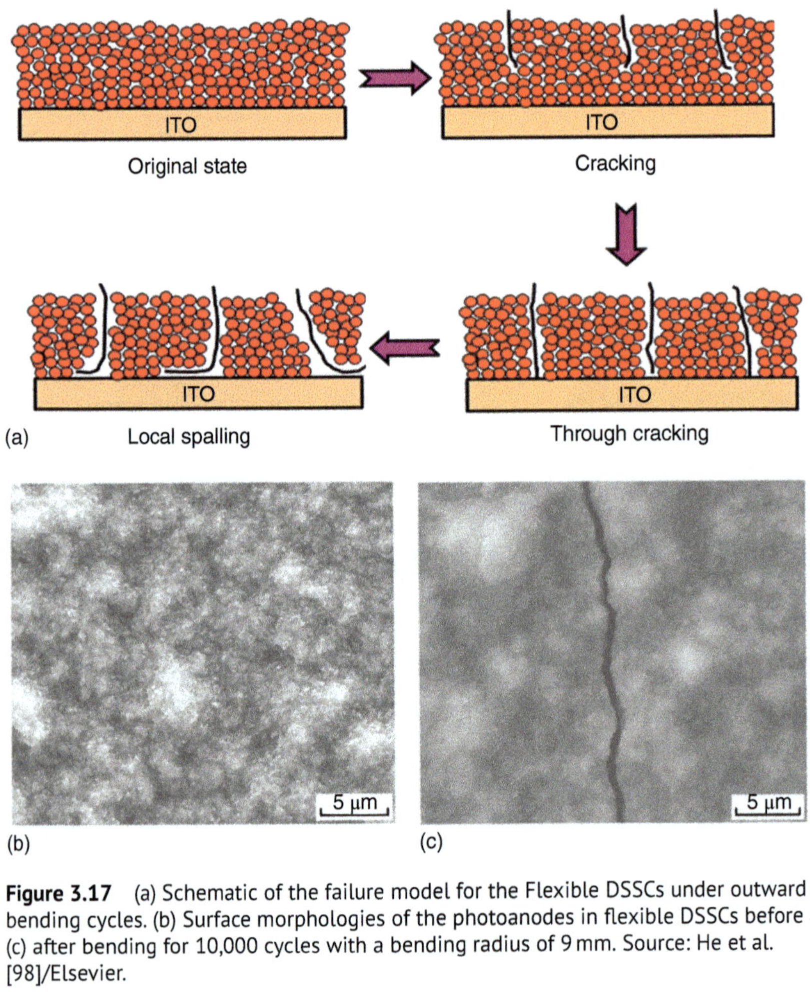

Figure 3.17 (a) Schematic of the failure model for the Flexible DSSCs under outward bending cycles. (b) Surface morphologies of the photoanodes in flexible DSSCs before and (c) after bending for 10,000 cycles with a bending radius of 9 mm. Source: He et al. [98]/Elsevier.

DSSCs. It mainly involves metal, carbon, and composite-based electrodes. Particularly, graphene, platinum, and metals/carbon composited materials are remarkable catalytic materials and have been widely used as counter electrodes by loading on flexible substrates. For example, Yin et al. exploited an electrostatic spinning method to fabricate flexible Ag@ZrO_2/C nanofiber film as the counter electrode. Such flexible plastic DSSCs achieved a PCE of 4.77% with enhanced mechanical stability [99]. Although the above materials show excellent catalytic properties, their poor mechanical stability has hindered the application of flexible DSSCs. In addition, the high cost and complex process restrict their practicability. Fortunately, conductive polymers with high conductivity, excellent doping capacity, and high mechanical stability were introduced as counter electrode materials to simplify manufacturing procedures. Ma et al. synthesized PEDOT on ITO/PEN as a flexible counter electrode through electrochemical methods and attained an IPCE of 7.18% [100]. Yue et al. fabricated nickel sulfide/polyaniline/titanium (NiS/PANI/Ti) as the counter electrode via electrodeposition [101]. All these electrodes were fabricated by simple process and demonstrated excellent flexibility and mechanical stability.

3.6 Flexible Perovskite Solar Cells

Perovskites have a general structure of ABX_3, wherein A is a monovalent cation, for example, methylammonium (MA), formamidinium (FA), or cesium (Cs). B is a divalent metallic cation, such as lead (Pb) or tin (Sn). And X is a halide anion like iodide (I^-), bromide (Br^-), or chloride (Cl^-). Selecting a numerous variety of elements can confer the modulated optical and electronic properties of perovskite structure, which involves a low- and gentle-energy direct bandgap, high absorption coefficient, together with long charge diffusion length [102]. More importantly, a low-crystallization energy barrier endows the possibility for flexible PSCs via solution-based manufacturing processes at low temperatures.

For PSCs, various structures have been developed, as shown in Figure 3.18 [103]. As an example, a layered structure of TCE/TiO_2/PEDOT-PSS/perovskite/PCBM/metal has been developed, in which a p-type PEDOT: PSS serves as HTM at the bottom, and an n-type PCBM layer functions as an ETM at the top. Perovskite material works as the intrinsic absorber. Under light illumination, electron–hole pairs generate and dissociate in the perovskite layer. Then, electrons and holes are extracted by the build-in potential at the ETL/perovskite and HTL/perovskite interfaces, followed by being collected at the anode and cathode.

Flexible TCE is the most important component for flexible PSCs. The doped indium oxide (In_2O_3), tin oxide (SnO_2), and zinc oxide (ZnO) possess higher transparency and conductivity, which are widely used as the base materials of flexible TCEs for PSCs. On the other hand, stable chemical properties of the substrates are required because the perovskite precursors are generally dissolved in organic solvents such as DMF and DMSO. Therefore, PET and PEN films with great resistance to various solvents are widely used as polymer substrates. However, these substrates exhibited weak resistance to oxygen and moisture, which accelerates the

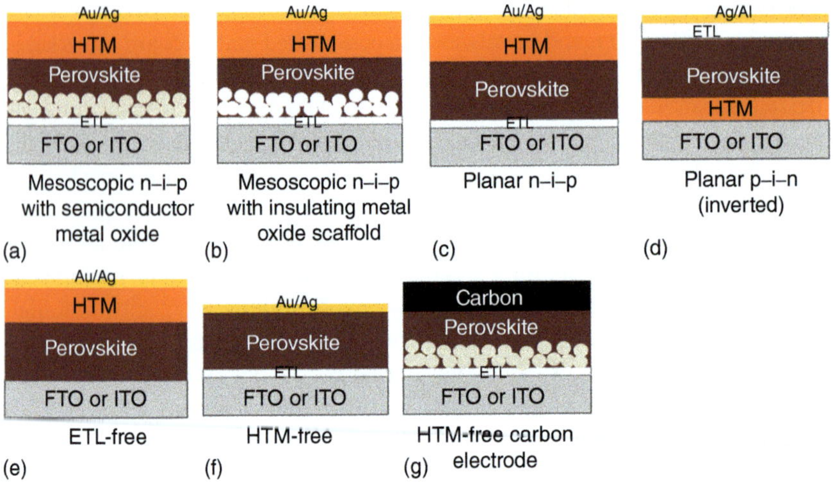

Figure 3.18 Variety of device architectures for perovskite solar cells. (a) Mesoscopic n-i-p with semiconductor metal oxide. (b) Mesoscopic n-i-p with insulating metal oxide scaffold. (c) Planar n-i-p. (d) Planar p-i-n (inverted). (e) ETL-free. (f) HTL-free. (g) HTL-free carbon electrode.

degradation of the device's performance. In addition, the surface roughness is a critical factor for fabricating flexible PSCs with large areas.

The perovskite precursor plays a decisive role in perovskite crystallization and film formation [104]. In general, the engineering of perovskite ink for the roll-to-roll process involves perovskite composition management, additive engineering, and solvent selection. For composition management, it can determine the physical characteristics of the deposited perovskite film, e.g. the energy bandgap, light absorption, and thermal stability. Wherein MA^+, FA^+, and mixed-cation-based flexible PSCs have widely been reported. For example, Bi et al. found that the adjustment of the perovskite precursor ratio (FAI/MABr) can ameliorate the morphology and optoelectronic performance of perovskite films and realize a high PCE of 18.1% for flexible PSCs [105].

Additive engineering is another effective method to reduce the defects of perovskite crystals existing at grain boundaries and surfaces, thus enhancing the PCE and long-term stability. Using dimethyl sulfide as an additive to react with Pb^{2+}, an intermediate phase can be formed. Feng et al. reported controllable growth of perovskite crystalline on flexible substrates, which possess larger grains and lower defect densities [106]. The fabricated PSCs thus exhibit a high PCE of 18.4%. Wu et al. reported that synergistic additives comprised of NMP and MACl have a concordance optimization of $FAPbI_3$ perovskites grown on flexible substrates [107]. The introduction of NMP could retard the crystallization process of perovskites through forming $FAI{\cdot}PbI_2{\cdot}NMP$ intermediate. Meanwhile, the MACl could promote the phase transition of black perovskites at a lower temperature.

Interfacial engineering between the functional and perovskite layers can effectively improve the PCE of flexible PSCs. Lu et al. reported the highest PCE (22.44%)

for flexible PSCs via the construction of the solution-processed HADI-SnO_2 ETL. The multifunctional organic ions (HADI) doping between SnO_2 and perovskite layers not only slightly lifted the SnO_2 conduction band to match perovskite but also bridged between the SnO_2 and perovskite layers for effective charge transfer. In addition, it also played a passivation effect on buried perovskite surfaces [108]. The coffee-ring effect usually occurs in the fabrication of printing flexible perovskite modules, which leads to inhomogeneity, poor interfacial contact, and pinholes that severely affect reliability and repeatability. Fan et al. introduced a bionic interface layer (Bio-IL) involving sticky functional groups (levodopa) into perovskite and NiO_x. As shown in Figure 3.19, the super hydrophilic Bio-IL resulted in pinning the three-phase contact line and meanwhile boosted the sedimentation rate of perovskite precursor due to electrostatic attraction and high evaporation rate of solvent. Those superiorities suppressed the lateral transport of perovskite colloidal particles to acquire uniform perovskite film on large-area substrates. Consequently, flexible PSC modules (14.63 cm^2) achieved a 16.87% PCE with a near 90% yield rate of reproducibility [109].

Although PET/ITO electrodes afford the best PCE for flexible PSCs to date, the natural brittle property of ITO remains a major hindrance to wearable applications. Numerous works have explored novel electrodes with better mechanical stability, which can help renovate the mechanical robustness of flexible PSCs. For instance, metal nanowires, metal grids, carbon-based materials, and PEDOT: PSS have been widely explored. Jeong et al. prepared a polyimide (PI)/Cu grid/graphene flexible electrode by utilizing graphene as a capping layer to prevent inter-diffusion between metal and halide, thus enhancing the device's long-term stability with a high PCE of 16.4% for flexible devices [110]. 94% of the initial PCE could be retained after 10,000 bending cycles at a radius of 5 mm. Upon coating highly conductive PEDOT: PSS (PH1000) on the Ag-mesh electrodes, an improved PCE of 14.0% could be achieved for flexible PSCs. After 5000 bending cycles at a curvature radius of 5 mm, the PCE maintained by 95%, demonstrating good stability under mechanical deformation [1].

The introduction of additives can also enhance the mechanical flexibility of flexible PSCs. As illustrated in Figure 3.20, Zheng et al. used ammonium formate ($HCOONH_4$) as an additive to pre-bury into the ETL layer, from which the integral modification, including ETL, perovskite layer, and their interface could be in situ realized *via* the bottom-up infiltration process. The $HCOONH_4$ treatment

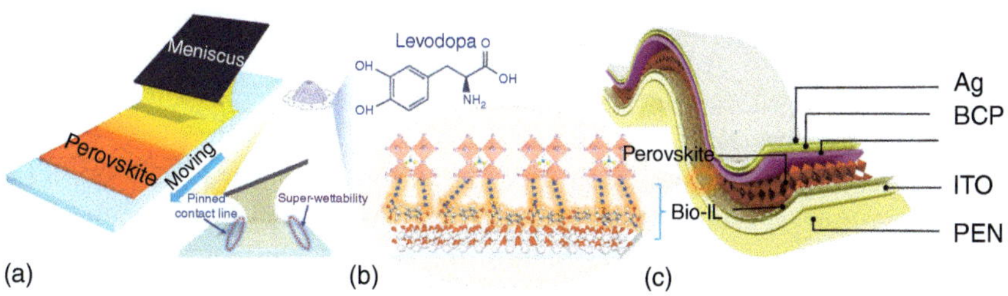

Figure 3.19 (a) Schematic of meniscus-coating for perovskite film. (b) Biomimetic mechanisms, and (c) perovskite device configuration.

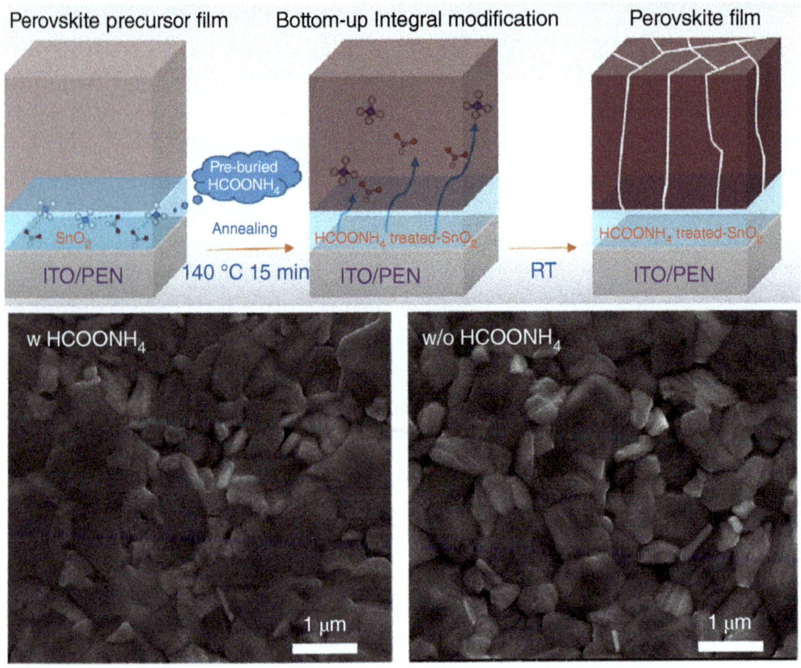

Figure 3.20 (a) Schematic illustration of the functionalization process of pre-buried $HCOONH_4$ additive. (b) SEM images of perovskite surfaces based on pristine SnO_2 and SnO_2 with $HCOONH_4$. Source: (b) Zheng et al. [111]/John Wiley & Sons, Inc.

promoted electron extraction in ETL, reducing the defect densities within these layers. Meanwhile, the residual strain and micro-strain in perovskite film could be relaxed. Finally, $HCOONH_4$-treated flexible devices achieved a certified PCE of 21.9%. Keeping over 90% of its initial PCE after 4000 bending cycles indicated the stable, flexible durability of the obtained device [111].

3.7 Summary and Perspective

In this chapter, we introduced the flexible TCE, fabrication technique, and typical flexible solar cells. For the flexible TCEs, conductivity and transparency are considered the key parameters, and the ideal materials, including metal oxides (ITO, IZO, etc.), conducting polymers (PEDOT: PSS and its derivatives), and carbon nanomaterials (CNTs and GO) have been discussed and compared in detail. Synthesis techniques, such as CVD and solution-based methods, are introduced. Although using TCOs electrodes can ensure high PCEs for flexible solar cells, the high-temperature manufacturing process, and cracking problem have hindered their practical use. Solution-processed transparent conducting polymers, particularly PEDOT and its derivatives, thus become promising candidates for flexible solar cells. The state-of-the-art PEDOT electrodes show comparable performance with vacuum-deposited ITO on flexible substrates, along with much higher mechanical flexibility.

Meanwhile, it enables a significantly lower manufacturing cost than TCOs. Moreover, a lower posttreatment temperature for the conducting polymer-based electrodes can be fully compatible with the roll-to-roll fabrication. However, the conductivity limitation of the polymers has remains an issue for large-area commercial solar modules. Carbon materials can be ideal candidates incorporated with high conductivity and optical transmittance. However, the synthesis of high-quality and repeating sizes of carbon materials should be further investigated.

For various fabrication techniques, we introduced solution-based coating and printing processes with the technology routes, parameters, and determinable factors about the quality of the film and discussed the development in PSC applications. Application scenarios determine the selection of these manufacturing methods. Generally, spin coating is widely used in lab-scale fabrication to deposit high-quality PSCs in small areas. Slot-die coating and screen-printing techniques are more compatible with the roll-to-roll process, which represents promising candidates for large-scale fabrication of solar modules. The gravure printing has also been reported for PSCs in recent years. Inkjet printing is a field-controlled method to form precise patterns on the substrate, which can avoid additional lithographic processes. It is worth noting that the roll-to-roll system can realize all-printed device architectures, which show the great commercial potential for PSCs.

At last, we discussed the common device structures and mechanisms for flexible OSCs, DSSCs, and PSCs. Also, we summarized the main achievements of those flexible film solar cells in terms of both photovoltaic performance and mechanical stability. For all the flexible solar cells, the flexible TCE plays an important role in the PCEs and mechanical stability. Particularly, solution-processed polymer, carbon, and metal not only show great potential for large-scale thin-film solar modules *via* the Roll-to-Roll manufacturing process but also demonstrate excellent mechanical robustness and adhesion to substrates for flexible solar cells.

Considering the practical and wearable application of flexible thin-film solar cells, several key bottlenecks must be solved, as detailed below. The first challenge comes from the long-term stability of the flexible solar cells. To meet the target of Levelized cost of energy (LCOE) of 0.06 USD/kWh (Sunshot 2020 utility-scale solar cost target) [112], a 30-year lifetime of perovskite solar modules with 15% PCE is necessary, which is far longer than the state-of-the-art PSCs. In fact, this instability is attributed to an intrinsic material property or the interface between different components of the solar cells. Although degradation of polymer electrodes can be partially solved by posttreatment or surface modification, the long-term stability still demands comprehensive and detailed studies in the future. In addition, effective device encapsulation is necessary to prevent the degradation derived from water and oxygen. For instance, a flexible encapsulation technique can be an ideal solution to extend the lifetime of the perovskites. How to overcome the natural rigidity of perovskite films? It might be a vital question that decides the fate of flexible PSCs. Although OSCs exhibit unique advantages such as being flexible and stable, their out-of-order molecular packing and low effective charge separation should be further improved towards high PCE and repeatability.

Another challenge lies in the precise control of the large-scale fabrication of flexible solar modules. Although Roll-to-Roll technology has been widely used for the preparation of transport layers and active layers, the obtained devices show relatively low PCEs. More efforts involving additive introduction, ink engineering, and interface modification should be made. Meanwhile, the theory of morphology and crystallography of functional or active layers should be systematically studied. In addition, flexible tandem solar cells can be a highly promising choice for flexible solar cells [113–116].

References

1 Li, Y., Meng, L., Yang, Y. et al. (2016). High-efficiency robust perovskite solar cells on ultrathin flexible substrates. *Nat. Commun.* 7 (1): 10214.

2 Chen, J., Huang, Y., Zhang, N. et al. (2016). Micro-cable structured textile for simultaneously harvesting solar and mechanical energy. *Nat. Energy* 1: 16138.

3 Zhang, K., Gao, K., Xia, R. et al. (2016). High-performance polymer tandem solar cells employing a new n-type conjugated polymer as an interconnecting layer. *Adv. Mater.* 28 (24): 4817–4823.

4 Li, Z., Klein, T.R., Kim, D.H. et al. (2018). Scalable fabrication of perovskite solar cells. *Nat. Rev. Mater.* 3 (4): 18017.

5 Li, X., Li, P., Wu, Z. et al. (2021). Review and perspective of materials for flexible solar cells. *Mater. Rep.: Energy* 1 (1): 100001.

6 Tang, G. and Yan, F. (2021). Recent progress of flexible perovskite solar cells. *Nano Today* 39: 101155.

7 Yang, D., Yang, R., Priya, S. et al. (2019). Recent advances in flexible perovskite solar cells: fabrication and applications. *Angew. Chem. Int. Ed.* 58 (14): 4466–4483.

8 Ellmer, K. (2012). Past achievements and future challenges in the development of optically transparent electrodes. *Nat. Photonics* 6 (12): 809–817.

9 Stadler, A. (2012). Transparent conducting oxides—an up-to-date overview. *Materials* 5 (4): 661–683.

10 George, J. and Menon, C.S. (2000). Electrical and optical properties of electron beam evaporated ITO thin films. *Surf. Coat. Technol.* 132 (1): 45–48.

11 Tang, C.W. (1986). Two-layer organic photovoltaic cell. *Appl. Phys. Lett.* 48 (2): 183–185.

12 Zardetto, V., Brown, T.M., Reale, A. et al. (2011). Substrates for flexible electronics: a practical investigation on the electrical, film flexibility, optical, temperature, and solvent resistance properties. *J. Polym. Sci., Part B: Polym. Phys.* 49 (9): 638–648.

13 Kim, M.-G., Kanatzidis, M.G., Facchetti, A. et al. (2011). Low-temperature fabrication of high-performance metal oxide thin-film electronics via combustion processing. *Nat. Mater.* 10 (5): 382–388.

14 Kawazoe, H., Yasukawa, M., Hyodo, H. et al. (1997). P-type electrical conduction in transparent thin films of $CuAlO_2$. *Nature* 389 (6654): 939–942.

15 Kanehara, M., Koike, H., Yoshinaga, T. et al. (2009). Indium tin oxide nanoparticles with compositionally tunable surface plasmon resonance frequencies in the near-IR region. *JACS* 131 (49): 17736–17737.

16 Chen, M.-F., Lin, K.-m., and Ho, Y.-S. (2012). Laser annealing process of ITO thin films using beam shaping technology. *Opt. Lasers Eng.* 50 (3): 491–495.

17 Lee, D., Pan, H., Ko, S.H. et al. (2012). Non-vacuum, single-step conductive transparent ZnO patterning by ultra-short pulsed laser annealing of solution-deposited nanoparticles. *Appl. Phys. A* 107 (1): 161–171.

18 Dauzon, E., Lin, Y., Faber, H. et al. (2020). Stretchable and transparent conductive PEDOT:PSS-based electrodes for organic photovoltaics and strain sensors applications. *Adv. Funct. Mater.* 30 (28): 2001251.

19 Kaltenbrunner, M., Adam, G., Głowacki, E.D. et al. (2015). Flexible high power-per-weight perovskite solar cells with chromium oxide–metal contacts for improved stability in air. *Nat. Mater.* 14 (10): 1032–1039.

20 Na, S.-I., Kim, S.-S., Jo, J. et al. (2008). Efficient and flexible ITO-free organic solar cells using highly conductive polymer anodes. *Adv. Mater.* 20 (21): 4061–4067.

21 Hu, X., Huang, Z., Li, F. et al. (2019). Nacre-inspired crystallization and elastic "brick-and-mortar" structure for a wearable perovskite solar module. *Energy Environ. Sci.* 12 (3): 979–987.

22 Hu, X., Meng, X., Zhang, L. et al. (2019). A mechanically robust conducting polymer network electrode for efficient flexible perovskite solar cells. *Joule* 3 (9): 2205–2218.

23 Xia, Y., Sun, K., and Ouyang, J. (2012). Solution-processed metallic conducting polymer films as transparent electrode of optoelectronic devices. *Adv. Mater.* 24 (18): 2436–2440.

24 Kim, N., Kee, S., Lee, S.H. et al. (2014). Highly conductive PEDOT:PSS nanofibrils induced by solution-processed crystallization. *Adv. Mater.* 26 (14): 2268–2272.

25 Song, W., Fan, X., Xu, B. et al. (2018). All-solution-processed metal-oxide-free flexible organic solar cells with over 10% efficiency. *Adv. Mater.* 30 (26): 1800075.

26 Zhang, Y., Ng, S.-W., Lu, X. et al. (2020). Solution-processed transparent electrodes for emerging thin-film solar cells. *Chem. Rev.* 120 (4): 2049–2122.

27 Gueye, M.N., Carella, A., Massonnet, N. et al. (2016). Structure and dopant engineering in PEDOT thin films: practical tools for a dramatic conductivity enhancement. *Chem. Mater.* 28 (10): 3462–3468.

28 Wang, X., Zhang, X., Sun, L. et al. (2018). High electrical conductivity and carrier mobility in oCVD PEDOT thin films by engineered crystallization and acid treatment. *Sci. Adv.* 4 (9): eaat5780.

29 Worfolk, B.J., Andrews, S.C., Park, S. et al. (2015). Ultrahigh electrical conductivity in solution-sheared polymeric transparent films. *Proc. Natl. Acad. Sci. U.S.A.* 112 (46): 14138–14143.

30 Petrosino, M., Vacca, P., Miscioscia, R. et al. (2007). *Effect of PEDOT:PSS Ratio on the Electrical and Optical Properties of OLEDs*. SPIE.

31 Chen, Y., Kang, K.S., Han, K.J. et al. (2009). Enhanced optical and electrical properties of PEDOT: PSS films by the addition of MWCNT-sorbitol. *Synth. Met.* 159 (17): 1701–1704.

32 Cruz-Cruz, I., Reyes-Reyes, M., Aguilar-Frutis, M.A. et al. (2010). Study of the effect of DMSO concentration on the thickness of the PSS insulating barrier in PEDOT:PSS thin films. *Synth. Met.* 160 (13): 1501–1506.

33 Wang, X.-Y., Narita, A., and Müllen, K. (2017). Precision synthesis versus bulk-scale fabrication of graphenes. *Nat. Rev. Chem.* 2 (1): 0100.

34 Coleman, J.N. (2013). Liquid exfoliation of defect-free graphene. *Acc. Chem. Res.* 46 (1): 14–22.

35 Ciesielski, A. and Samorì, P. (2014). Graphene via sonication assisted liquid-phase exfoliation. *Chem. Soc. Rev.* 43 (1): 381–398.

36 Li, X., Cai, W., An, J. et al. (2009). Large-area synthesis of high-quality and uniform graphene films on copper foils. *Science* 324 (5932): 1312–1314.

37 Boyd, D., Lin, W.-H., Hsu, C.-C. et al. (2015). Single-step deposition of high-mobility graphene at reduced temperatures. *Nat. Commun.* 6 (1): 1–8.

38 Wang, X., Zhi, L., and Müllen, K. (2008). Transparent, conductive graphene electrodes for dye-sensitized solar cells. *Nano Lett.* 8 (1): 323–327.

39 Reina, A., Jia, X., Ho, J. et al. (2009). Large area, few-layer graphene films on arbitrary substrates by chemical vapor deposition. *Nano Lett.* 9 (1): 30–35.

40 Wang, Y., Chen, X., Zhong, Y. et al. (2009). Large area, continuous, few-layered graphene as anodes in organic photovoltaic devices. *Appl. Phys. Lett.* 95 (6): 209.

41 Sung, H., Ahn, N., Jang, M.S. et al. (2016). Transparent conductive oxide-free graphene-based perovskite solar cells with over 17% efficiency. *Adv. Energy Mater.* 6 (3): 1501873.

42 Bae, S., Kim, H., Lee, Y. et al. (2010). Roll-to-roll production of 30-inch graphene films for transparent electrodes. *Nat. Nanotechnol.* 5 (8): 574–578.

43 Graham, A., Duesberg, G., Hoenlein, W. et al. (2005). How do carbon nanotubes fit into the semiconductor roadmap? *Appl. Phys. A* 80 (6): 1141–1151.

44 Yamada, T., Namai, T., Hata, K. et al. (2006). Size-selective growth of double-walled carbon nanotube forests from engineered iron catalysts. *Nat. Nanotechnol.* 1 (2): 131–136.

45 Pasquier, A.D., Unalan, H.E., Kanwal, A.P. et al. (2005). Conducting and transparent single-wall carbon nanotube electrodes for polymer-fullerene solar cells. *Appl. Phys. Lett.* 87: 203511.

46 Tenent, R.C., Barnes, T.M., Bergeson, J.D. et al. (2009). Ultrasmooth, large-area, high-uniformity, conductive transparent single-walled-carbon-nanotube films for photovoltaics produced by ultrasonic spraying. *Adv. Mater.* 21 (31): 3210–3216.

47 Zhang, D., Ryu, K., Liu, X. et al. (2006). Transparent, conductive, and flexible carbon nanotube films and their application in organic light-emitting diodes. *Nano Lett.* 6 (9): 1880–1886.

48 Geng, H.-Z., Kim, K.K., So, K.P. et al. (2007). Effect of acid treatment on carbon nanotube-based flexible transparent conducting films. *JACS* 129 (25): 7758–7759.

49 Lee, J.-W., Jeon, I., Lin, H.-S. et al. (2018). Vapor-assisted ex-situ doping of carbon nanotube toward efficient and stable perovskite solar cells. *Nano Lett.* 19 (4): 2223–2230.

50 Zhou, Y., Yin, X., Luo, Q. et al. (2018). Efficiently improving the stability of inverted perovskite solar cells by employing polyethylenimine-modified carbon nanotubes as electrodes. *ACS Appl. Mater. Interfaces* 10 (37): 31384–31393.

51 Yang, Z., Ren, J., Zhang, Z. et al. (2015). Recent advancement of nanostructured carbon for energy applications. *Chem. Rev.* 115 (11): 5159–5223.

52 Xue, P., Cheng, P., Han, R.P. et al. (2022). Printing fabrication of large-area non-fullerene organic solar cells. *Mater. Horiz.* 9 (1): 194–219.

53 Zabihi, F., Xie, Y., Gao, S. et al. (2015). Morphology, conductivity, and wetting characteristics of PEDOT: PSS thin films deposited by spin and spray coating. *Appl. Surf. Sci.* 338: 163–177.

54 Krebs, F.C. (2009). Fabrication and processing of polymer solar cells: a review of printing and coating techniques. *Sol. Energy Mater. Sol. Cells* 93 (4): 394–412.

55 Wu, W.-Q., Wang, Q., Fang, Y. et al. (2018). Molecular doping enabled scalable blading of efficient hole-transport-layer-free perovskite solar cells. *Nat. Commun.* 9 (1): 1625.

56 Yu, Y., Zhang, F., Hou, T. et al. (2021). A review on gas-quenching technique for efficient perovskite solar cells. *Sol. RRL* 5 (10): 2100386.

57 Gao, L.-L., Li, C.-X., Li, C.-J. et al. (2017). Large-area high-efficiency perovskite solar cells based on perovskite films dried by the multi-flow air knife method in air. *J. Mater. Chem. A* 5 (4): 1548–1557.

58 Deng, Y., Van Brackle, C.H., Dai, X. et al. (2019). Tailoring solvent coordination for high-speed, room-temperature blading of perovskite photovoltaic films. *Sci. Adv.* 5 (12): eaax7537.

59 Deng, Y., Zheng, X., Bai, Y. et al. (2018). Surfactant-controlled ink drying enables high-speed deposition of perovskite films for efficient photovoltaic modules. *Nat. Energy* 3 (7): 560–566.

60 Bu, T., Li, J., Zheng, F. et al. (2018). Universal passivation strategy to slot-die printed SnO_2 for hysteresis-free efficient flexible perovskite solar module. *Nat. Commun.* 9 (1): 4609.

61 Yang, Z., Chueh, C.-C., Zuo, F. et al. (2015). High-performance fully printable perovskite solar cells via blade-coating technique under the ambient condition. *Adv. Energy Mater.* 5 (13): 1500328.

62 Heo, Y.-J., Kim, J.-E., Weerasinghe, H. et al. (2017). Printing-friendly sequential deposition via intra-additive approach for roll-to-roll process of perovskite solar cells. *Nano Energy* 41: 443–451.

63 Hwang, K., Jung, Y.-S., Heo, Y.-J. et al. (2015). Toward large scale roll-to-roll production of fully printed perovskite solar cells. *Adv. Mater.* 27 (7): 1241–1247.

64 Qin, T., Huang, W., Kim, J.-E. et al. (2017). Amorphous hole-transporting layer in slot-die coated perovskite solar cells. *Nano Energy* 31: 210–217.

65 Wang, Q., Eslamian, M., Zhao, T. et al. (2018). Achieving fully blade-coated ambient-processed perovskite solar cells by controlling the blade-coater temperature. *IEEE J. Photovoltaics* 8 (6): 1662–1669.

66 Galagan, Y., Di Giacomo, F., Gorter, H. et al. (2018). Roll-to-roll slot die coated perovskite for efficient flexible solar cells. *Adv. Energy Mater.* 8 (32): 1801935.

67 Aleeva, Y. and Pignataro, B. (2014). Recent advances in upscalable wet methods and ink formulations for printed electronics. *J. Mater. Chem. C* 2 (32): 6436–6453.

68 Larsen-Olsen, T.T., Søndergaard, R.R., Norrman, K. et al. (2012). All printed transparent electrodes through an electrical switching mechanism: a convincing alternative to indium-tin-oxide, silver and vacuum. *Energy Environ. Sci.* 5 (11): 9467–9471.

69 Zhou, L., Yu, M., Chen, X. et al. (2018). Screen-printed poly(3,4-ethylenedioxythiophene):poly(styrenesulfonate) grids as ITO-free anodes for flexible organic light-emitting diodes. *Adv. Funct. Mater.* 28 (11): 1705955.

70 Overgaard, M.H., Kühnel, M., Hvidsten, R. et al. (2017). Highly conductive semitransparent graphene circuits screen-printed from water-based graphene oxide ink. *Adv. Mater. Technol.* 2 (7): 1700011.

71 Ramasamy, E., Lee, W.J., Lee, D.Y. et al. (2007). Portable, parallel grid dye-sensitized solar cell module prepared by screen printing. *J. Power Sources* 165 (1): 446–449.

72 Park, J., Shin, K., and Lee, C. (2016). Roll-to-roll coating technology and its applications: a review. *Int. J. Precis. Eng. Manuf.* 17 (4): 537–550.

73 Lee, C., Kang, H., Kim, C. et al. (2010). A novel method to guarantee the specified thickness and surface roughness of the roll-to-roll printed patterns using the tension of a moving substrate. *J. Microelectromech. Syst.* 19 (5): 1243–1253.

74 Kim, Y.Y., Yang, T.-Y., Suhonen, R. et al. (2019). Gravure-printed flexible perovskite solar cells: toward roll-to-roll manufacturing. *Adv. Sci.* 6 (7): 1802094.

75 Kim, Y.Y., Yang, T.-Y., Suhonen, R. et al. (2020). Roll-to-roll gravure-printed flexible perovskite solar cells using eco-friendly antisolvent bathing with wide processing window. *Nat. Commun.* 11 (1): 5146.

76 Fukuda, K., Yu, K., and Someya, T. (2020). The future of flexible organic solar cells. *Adv. Energy Mater.* 10 (25): 2000765.

77 Bi, S., Leng, X., Li, Y. et al. (2019). Interfacial modification in organic and perovskite solar cells. *Adv. Mater.* 31 (45): 1805708.

78 Kumari, T., Lee, S.M., Kang, S.-H. et al. (2017). Ternary solar cells with a mixed face-on and edge-on orientation enable an unprecedented efficiency of 12.1%. *Energy Environ. Sci.* 10 (1): 258–265.

79 Cui, Y., Xu, Y., Yao, H. et al. (2021). Single-junction organic photovoltaic cell with 19% efficiency. *Adv. Mater.* 33 (41): 2102420.

80 Li, C., Zhou, J., Song, J. et al. (2021). Non-fullerene acceptors with branched side chains and improved molecular packing to exceed 18% efficiency in organic solar cells. *Nat. Energy* 6 (6): 605–613.

81 Yan, T., Song, W., Huang, J. et al. (2019). 16.67% rigid and 14.06% flexible organic solar cells enabled by ternary heterojunction strategy. *Adv. Mater.* 31 (39): 1902210.

82 Chen, Z., Song, W., Yu, K. et al. (2021). Small-molecular donor guest achieves rigid 18.5% and flexible 15.9% efficiency organic photovoltaic via fine-tuning microstructure morphology. *Joule* 5 (9): 2395–2407.

83 Lin, Z., Guan, W., Cai, W. et al. (2021). Effect of anode interfacial modification on the performance of laminated flexible ITO-free organic solar cells. *Energy Sci. Eng.* 9 (4): 502–508.

84 Cairns, D.R., Paine, D.C., and Crawford, G.P. (2001). The mechanical reliability of sputter-coated indium tin oxide polyester substrates for flexible display and touchscreen applications. *MRS Proc.* 666 (F3): 24.

85 Deng, B., Hsu, P.-C., Chen, G. et al. (2015). Roll-to-roll encapsulation of metal nanowires between graphene and plastic substrate for high-performance flexible transparent electrodes. *Nano Lett.* 15 (6): 4206–4213.

86 Kayser, L.V. and Lipomi, D.J. (2019). Stretchable conductive polymers and composites based on PEDOT and PEDOT:PSS. *Adv. Mater.* 31 (10): 1806133.

87 Zhang, Y.-X., Fang, J., Li, W. et al. (2019). Synergetic transparent electrode architecture for efficient non-fullerene flexible organic solar cells with >12% efficiency. *ACS Nano* 13 (4): 4686–4694.

88 Seo, K.-W., Lee, J., Jo, J. et al. (2019). Highly efficient (>10%) flexible organic solar cells on PEDOT-free and ITO-free transparent electrodes. *Adv. Mater.* 31 (36): 1902447.

89 Zeng, G., Chen, W., Chen, X. et al. (2022). Realizing 17.5% efficiency flexible organic solar cells via atomic-level chemical welding of silver nanowire electrodes. *JACS* 144 (19): 8658–8668.

90 Root, S.E., Savagatrup, S., Printz, A.D. et al. (2017). Mechanical properties of organic semiconductors for stretchable, highly flexible, and mechanically robust electronics. *Chem. Rev.* 117 (9): 6467–6499.

91 Xiong, S., Hu, L., Hu, L. et al. (2019). 12.5% flexible nonfullerene solar cells by passivating the chemical interaction between the active layer and polymer interfacial layer. *Adv. Mater.* 31 (22): 1806616.

92 Xu, Z., Han, Y., Bai, Y. et al. (2022). Thermoplastic elastomer enhanced interface adhesion and bending durability for flexible organic solar cells. *npj Flexible Electron.* 6 (1): 56.

93 Song, W., Yu, K., Ge, J. et al. (2022). Entangled structure morphology by polymer guest enabling mechanically robust organic solar cells with efficiencies of over 16.5%. *Matter* 5 (6): 1877–1889.

94 Li, G., Sheng, L., Li, T. et al. (2019). Engineering flexible dye-sensitized solar cells for portable electronics. *Sol. Energy* 177: 80–98.

95 Liu, W., Zhang, H., Wang, H.-g. et al. (2017). Titanium mesh supported TiO_2 nanowire arrays/upconversion luminescence Er^{3+}-Yb^{3+} codoped TiO_2 nanoparticles novel composites for flexible dye-sensitized solar cells. *Appl. Surf. Sci.* 422: 304–315.

96 Song, L., Zhai, J., Du, P. et al. (2018). A novel bilayer photoanode made of carbon nanotubes incorporated TiO_2 nanorods and Mg^{2+} doped TiO_2 nanorods for flexible dye-sensitized solar cells. *Thin Solid Films* 646: 44–52.

97 Miclau, M., Dabici, A., Vajda, M. et al. (2018). CrOOH as high-performance surface passivation material for dye-sensitized solar cell. *Mater. Lett.* 216: 119–122.

98 He, X.-L., Yang, G.-J., Li, C.-J. et al. (2015). Failure mechanism for flexible dye-sensitized solar cells under repeated outward bending: cracking and spalling off of nano-porous titanium dioxide film. *J. Power Sources* 280: 182–189.

99 Yin, X., Xie, X., Song, L. et al. (2018). Enhanced performance of flexible dye-sensitized solar cells using flexible Ag@ ZrO_2/C nanofiber film as low-cost counter electrode. *Appl. Surf. Sci.* 440: 992–1000.

100 Ma, J., Qingfeng, S., Fengbao, Z. et al. (2018). Improvement on the catalytic activity of the flexible PEDOT counter electrode in dye-sensitized solar cells. *Mater. Res. Bull.* 100: 213–219.

101 Motlagh, M.S. and Mottaghitalab, V. (2017). The charge transport characterization of the polyaniline coated carbon fabric as a novel textile based counter electrode for flexible dye-sensitized solar cell. *Electrochim. Acta* 249: 308–317.

102 Dirin, D.N., Protesescu, L., Trummer, D. et al. (2016). Harnessing defect-tolerance at the nanoscale: highly luminescent lead halide perovskite nanocrystals in mesoporous silica matrixes. *Nano Lett.* 16 (9): 5866–5874.

103 Jena, A.K., Kulkarni, A., and Miyasaka, T. (2019). Halide perovskite photovoltaics: background, status, and future prospects. *Chem. Rev.* 119 (5): 3036–3103.

104 Kim, C.M.T., Atourki, L., Ouafi, M. et al. (2021). A synopsis of progressive transition in precursor inks development for metal halide perovskites-based photovoltaic technology. *J. Mater. Chem. A* 9 (47): 26650–26668.

105 Bi, C., Chen, B., Wei, H. et al. (2017). Efficient flexible solar cell based on composition-tailored hybrid perovskite. *Adv. Mater.* 29 (30): 1605900.

106 Feng, J., Zhu, X., Yang, Z. et al. (2018). Record efficiency stable flexible perovskite solar cell using effective additive assistant strategy. *Adv. Mater.* 30 (35): 1801418.

107 Wu, C., Wang, D., Zhang, Y. et al. (2019). $FAPbI_3$ flexible solar cells with a record efficiency of 19.38% fabricated in air via ligand and additive synergetic process. *Adv. Funct. Mater.* 29 (34): 1902974.

108 Yang, L., Feng, J., Liu, Z. et al. (2022). Record-efficiency flexible perovskite solar cells enabled by multifunctional organic ions interface passivation. *Adv. Mater.* 34 (24): 2201681.

109 Fan, B., Xiong, J., Zhang, Y. et al. (2022). A bionic interface to suppress the coffee-ring effect for reliable and flexible perovskite modules with a near-90% yield rate. *Adv. Mater.* 34 (29): 2201840.

110 Jeong, G., Koo, D., Seo, J. et al. (2020). Suppressed interdiffusion and degradation in flexible and transparent metal electrode-based perovskite solar cells with a graphene interlayer. *Nano Lett.* 20 (5): 3718–3727.

111 Zheng, Z., Li, F., Gong, J. et al. (2022). Pre-buried additive for cross-layer modification in flexible perovskite solar cells with efficiency exceeding 22%. *Adv. Mater.* 34 (21): 2109879.

112 Song, Z., McElvany, C.L., Phillips, A.B. et al. (2017). A technoeconomic analysis of perovskite solar module manufacturing with low-cost materials and techniques. *Energy Environ. Sci.* 10 (6): 1297–1305.

113 Yang, Z., Rajagopal, A., Chueh, C.-C. et al. (2016). Stable low-bandgap Pb–Sn binary perovskites for tandem solar cells. *Adv. Mater.* 28 (40): 8990–8997.

114 Rajagopal, A., Yang, Z., Jo, S.B. et al. (2017). Highly efficient perovskite–perovskite tandem solar cells reaching 80% of the theoretical limit in photovoltage. *Adv. Mater.* 29 (34): 1702140-n/a.

115 Yang, Z., Yu, Z., Wei, H. et al. (2019). Enhancing electron diffusion length in narrow-bandgap perovskites for efficient monolithic perovskite tandem solar cells. *Nat. Commun.* 10 (1): 4498.

116 Yu, Z., Yang, Z., Ni, Z. et al. (2020). Simplified interconnection structure based on $C60/SnO_{2-x}$ for all-perovskite tandem solar cells. *Nat. Energy* 5 (9): 657–665.

4

Flexible Fiber Electrodes

4.1 Introduction

Conventional bulky and planar solar cells cannot fully meet the requirements of wearable electronics attributed to their poor flexibility and breathability. As a result, solar cells in a novel fiber format have been intensively pursued in the past decade due to their superior flexibility, weavability, and wearability, which are fully compatible with daily clothes and tolerant of complex deformations such as bending, twisting, and stretching [1–3].

Importantly, fiber electrodes are considered as the basic building blocks of fiber-shaped solar cells. Owing to the unique one-dimensional configuration, the requirements of fiber electrodes are very different from their planar counterparts [4–6]. For instance, a high electrical conductivity is quite important to reduce the inner resistances of fiber electrodes that demonstrate much longer pathways of charge transport than planar ones. Highly conductive fiber electrodes can guarantee the stable performance of fiber-shaped devices with lengths in meter or even kilometer levels. In addition, strong and flexible fiber electrodes are desirable considering the frequent deformations during fabrication and application, which are particularly important for the weavability into textiles. Furthermore, the highly curved fiber electrodes pose a grand challenge for effective contact between different components. Rational optimization of the interface between the photoactive material and semiconductor skeleton is thus key to improving the charge separation and transport for dye-sensitized solar cells (DSSCs).

To date, a variety of conductive wires/fibers have been explored as potential electrodes for fiber-shaped solar cells, including metal, polymer, carbon, and their composites [7–10]. The stiffness and rigidity of the metal wires create local stress under varying deformations, therefore impeding the flexibility of the produced devices. However, high electrical conductivity of metal makes them suitable for the production of long fiber-shaped solar cells. Polymer-based fibers are flexible and lightweight, but their electrical conductivities are highly dependent on the incorporation of conductive components. To this end, a series of fiber electrodes based on carbon nanomaterials such as multi-walled carbon nanotube (MWCNT) and graphene have been developed, aiming to achieve high electrical conductivity, mechanical strength, and flexibility. In addition, excellent electrochemical activity

Wearable Solar Cells: Mechanisms, Materials, and Devices, First Edition.
Hao Sun, Zhibin Yang, Longbin Qiu, and Huisheng Peng.

originated from their high specific surface offers great prospects for applications in fiber-shaped solar cells.

4.2 Metal Wires

Metal wires have been widely recognized as important fiber electrodes owing to their remarkable electrical conductivities and low-cost features. The basic properties of common metal wires are listed in Table 4.1. For the preparation of fiber-shaped solar cells, titanium (Ti), platinum (Pt), and stainless-steel wires have been used as the electrodes. For instance, Ti wire is widely used as the working electrode of fiber-shaped DSSCs, attributed to the high compatibility with titanium dioxide (TiO_2) nanorods, nanowires, and nanotubes. On the other hand, Pt wire demonstrates high electrical conductivity, catalytic activity, and stability, thus making it an ideal counter electrode in fiber-shaped DSSC to catalytic redox reactions of triiodide.

4.2.1 Titanium

As briefly mentioned above, the low density, high mechanical strength, and high electrochemical stability of Ti make it an ideal candidate for solar cell applications. In addition, TiO_2 can be incorporated on the surface of Ti wire via hydrothermal, dip-coating, and anodization methods, and serve as the working electrodes of dye-sensitized polymer and perovskite solar cells. The morphologies of TiO_2/Ti electrodes have been proven to be critical for the photovoltaic conversion process of the obtained solar cells. For TiO_2 nanoparticles, the numerous grain boundaries are prone to result in high electrical resistances that restrain the charge transport from semiconductor to electrode [11]. In addition, TiO_2 nanoparticles might peel off from the Ti substrate under deformation. Therefore, one-dimensional TiO_2 nanorods, nanowires, and nanotubes have been introduced to replace TiO_2 nanoparticles, which are expected to provide shortened charge transport pathways. For example,

Table 4.1 Properties of common metal materials.

Name	Density (g/cm^3)	Young's modulus (GPa)	Conductivity (10^4 S/cm)
Copper (Cu)	8.9	110–128	58.8
Aluminum (Al)	2.7	70	35.5
Stainless steel (Type 304)	7.8	211	10.4
Titanium (Ti)	4.5	116	2.4
Nickel (Ni)	8.9	200	14.4
Platinum (Pt)	21.4	168	9.5
Gold (Au)	19.3	79	45.1
Silver (Ag)	10.5	83	63.0

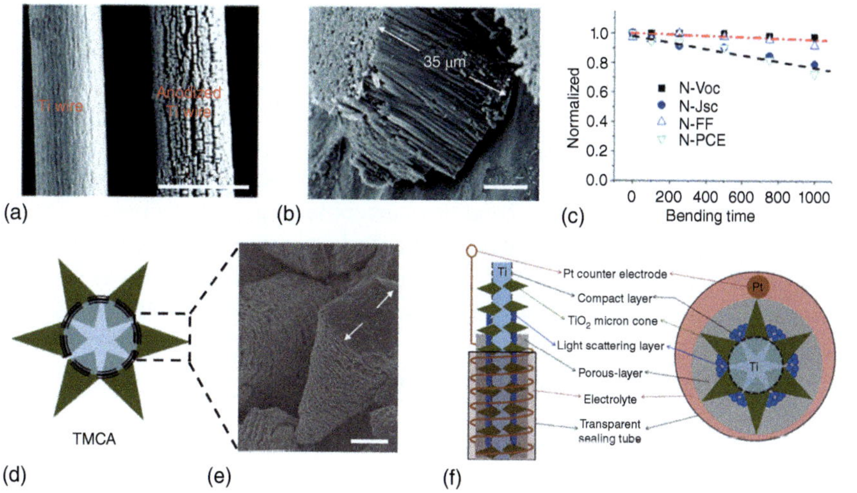

Figure 4.1 (a) Morphological comparison before and after anodic oxidization of Ti wires. (b) Side view of TiO_2 nanotube array on Ti wire. (c) Relationship between the performances of DSSC and bending times, including open circuit voltage (V_{OC}), J_{SC}, fill factor (FF), and PCE. Source: Lv et al. [12], © 2012/Royal Society of Chemistry. (d) Top view of Ti/TiO_2 micron-cone array. (e) Scanning electron microscopy (SEM) images of the micron cone, the white arrows pointing to the embedded part into Ti substrate. (f) Schematic illustrations of DSSC from the side and top views. Source: Liu et al. [13], © 2015/ Elsevier. Scale bar: (a), 300 μm; (b), 10 μm; (e), 10 μm.

fiber-shaped DSSC was prepared using a Ti wire covered by TiO_2 nanotube arrays as the fiber photoanode (Figure 4.1a) [12]. Vertically grown TiO_2 nanotube arrays were stably attached on Ti wire via anodic oxidization. The length of the TiO_2 nanotubes could be controlled by the reaction time (Figure 4.1b). The enhanced interaction between the TiO_2 nanotube and Ti wire promoted the structure stability under deformation. The large surface area further benefited the adsorption of photoactive materials, and the vertically oriented structure facilitated charge separation and transport. Therefore, the short-circuit current (J_{SC}) and power conversion efficiency (PCE) of the resulted solar cells could remain at 80% and 73% of the initial values (6.93% and 15.42 mA/cm^2) after bending for 1000 times (Figure 4.1c).

The pursuit of higher PCEs has driven the optimization of the TiO_2 layer. For instance, Ti/TiO_2 micron-cone array (TMCA) was developed via synergy of chemical and electrochemical reactions, serving as the supportive frame for photoanode and electron transfer channel (Figure 4.1d) [13]. Meanwhile, one-third of the micron-cone array layer could grow into the Ti wire, which further enhanced the binding force between TiO_2 micron cone and Ti wire (Figure 4.1e). A three-layered structure of TiO_2 with compact, light-scattering, and porous layers was then introduced into the working electrode (Figure 4.1f). The optimal light scattering layer located between the porous and compact layers could improve the light-absorbing efficiency and J_{SC}. Therefore, the PCE of the resulted DSSC could reach 8.07% with a decent V_{OC}, J_{SC}, and FF of 702 mV, 16.036 mA/cm^2, and 0.717, respectively.

After bending for 100 cycles, the PCE could be maintained by 92.01%, indicating the good flexibility of the obtained devices.

4.2.2 Stainless Steel

Stainless steel is a type of steel alloy with chromium doping, which exhibits excellent corrosion resistance in most conditions except in saline environments. When exposed to oxygen, stainless steel can form a passivation film comprised of Cr_xO_y, which prevents penetration into stainless steel, thereby inhibiting further surface corrosion. Electrochemically active components such as TiO_2, ZnO, Pt, and Ag can be decorated on the surface of stainless-steel wire via dip coating. And the obtained wires with enhanced surface area and catalytic properties can serve as working electrodes of fiber-shaped solar cells.

For example, stainless-steel wires with rational modification can serve as both working and counter electrodes of fiber-shaped solar cells [14]. A thin layer of TiO_x was coated on a stainless-steel wire as an electron transport layer via drawing through a coating cup containing an isopropanol solution of tetra(*n*-butyl)titanate. Subsequently, poly(3-hexyl-2,5-thiophene) and (6,6)-phenyl-C_{61} butyric acid methyl ester were placed as the hole transport layer. In addition, Ag paste was coated on another stainless-steel wire serving as the counter electrode (Figure 4.2a). The extremely rough surface of Ti wire could shelter from the submicron coatings of the organic photovoltaic layers, which was detrimental for the PCEs (Figure 4.2b). As a result, the stainless-steel wire with a smoother surface was used, which well matched the thicknesses (~250 nm) of the photovoltaic material (Figure 4.2c). The assembled fiber-shaped photovoltaic devices showed a PCE of 3.27% with good flexibility.

4.2.3 Other Metals

Noble metal wires (Pt, Ag, and Au) have also attracted extensive attention as the fiber electrodes due to their high electrical conductivity and electrochemical catalytic activity [15–17]. For instance, Pt is a common counter electrode in DSSCs due to its excellent catalytic capability. A DSSC was obtained by twisting a Pt wire

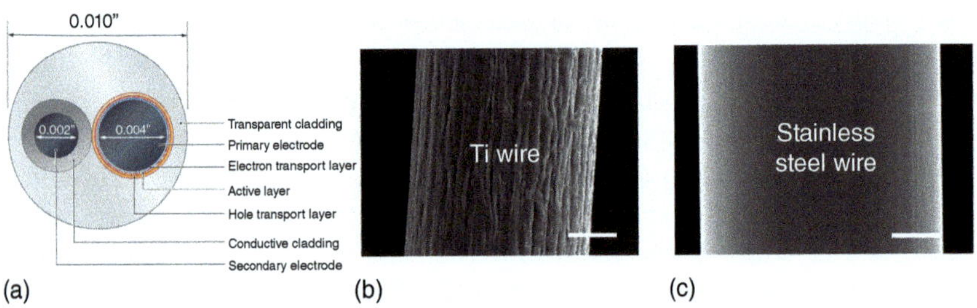

Figure 4.2 (a) Top schematic view of fiber-shaped solar cell. SEM images of the surfaces of Ti wire (b) and stainless-steel wire (c). Source: Lee et al. [14], © 2009/American Association for the Advancement of Science. Scale bar: (b) and (c), 20 μm.

and a TiO_2-modified stainless-steel wire [18]. With the TiO_2 layer separating the working electrode and counter electrode, the device could work normally without short circuit. During the twisting process, local stress concentration on certain parts could be produced between electrodes due to the intrinsic rigid of the Pt and stainless-steel wires, where the carrier's back-transfer rate increased, resulting in the reduction of voltage and current. A polymer protective layer was thus coated on the surface of Pt electrode for relieving the local stress. A DSSC with a length of a few centimeters could produce a V_{OC} of 0.61 V and a PCE of 0.27%. The low PCE was attributed to the insufficient charge transmission process at the interfaces between Pt and stainless-steel wires. The high price, mass density, and intrinsic rigidity of the Pt wires have hindered their large-scale applications. Efforts have thus been devoted to the development of Pt-free counter electrodes with comparable catalytic activity for fiber-shaped DSSCs with low cost, high flexibility and excellent stability.

4.3 Polymer Fibers

Polymer fibers such as nylon, cotton, and polyester are promising candidates for wearable solar cells due to their low density, flexibility, and chemical stability. Conductive and electrochemically active components can be incorporated into polymer fibers to overcome their intrinsic insulation and electrochemical inertia.

Incorporation of metal, carbon, and conductive polymers into polymer fibers has been widely used [19–23]. For instance, polymer fiber electrodes were reported by coating poly(3,4-ethylenedioxythiophene)/poly(styrenesulfonate) (PEDOT:PSS) on the surface of cotton yarns (Figure 4.3a) [21]. The conductive PEDOT:PSS layer could tightly bind with the monofilaments with well-controlled loading by varying the dip-coating times. The electrical conductivity reached 109 S/cm when the loading of PEDOT:PSS increased to 0.316 mg/cm. Two pairs of redox peaks of the cyclic voltammetry curves indicated the high catalytic activity of the composite

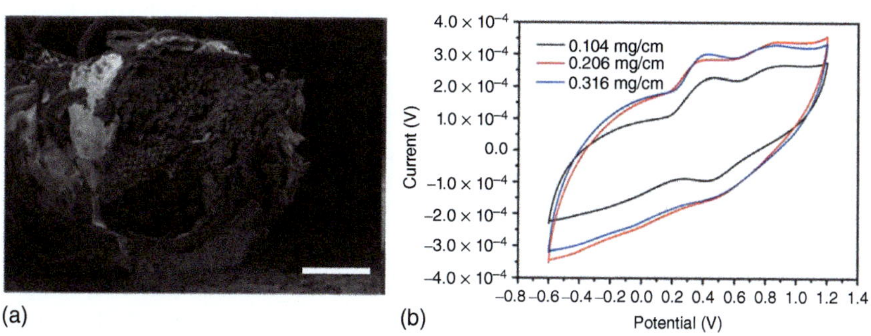

Figure 4.3 (a) The cross-section SEM images of the flexible conductive cotton yarn electrodes. (b) Cyclic voltammetry curves of the flexible conductive fiber electrodes with different PEDOT:PSS mass loadings. Source: Hou et al. [21], © 2012/Royal Society of Chemistry. Scale bar: (a), 25 μm.

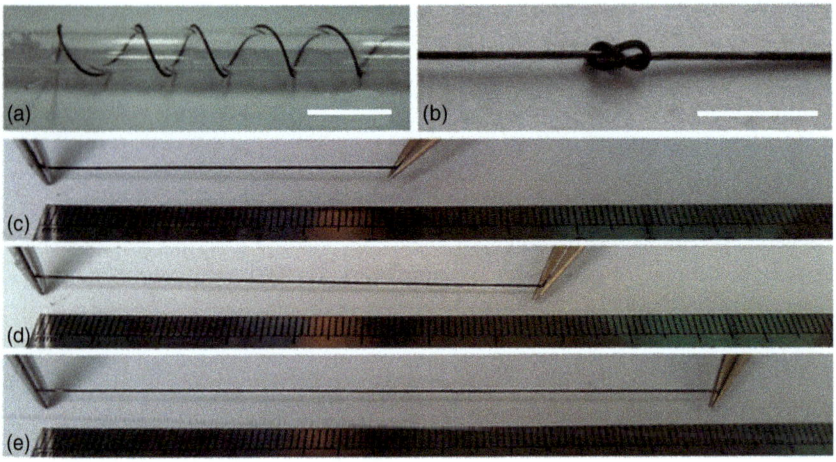

Figure 4.4 (a, b) Photographs of the stretchable electrodes with different shapes. (c–e) A stretchable electrode before and after stretching by 50% and 100%, respectively. Source: Yang et al. [28], © 2014/John Wiley & Sons. Scale bar: (a), 2 cm; (b), 1 cm.

fibers (Figure 4.3b), which were used as the counter electrodes of DSSCs with a PCE of 4.8%. In addition, conductive materials such as MXene and carbon black have been coated onto a variety of polymer fibers via electrospinning [24–26]. The created nanostructures with increased surface areas improved the electrochemical performance of the resulted fiber electrodes.

However, the weak interaction between the conductive component and polymer fiber substrate generally leads to inferior structure stability under deformations. Therefore, enhanced adhesion between the aligned MWCNT films and polymer fibers could significantly promote structure stability [27]. For instance, a new family of aligned MWCNT/rubber composite fibers was produced by wrapping aligned MWCNT films on an elastic rubber fiber using a "rotation-translation" method [28]. The conductive fiber electrodes could be wound on a glass tube or knotted into multiple shapes (Figure 4.4a,b). No observable structural changes were found after repeated stretching at a high strain of 100% (Figure 4.4c–e). Such stretchable composite fiber could serve as the counter electrode for the realization of excellent stretchability of fiber-shaped DSSCs [28, 29].

Besides coating on polymer fibers, conductive components can also be mixed with polymer and extruded to produce a homogeneous composite fiber [30]. A spinnable mixture composed of Ag particles, modified MWCNTs, 1-butyl-4-methylpyridinium tetrafluoroborate, and poly(vinylidene fluoride-*co*-hexafluoropropylene) was extruded into the hexane and then hot rolled to obtain conductive stretchable fibers (Figure 4.5a). The strain and initial conductivity of the resulted fibers could reach 490% and 1.7×10^4 S/cm, respectively, through changing the concentration of Ag particles. Importantly, the single fiber and fabric could be stretched for 300 cycles at a strain of 100% without significant change in the electrical resistance (Figure 4.5b,c), delivering excellent structural stability upon frequent deformations.

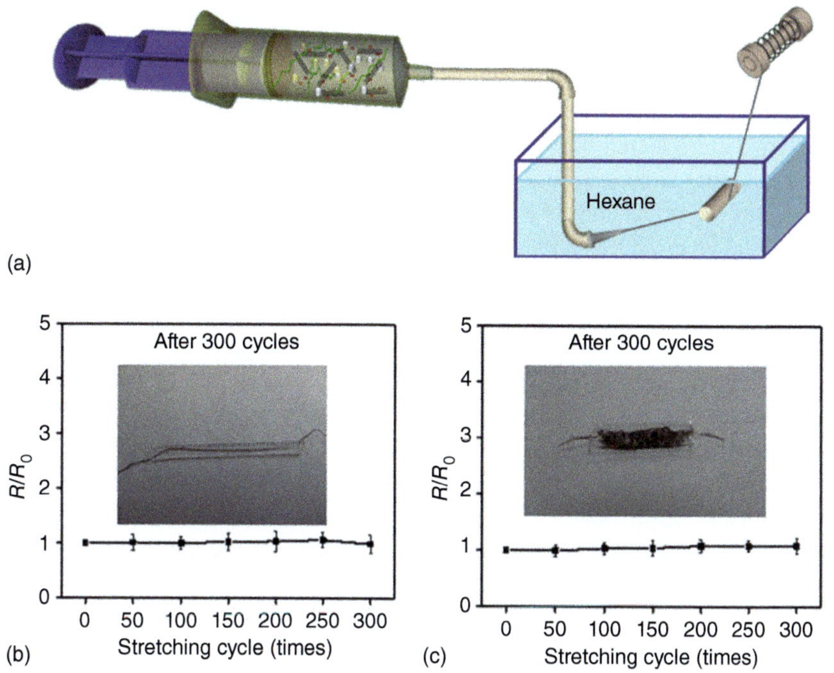

Figure 4.5 (a) Schematic illustration of the wet-spinning process. The single fiber (b) and fabric (c) are repeatedly stretched and released at a tensile strain of 100%. Source: Ma et al. [30], © 2014/American Chemical Society.

4.4 Carbon-Based Fiber

As mentioned above, the unstable interfaces and limited tensile strengths of the polymer-based electrodes could cause the structure failure of the obtained devices. In the past decade, carbon-based fibers such as MWCNT, graphene, and carbon fibers have been explored to solve the above challenge, owing to their excellent mechanical properties, high electrical conductivities, and large specific surface areas. Herein, the preparation and key properties of carbon fibers, MWCNT, and graphene fibers will be introduced, focusing on the applications in fiber-shaped solar cells.

4.4.1 Carbon Fiber

Carbon fibers are mainly prepared from several precursors, including polyacrylonitrile, rayon, acrylic acid, and pitch via a preferred wet-spinning method. Bare carbon fibers can be used as the counter electrodes, but their low specific surface areas have hindered the catalytic activities, resulting in a low PCE of less than 2% [31]. Therefore, modified carbon fibers by transition metal compounds (e.g. metal nitrides [32] and chalcogens [33]) have been developed as novel counter electrodes to improve the electrocatalytic properties. For example, $NiCo_2S_4$ nanosheets *in-situ* grown on a carbon fiber ($NiCo_2S_4$-CF) have served as the counter electrodes of fiber-shaped DSSCs to improve the photovoltaic conversion performance (Figure 4.6a–c) [33]. $NiCo_2S_4$ nanosheets demonstrated comparable catalytic activity as Pt. Besides, the large voids among nanosheets provided direct transport tunnels for ion diffusion, and the two-dimensional morphology afforded larger contact areas between the catalytic material and electrolyte. As a result, the $NiCo_2S_4$-CF counter electrode improved the redox reactions of triiodide, achieving a high PCE of 6.31% compared with 6.03% based on bare Pt wire.

Modified carbon fiber electrodes can also be employed as the photoanodes of fiber-shaped DSSCs. For example, ZnO nanowires and TiO_2 nanoparticles were deposited on carbon fibers via dip coating, and the resulted fiber-shaped DSSCs showed low PCEs of 0.006% and 1.9%, respectively. This could be attributed to the limited charge transport and severe electron recombination within the thick photovoltage layers [35, 36]. In order to improve the photovoltaic conversion performance, the morphology of the photoactive layer and the device configuration of the DSSCs required further optimization. For example, controllable synthesis of vertically aligned and hierarchical TiO_2 nanowire/nanoparticles on the surface of carbon fiber was presented (Figure 4.6d) [34]. The optimized configuration enhanced the loading capability for photoactive materials (N719 dye) and suppressed the charge recombination at the carbon fiber/TiO_2 interface. Therefore, an optimal PCE of 2.46% was achieved using a Pt-modified carbon fiber counter electrode. The nanostructured TiO_2 and enhanced interfacial contact with carbon fiber were crucial for decreasing the charge recombination during photovoltaic conversion.

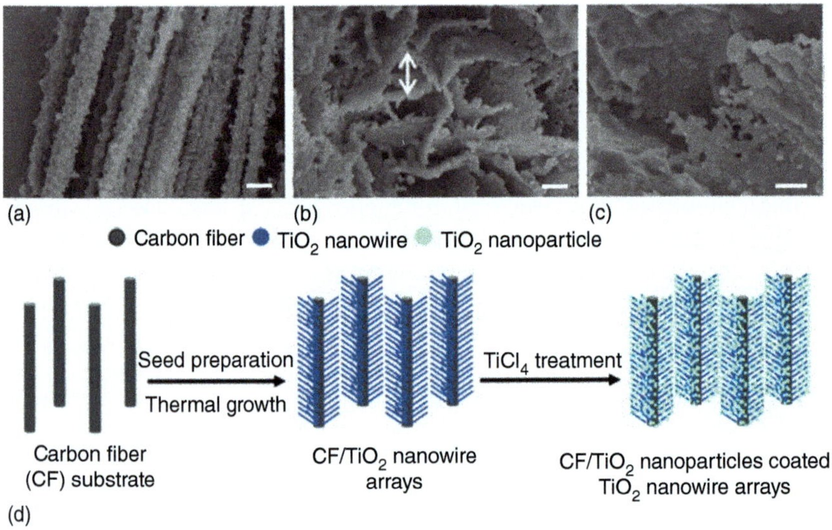

Figure 4.6 (a–c) SEM images of $NiCo_2S_4$-CF in different magnifications. Source: Chi et al. [33], © 2017/Springer Nature. (d) Preparation route of vertically aligned and hierarchical TiO_2 nanowire/nanoparticle grown on carbon fiber substrates. Source: Cai et al. [34], © 2014/John Wiley & Sons. Scale bar: (a), 3 μm; (b), 600 nm; (c), 300 nm.

4.4.2 Carbon Nanotube Fiber

Carbon nanotube (CNT) fibers have attracted extensive attention as the working or counter electrodes to fabricate fiber-shaped solar cells due to their remarkable tensile strength (5.5 GPa) [37], electrical conductivity (1.09×10^3 S/cm) [38] and excellent mechanical flexibility. As the basic building blocks of the CNT fiber, CNT is well recognized as a high-performing material with remarkable electrical conductivity (10^6 S/cm) and tensile strength (100 GPa) since its first discovery in 1991 [39–42]. However, the rational assembling of CNTs plays a critical role in the physical and chemical properties of CNT fibers.

Wet-spinning represents a widely used method for the preparation of CNT fibers. Generally, CNTs were dispersed in solvents (e.g. water, fuming sulfuric acid, chlorosulfuric acid, and dimethyl sulfoxide) by ultrasonication or shear mixing as the spinning dopes [43–45]. Subsequently, the dispersions were injected via a spinneret/needle into a liquid coagulation bath, e.g. aqueous polyvinyl alcohol solution, dilute sulfuric acid, acetone, and water, and the extruded flows were solidified into fibers by aggregation of the CNTs upon drawing forces [44–46]. Post-treatment, such as annealing, could remove organic impurities and increase the conductivity of the CNT fiber. To date, approaches based on surfactant-stabilized and superacid-aid spinning have been reported.

A pioneer work on the wet-spinning of CNT fibers was reported in 2000, using sodium dodecyl sulfate as a surfactant to facilitate the dispersion of single-walled carbon nanotubes (SWCNTs) [46]. The obtained dispersion was injected into an aqueous coagulation bath containing 5 wt% of polyvinyl alcohol to form a ribbon. In this process, flow-induced alignment of the nanotubes took place due to the shear force generated by the rotating stage. A SWCNT fiber with helical structure could be obtained when the polyvinyl alcohol solution was pumped out from container (Figure 4.7a,c). This optimized wet-spinning method resulted in a tensile strength of 150 MPa and electrical conductivity of ~10 S/cm. The tying knots revealed the high flexibility and resistance to torsion of the resulted SWCNT fibers (Figure 4.7b). The concentration of surfactant was crucial to obtain uniformly dispersed SWCNTs. Low-concentration surfactant failed to afford the stabilization of SWCNT/water dispersion, which induced electrostatic repulsion to compromise intertube van der Waals attractions. At a high concentration, encasing the SWCNTs was difficult due to the aggregates of surfactant. Only appropriate concentrations of surfactants, e.g. an optimum at 1.0 wt% sodium dodecyl sulfate for 0.35 wt% nanotubes in water, could assist SWCNTs in forming homogeneous water dispersion. Unfortunately, the properties of the resulted SWCNT fibers are largely affected by surfactants (Figure 4.7d).

To overcome the negative effects of surfactants, superacids were used for the stabilization of SWCNT liquid crystal dispersion to obtain highly oriented SWCNT fibers. SWCNTs were protonated in fuming sulfuric acid, and the individual charged SWCNT was surrounded by a large number of sulphonic acid groups [44]. The formed SWCNT/acid charge-transfer complexes were highly mobile, dispersible, and easily aligned into ordered mesophases (Figure 4.8a,b), resulting in an enhanced Young's modulus of 120 GPa, tensile strength of 116 MPa, and high

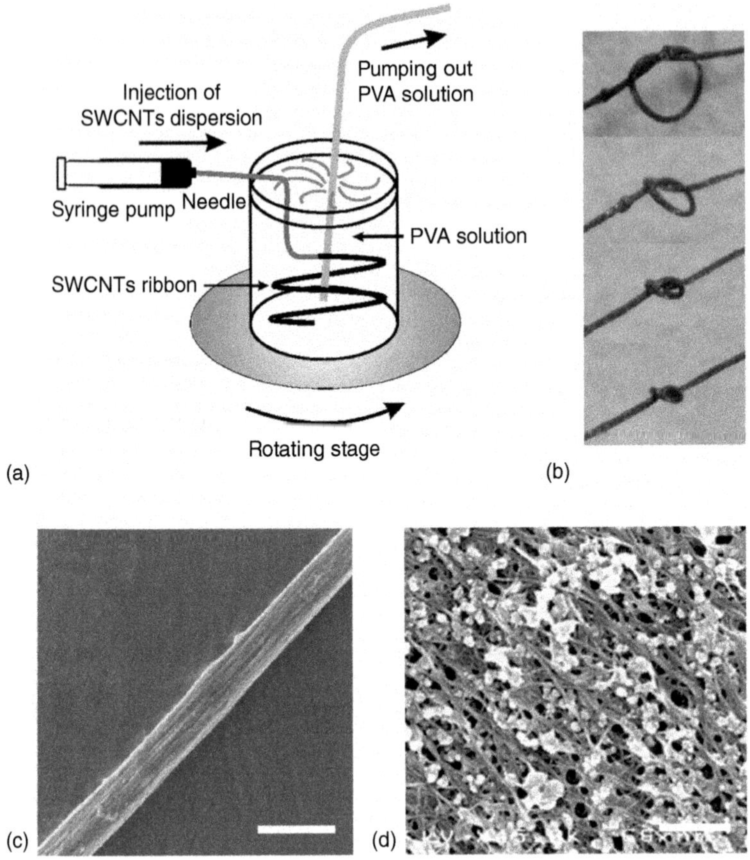

Figure 4.7 (a) Schematic illustration of the experimental setup for fabricating SWCNT fibers. (b) Tying knots of the SWCNT fibers. (c) SEM image of a SWCNT fiber. (d) SEM image showing the presence of numerous spherical carbon impurities. Source: Vigolo et al. [46], © 2000/American Association for the Advancement of Science. Scale bar: (c), 25 μm; (d), 300 nm.

electrical conductivity of 5000 S/cm. In addition, the residual acid could be removed by subsequent washing, and the obtained SWCNT fiber exhibited a smooth surface without contaminants or impurities (Figure 4.8c,d).

Chlorosulfuric acid could also benefit the dispersion of SWCNTs at a relatively high concentration of 0.5 wt% (Figure 4.9a) [47]. The high-concentration SWCNT was important for the formation of liquid crystal phases, which improved the alignment of SWCNTs. The SWCNT fibers prepared by wet spinning showed a tensile strength of 50–150 MPa and electrical conductivity of 8300 S/cm (Figure 4.9b). Although the wet-spinning method is very appealing for industrial production of SWCNT fibers, improvements are needed to further optimize the preparation process. For instance, a minimal moisture would result in phase separation and precipitation of needle-like crystal solvates. Additionally, the complete removal of residual strong acids during the purification process increased the production

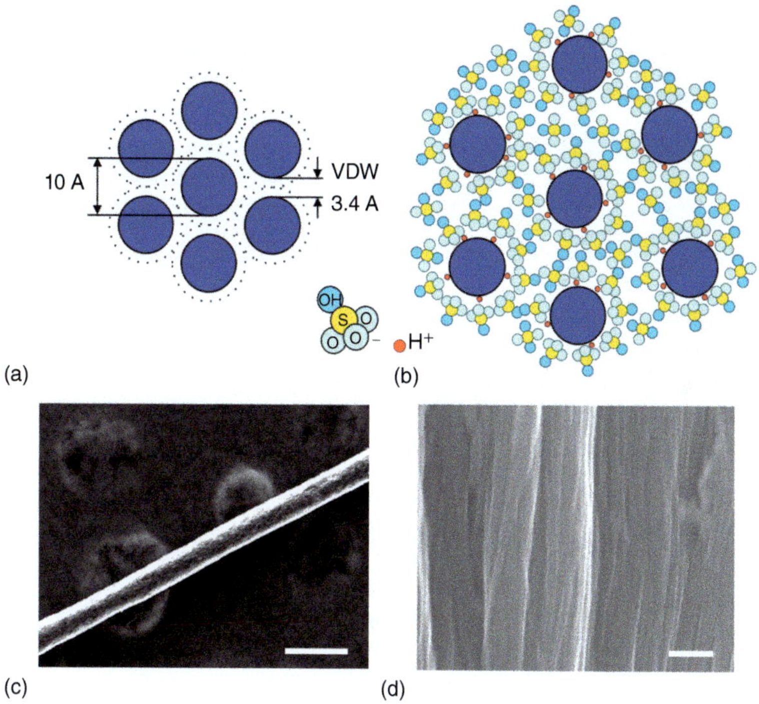

Figure 4.8 (a) Cartoon of SWCNTs with van der Waals contact. (b) Formed charge-transfer complex after SWCNT fibers dispersed in sulfuric acid. (c) An annealed neat SWCNT fiber spun from sulfuric acid dispersion. (d) SEM image of the fiber surface showing a smooth appearance without any surface contaminants or impurities. Source: Ericson et al. [44], © 2004/American Association for the Advancement of Science. Scale bar: (c), 200 μm; (d), 100 nm.

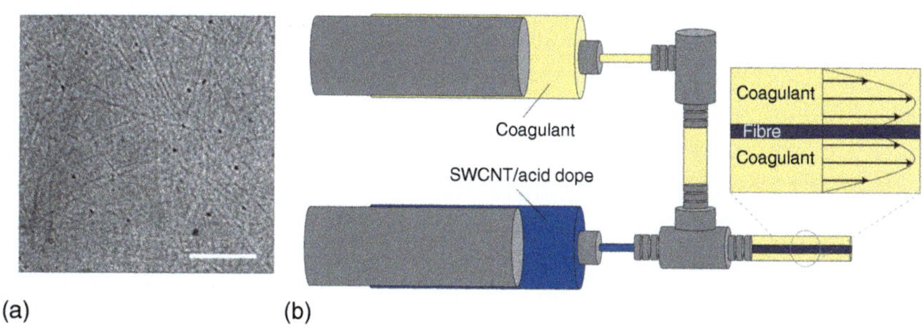

Figure 4.9 (a) TEM image of SWCNTs in CSA. (b) The wet-spinning process of fibers from high-concentration SWCNTs solutions. Source: Davis et al. [47], © 2009/Springer Nature. Scale bar: (a), 100 nm.

complexity. More importantly, the spinning of MWCNTs in superacid is still challenging due to the limited solubility.

Compared with the wet-spinning method, the dry-spinning method avoids the use of surfactants or superacids for the enhancement of MWCNT dispersion. The typical dry-spinning method is based on spinnable MWCNT arrays, in which

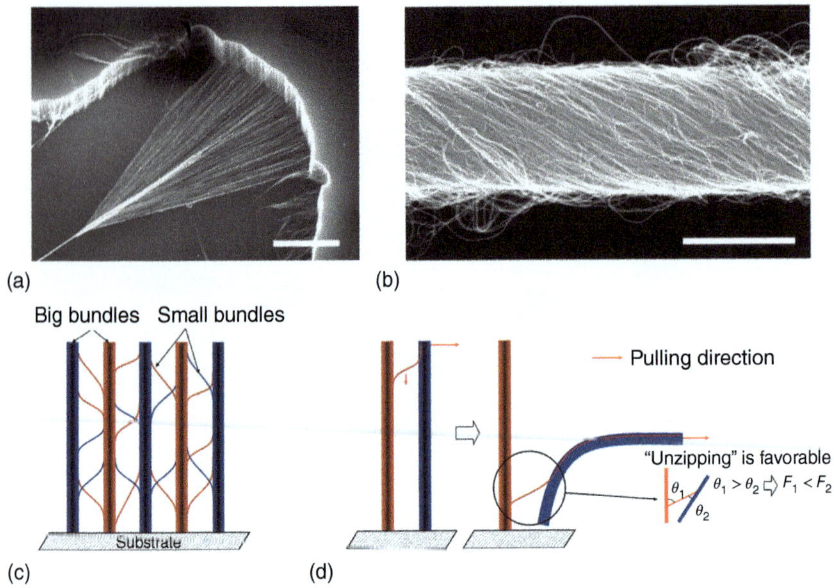

Figure 4.10 (a) SEM images of the twist-drawing spinning process. (b) SEM images of the obtained single MWCNT fiber. Source: Zhang et al. [48], © 2004/American Association for the Advancement of Science. Schematic illustration of the two spinning models by interconnection (c) and van der Waals interactions (d). Source: Kuznetsov et al. [49], © 2011/American Chemical Society. Scale bar: (a), 200 μm; (b), 2 μm.

the MWCNTs are vertically aligned like the forest. Zhang et al. prepared the first aligned MWCNT fiber based on a spinnable MWCNT array. An aligned MWCNT sheet was drawn from the array, followed by twisting into a fiber using a rotating spindle (Figure 4.10a) [48]. The MWCNT bundles showed good alignment that resulted in a high tensile strength and electrical conductivity of 150–300 MPa and ~300 S/cm, respectively. Abrasion and overhand knots were easily achieved without breakage (Figure 4.10b). The continuous spinnability originated from the strong van der Waals interactions among MWCNT bundles (Figure 4.10c,d) [49].

More efforts have been made to improve the strength of aligned MWCNT fibers toward real-world applications. Zhang et al. treated the aligned MWCNT fiber using ethanol for densification [50]. The surface tension of organic solvents with the meniscus shape played a critical role in improving the density of the obtained fibers (Figure 4.11a). It showed a high tensile strength of ~600 MPa, along with excellent flexibility and shape-memory capability (Figure 4.11b). In addition, the mechanical (tensile strength and Young's modulus) and electrical properties could be enhanced through the optimization of MWCNT arrays [52] and other post-treatment strategies such as mechanical compression [53], chemical treatment [54, 55], plying [48], and laser heating under traction force [56]. Some inherent advantages are unique for forest spinning over other spinning methods, such as easy control of the process variables, fewer impurities, and enhanced contact among MWCNT bundles. However, the spinnable MWCNT forests require strict

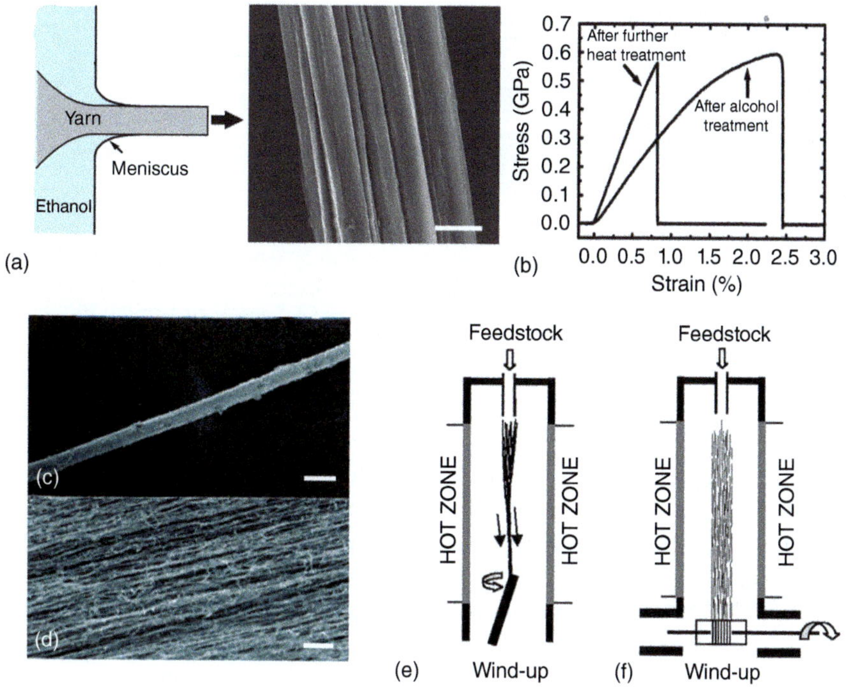

Figure 4.11 (a) Pulling a MWCNT film through ethanol into a tight fiber. (b) Stress–strain curves of ethanol-treated and further heat-treated MWCNT fibers. Source: Zhang et al. [50], © 2006/John Wiley & Sons. (c, d) SEM images of a fiber that consists of well-aligned MWCNTs at different magnifications. (e) Schematic of the direct-spinning process. (f) Schematic of the wind-up assembly outside the furnace hot zone. Source: Li et al. [51], © 2004/American Association for the Advancement of Science. Scale bar: (a), 10 μm; (c), 50 μm; (d), 2 μm.

preparation conditions such as catalyst, gas purity, and temperature control. In general, highly aligned MWCNTs with rather clean surfaces and strong van der Waals interaction are the prerequisites for the continuous spinning of MWCNT fibers.

Direct spinning of MWCNT fibers was developed by Li et al. to produce continuous MWCNT aerogel fibers in one single step (Figure 4.11c,d) [51]. MWCNT aerogel was synthesized by chemical vapor deposition using an ethanol feedstock (including small quantities of ferrocene and thiophene) and H_2 carrier gas. The formed aerogel was captured and wound out of the hot zone to form a continuous fiber (Figure 4.11e,f). In this process, the rapid growth of high-purity MWCNT aerogel was vital for continuous fiber spinning. The highly aligned MWCNTs along the axial direction contributed to a remarkable tensile strength and electrical conductivity of 0.1–1.0 GPa and 8300 S/cm, respectively. Further studies on mechanical and electrical property improvement of MWCNT fibers involve fabrication process optimization, such as the synthesis of long and large-diameter MWCNTs [57], densification using acetone [58], and variation of the winding rate [59]. The reduced production cost of the direct-spinning method is a highly appealing feature.

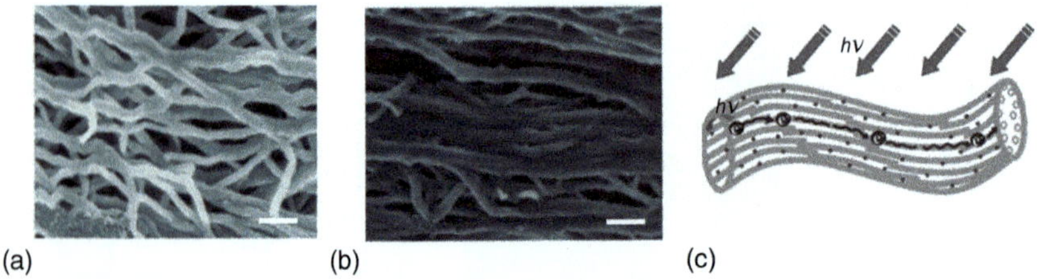

Figure 4.12 SEM images of nanotubes before (a) and after (b) incorporation of N719. (c) Diagram of the production and transportation of photoelectrons in a nanotube/N719 composite fiber. Source: Chen et al. [60], © 2011/John Wiley & Sons. Scale bar: (a), 100 nm; (b), 100 nm.

However, the removal of the remaining impurities, such as catalysts, requires extra treatment to better meet the application requirement.

The above progress on MWCNT fibers made them promising candidates as the photoanodes and counter electrodes of fiber-shaped dye-sensitized, polymer, and perovskite solar cells. A pioneer work was proposed by Chen et al. in 2011, which employed modified aligned MWCNT fibers as the photoanodes in fiber-shaped DSSCs [60]. The N719 dye molecules were well distributed among MWCNTs based on the π–π interaction and van der Waals force (Figure 4.12a,b), which facilitated the electron injection from excited N719 molecules into the conduction band of nanotubes (Figure 4.12c). The alignment of MWCNTs enhanced the charge transport based on the 3D hopping mechanism, consequently improving the separation of electrons and holes. Coupling with a Pt wire as the counter electrode, the achieved DSSC in a planar format showed a PCE of 2.2%. This work opened the avenue of aligned MWCNT fibers in wearable solar cells.

To further improve the PCEs, aligned MWCNT fibers were further optimized by coating a TiO_2 nanoparticle layer serving as the photoanodes of fiber-shaped DSSCs (Figure 4.13a) [61]. The electrons were injected from the excited dye molecules into the conduction band of TiO_2 nanoparticles, and were separated and transported along the MWCNT fiber (Figure 4.13b). The TiO_2 nanoparticles could enhance the adsorption of dye molecules to improve the photoinduced charge separation and transport. In addition, the aligned MWCNT fibers with high mechanical and electrical properties were used as the counter electrode, resulting in all-carbon-based fiber-shaped DSSCs. A high PCE of 2.94% was thus achieved with good flexibility and weavability (Figure 4.13c).

Counter electrodes of fiber-shaped polymer solar cells using rigid metal wires might easily penetrate the ultrathin polymer layers, leading to the occurrence of a short circuit during the assembling process. To this end, the highly flexible MWCNT fibers can be ideal counter electrodes of fiber-shaped polymer solar cells. For instance, Ti wires with aligned TiO_2 nanotubes were coated by photoactive materials, including the poly(3-hexyl-2,5-thiophene) and (6,6)-phenyl-C_{71} butyric acid methyl ester as the photoanode, coupling with an aligned MWCNT fiber as the counter electrode to produce a fiber-shaped polymer solar cell (Figure 4.14a) [62]. A decent PCE of 1.78% was achieved, compared with only 0.48% based on an

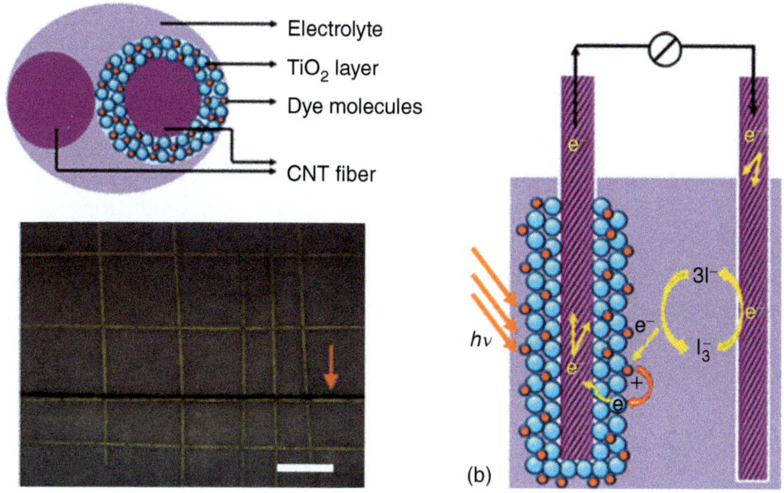

Figure 4.13 Top view (a) and working mechanism (b) of the all carbon-based DSSC fabricated from two MWCNT fibers. (c) A fiber-shaped solar cell is woven into a textile composed of aramid fibers. Source: Chen et al. [61], © 2012/American Chemical Society. Scale bar: (c), 100 μm.

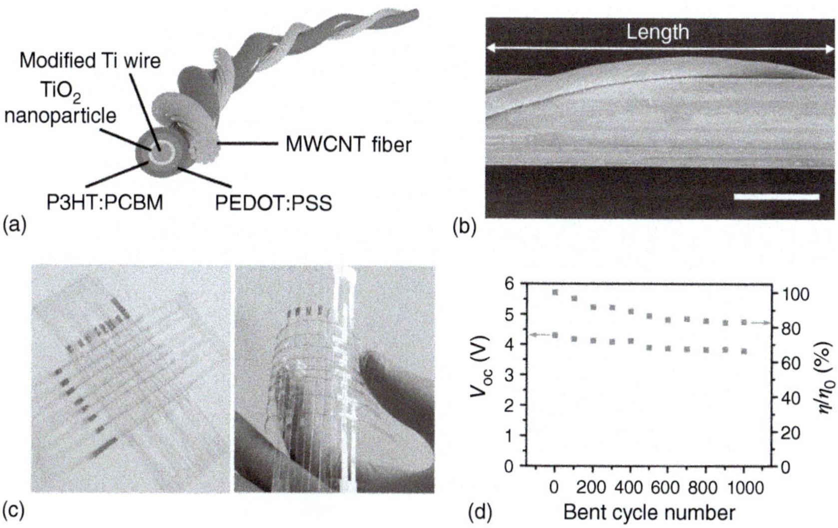

Figure 4.14 (a) Schematic illustration of the fiber-shaped polymer solar cells. (b) SEM image of the twisted length. (c) Polymer solar cell textiles are bent into different structures. (d) Dependence of PCE and voltage on bent cycle number in a polymer solar cell textile. η_0 and η correspond to the PCE before and after bending, respectively. Source: Zhang et al. [62], © 2014/John Wiley & Sons. Scale bar: (b), 150 μm.

Ag wire counter electrode, owing to the low work function of 4.3 eV for Ag metal. MWCNT fiber with a high work function of 4.8–5.1 eV enabled more effective hole collection, thus more suitable for serving as the counter electrode. Increased twisted lengths and diameters of MWCNT fibers led to decreased PCEs owing to the severe light shadowing (Figure 4.14b). The obtained polymer solar cells could be woven into various flexible structures such as textiles (Figure 4.14c), and the PCE could be maintained by 85% after bending for 1000 cycles (Figure 4.14d), indicating the high flexibility toward wearable applications.

The MWCNT fibers can also replace the noble metal wires in fiber-shaped perovskite solar cells, owing to their outstanding flexibility, lightweight, and high accessibility for enhanced catalytic activity. For instance, a MWCNT fiber electrode was sequentially coated with a thin layer of compact n-type TiO_2, mesoporous TiO_2, $CH_3NH_3PbI_{3-x}Cl_x$, poly(3-hexylthiophene) (P3HT)/SWCNT and silver nanowire, which was then twisted with a pristine MWCNT fiber to form a double-twisted architecture (Figure 4.15a) [63]. The n-type TiO_2 layer was used to prevent electron–hole recombination toward high PCE, and its thickness was controlled by changing the dip-coating times. The obtained fiber-shaped perovskite solar cell showed a PCE of 3.3% (Figure 4.15b). The mesoporous TiO_2 layer was of importance for the uniform and dense deposition of perovskite resulted from physical adsorption, which contributed to excellent structural stability under deformations. The PCE remained almost unchanged after 1000 bending cycles (Figure 4.15c), and

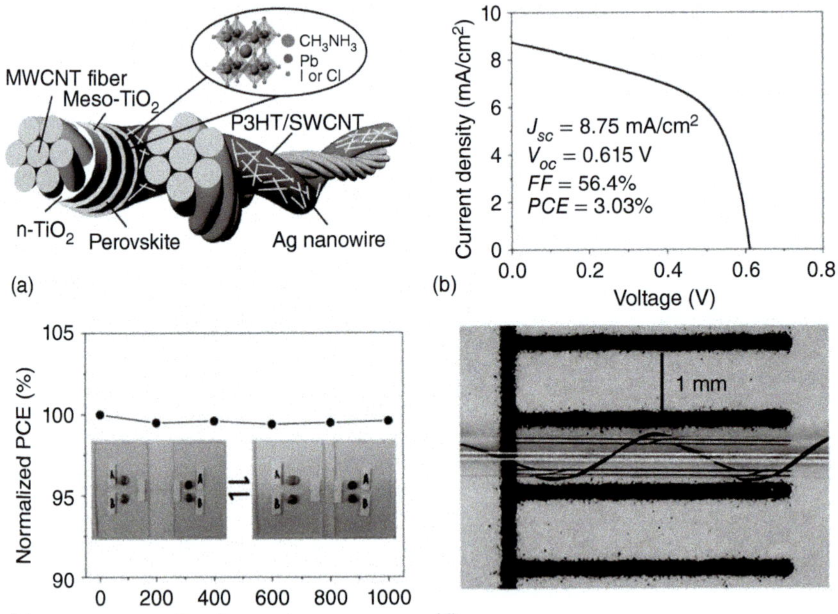

Figure 4.15 Schematic illustration (a) and *J*–*V* curve (b) of the double-twisted fiber-shaped perovskite solar cell. (c) The PCE at different bending cycles. (d) The photograph of the perovskite solar cell wrapped onto a glassy tube. Source: Li et al. [63], © 2015/John Wiley & Sons.

the solar cell could be readily wrapped on a glass tube without structure damage or performance degradation (Figure 4.15d).

4.4.3 Graphene Fiber

Individual graphene sheet exhibits remarkably high mechanical strength (130 GPa), modulus (1100 GPa) [64], and high electrical conductivity (1×10^4 S/cm) [65], which has attracted widespread attention since first prepared via hand peeling by Novoselov et al. [66]. Graphene can be assembled into one-dimensional macroscopic fibers, which opens up their application in fiber electronics [67].

Twisting aligned graphene/graphene oxide films into a fiber electrode represents a simple approach, which represents an early trial of this field. For example, Cruz-Silva et al. fabricated a large-area graphene oxide film by using a graphene oxide aqueous dispersion (Figure 4.16a) [68]. The obtained film was then twisted into a graphene oxide fiber, which exhibited high toughness and elongation of 17 J/m^3 and 76%, respectively. The well-ordered and concentrically packed layers provided high axial tensile strengths of the fibers, and the void spaces between the layers were responsible for toughness enhancement. In addition, no obvious damage was observed upon knotting and over-twisting (Figure 4.16c). Thermal treatment at 2800 °C was subsequently performed to obtain electrically conducting reduced graphene oxide fibers (electrical conductivity 416 S/cm) (Figure 4.16b). The reduced graphene oxide fibers could be used as electron field emitters at a low turn-on field of 0.48 V/μm and high current densities (5.3 A/cm^2), which outperformed chemical vapor deposition-grown graphene fibers. The remarkable field emission properties were attributed to the orientation of the reduced graphene

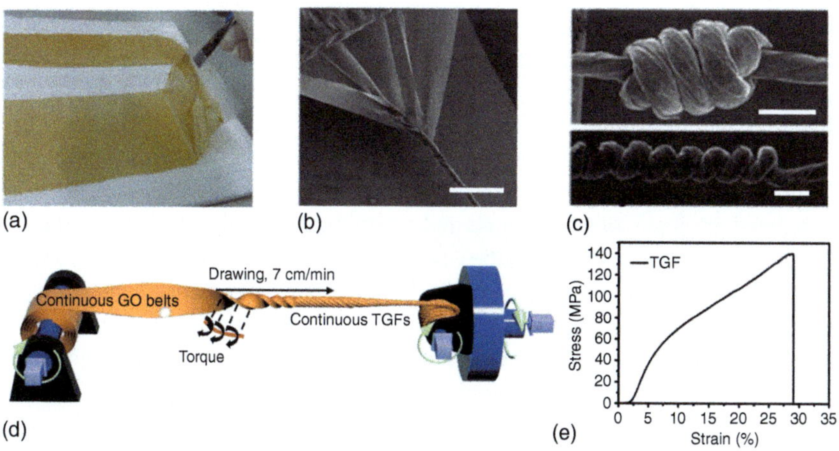

Figure 4.16 (a) Optical photograph of the thin graphene oxide film. (b) SEM image of the graphene oxide film being scrolled into a fiber. Source: Cruz-Silva et al. [68], © 2014/American Chemical Society. (c) Quadruple overhand knot and self-coiled homochiral fiber. (d) Scheme of the drawing-twisting process to obtain continuous graphene fibers. (e) Stress–strain curves of the twisted graphene fibers (TGF). Source: Fang et al. [69], © 2019/Royal Society of Chemistry. Scale bar: (b), 1 mm; (c), 500 μm.

oxide sheets in the fiber. However, the length of reduced graphene oxide fibers was severely hindered by the size of the graphene oxide films, and the discontinuity of the twisting process resulted in rather limited scalability.

A continuous production process was developed by Fang et al. connecting the liquid crystalline spinning of graphene oxide belts and the twisting step [69]. Graphene oxide belts were continuously obtained via a wet-spinning method based on a microfluidic system. Then, twisting the graphene oxide belts produced spirally oriented graphene fibers by using a twist-drawing apparatus (Figure 4.16d). Importantly, the rotation direction was adjustable; thus, both right-handed and left-hand graphene fibers with mechanical robustness (tensile strength of 130 MPa at a strain of 29%) and good flexibility could be acquired as controllable solvent actuators (Figure 4.16e). They exhibited the application potential in smart fabrics. Besides the twisting method, other strategies have been proposed to fabricate graphene fibers, such as the shape-confined hydrothermal method [70], chemical vapor deposition [71], and chemical reduction-induced self-assembly [72].

The graphene fibers produced by the above-mentioned methods usually exhibit low tensile strengths of less than 150 MPa because of the large interlayer distance and weak interaction between the graphene sheets. Wet spinning represents an efficient method for preparing high-quality graphene fibers. Graphene oxide sheets with numerous oxygen-containing functional groups can be dispersed into many polar solvents to form lyotropic nematic liquid crystals, serving as the basic building blocks of the graphene fibers. During fiber spinning, the orientation of graphene oxide sheets is enhanced, and the liquid crystalline defects decrease under shearing at the spinneret, which contributes to improved mechanical properties. Further chemical or thermal reduction treatment can reduce graphene oxide fibers into reduced graphene oxide (rGO) fibers with significantly promoted electrical conductivities. The tensile strengths can also be enhanced, because the reduction process decreases the interlayer distance of microcrystals, thus strengthening the interlayer interaction.

First, dry-spun graphene fibers were prepared by Xu et al. in 2011 by extruding graphene oxide liquid crystal dispersions into a coagulation bath to obtain graphene oxide fibers, followed by chemical reduction by hydroiodic acid (Figure 4.17a). The obtained graphene fiber demonstrated a remarkable tensile strength, Young's modulus, and electrical conductivities of 140 MPa, 7.7 GPa, and 2.5×10^2 S/cm, respectively [73]. In addition, they could be made into tight knots and woven into cotton textiles (Figure 4.17b,c). Many other strategies have been developed to improve the performance of graphene fibers, e.g. using large-size graphene oxide sheets to decrease boundaries and voids [75], flow-stretching in the rotating bath and stretching tension to enhance the orientation [76, 77], and mixing graphene oxide sheets with different sheet sizes to optimize the compactness of graphene fibers [78]. For instance, Xu et al. reported a type of ultra-strong graphene fiber with ultrahigh stiffness, tensile strength, and high electrical conductivity of 282 GPa, 1.45 GPa, and 8×10^3 S/cm, respectively (Figure 4.17d) [74]. Giant graphene oxide sheets with 24 μm average width were prepared via chemical oxidation and graphene oxide/*N*,*N*-dimethylformamide spinning dopes were subsequently

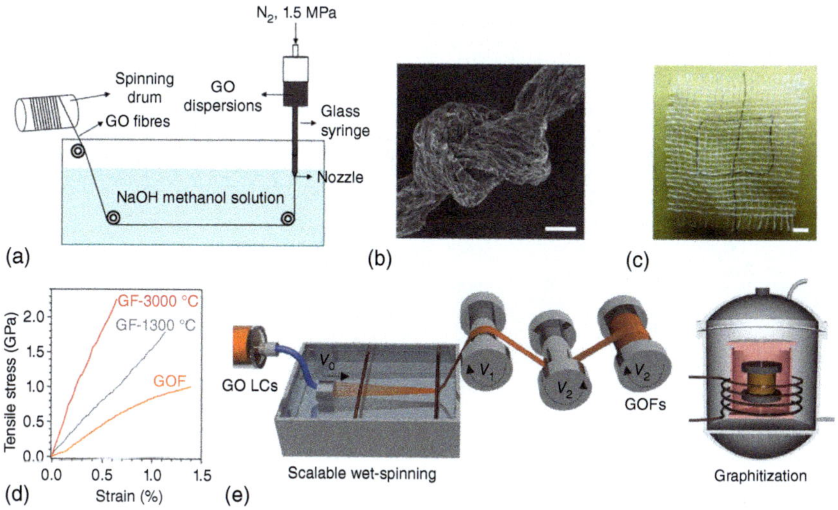

Figure 4.17 (a) Continuous wet-spinning of graphene oxide fibers from aqueous graphene oxide liquid crystals. (b) The tightened knot of a graphene fiber. (c) Several graphene fibers (horizontal) are knitted with cotton threads (vertical). Source: Xu et al. [73], © 2020/Springer Nature. (d) Typical stress–strain curves of graphene fibers. (e) Schematic illustration of wet-spinning of graphene oxide fibers with continuous stretching in two stages, followed by graphitization at 3000 °C to transform graphene oxide fibers into graphene fibers. Source: Xu et al. [74], © 2016/John Wiley & Sons. Scale bar: (b), 50 μm; (c), 2 mm.

extruded into a coagulation bath. A continued stretching tension was employed to enhance the alignment of graphene oxide sheets and densify the microstructure of graphene oxide fibers. The combination of chemical and thermal reduction using hydrogen iodide and 3000 °C annealing was performed to complete the reduction of graphene oxide fibers (Figure 4.17e). Besides the dense graphene fibers for high-performance applications, hollow [79], porous [80], and belt-like [81] counterparts have been produced by wet-spinning for various applications, such as flexible supercapacitors, electrochemical sensors, and absorbents.

Owing to its excellent physical and chemical properties, the graphene fibers have been widely used in fiber-shaped solar cells. For example, Chen et al. drew a hollow graphene tube/sheet from water to prepare a porous graphene fiber [71]. It exhibited a high electrical conductivity of more than 120 S/cm with high flexibility. The obtained graphene fiber was used as a counter electrode to replace the conventional Pt electrode in DSSCs toward low cost and high electrochemical properties. A fiber-shaped DSSC was fabricated using the graphene fiber and TiO_2 nanotubes/Ti wire as the counter and working electrode, respectively (Figure 4.18a), demonstrating a PCE of 3.25% compared with 2.14% of Pt wire (Figure 4.18b). This could be originated from the higher surface area of porous graphene fiber that showed higher electrochemical properties. In order to further improve PCEs, a graphene/Pt composite fiber was prepared by Yang et al., which uniformly electrodeposited Pt nanoparticles on the surface of graphene fiber using a double potential step electrochemical method (Figure 4.18c) [82]. The weight

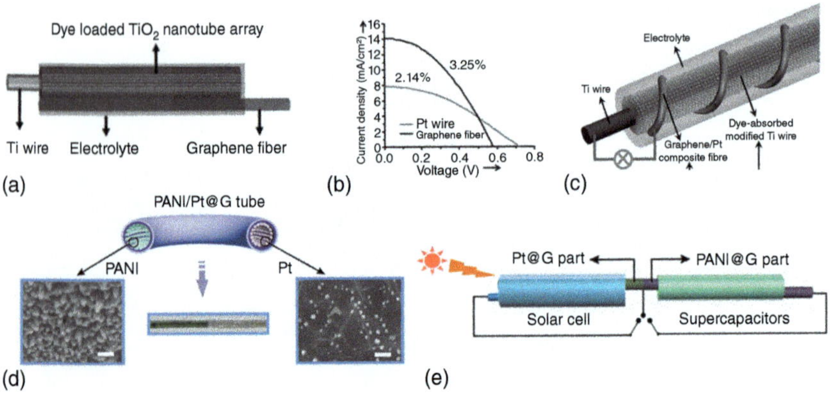

Figure 4.18 (a) Schematic illustration of the DSSC using the graphene fiber as the counter electrode. (b) *J–V* curves of the fiber-shaped DSSC based on the graphene fiber or Pt wire counter electrode. Source: Chen et al. [71], © 2015/John Wiley & Sons. (c) Schematic illustration of the DSSC prepared by using a graphene/Pt composite fiber as the counter electrode. Source: Yang et al. [82], © 2013/John Wiley & Sons. (d) Schematic of PANI//Pt@graphene tube. Inset: the SEM images of PANI@graphene tube and Pt@graphene tube. (e) Schematic illustration of the integrated device containing DSSC and supercapacitor. Source: Yao et al. [83], © 2020/Elsevier. Scale bar: (d) 200 nm.

percentage of Pt nanoparticles could be controlled by varying the deposition time. This enhanced the interaction between the graphene and Pt nanoparticles, contributing to improved transport of electrons across the interface. According to the cyclic voltammetry profile, graphene/Pt composite fibers demonstrated higher peak current densities and lower peak-to-peak voltage separations (<0.3 V) compared with Pt wire. This indicated that graphene/Pt composite fibers had a higher catalytic capability for the reduction of I_3^-, which was attributed to the higher surface areas of Pt nanoparticles. A higher PCE of 8.45% was achieved for the obtained DSSC based on the composite fiber electrode compared with 7.31% of Pt wire.

The comprehensively high mechanical, electrical, and electrochemical properties of graphene fibers make them ideal electrodes of integrated energy conversion and storage devices. For instance, a graphene fiber with a hollow structure was produced by assembling graphene tubes, which provided a novel platform for the functionalization of fiber electrodes [83]. The unique hollow tube structure enabled one graphene fiber to be selectively functionalized on the inner surface of the tube with different active materials, which could be used as a common electrode for integration of both energy conversion and storage in series. Polyaniline nanowires could selectively be grown inside of the left half segment of the hollow tube, serving as a high-performce counter electrode of supercapacitor, and Pt nanoparticles could be deposited inside of the right half segment as counter electrodes for DSSC (Figure 4.18d). With the two parts connected by an external wire, the electricity generated by the DSSC part under sunlight could be stored to the supercapacitor part in real-time. The integration of the common electrode avoided the energy loss generated by the external conducted wire of the traditional integration method.

As a result, the integrated device demonstrated a high PCE (4.73%), specific capacitance (284 mF/cm^2), and total energy conversion and storage efficiency (3.07%) (Figure 4.18e).

4.5 Functional Fibers

To better accommodate the complex and harsh application scenarios, wearable solar cells are expected to be functional. As a result, a variety of functional fiber electrodes have been developed for the functionalization of fiber-shaped solar cells. For instance, a stretchable electrode was produced by Yang et al. to accommodate the frequent strain and stress for wearable applications [28]. Aligned MWCNT sheets were wrapped on a rubber fiber using a rotary motor (Figure 4.19a). The resulted composite fibers demonstrated stable electrical properties under repeated stretching and releasing. The Ti wire with perpendicularly aligned TiO_2 nanotubes was made into a spring-like shape as the working electrode, and the stretchable conductive fiber was inserted into the working electrode as the counter electrode. The resulted fiber-shaped DSSC could be woven into a stretchable textile, which retained 90% of the original PCE after 50 stretching cycles at a strain of 20%.

Incorporation of functional guests into fiber electrodes without sacrificing the electrical conductivity is important while remaining challenging. Lima et al. developed a biscrolling process for incorporation of various functional guests (such as

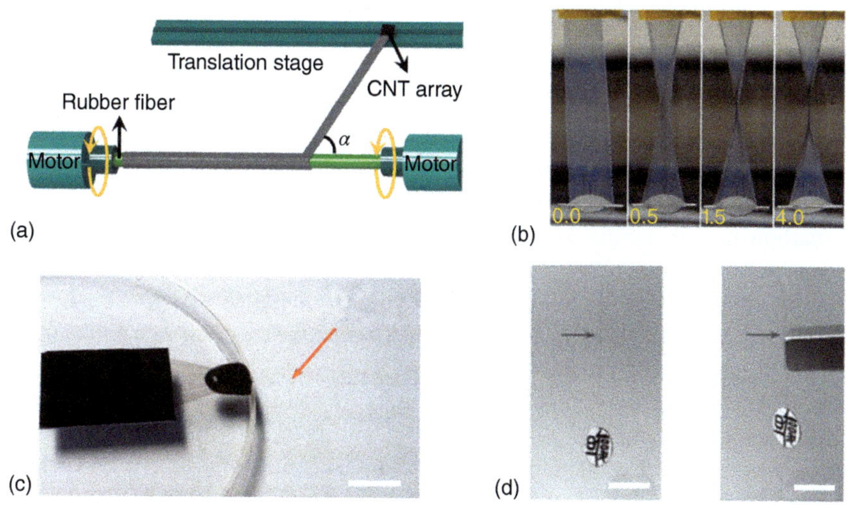

Figure 4.19 (a) Schematic illustration for fabrication of stretchable conductive fibers. Source: Yang et al. [28], © 2014/John Wiley & Sons. (b) Preparation of functional fiber electrode using a biscrolling method. The numbers at the left bottom corner represent the twisting turns. Source: Lima et al. [84], © 2011/American Association for the Advancement of Science. (c) A continuous process for the production of Fe_3O_4/MWCNT fibers with magnetic response. (d) Optical photographs of magnetic Fe_3O_4/MWCNT fibers before and after being attracted to a magnet. Source: Sun et al. [85], © 2013/John Wiley & Sons. Scale bar: (c), 1 cm; (d) 1 cm.

Si_3N_4, TiO_2, and $LiFePO_4$) into MWCNT fibers (Figure 4.19b) [84]. A high loading of up to 95% and good electrical conductivity were obtained by this method, providing an excellent platform for producing functional fiber electrodes. However, the fixation of both ends of MWCNT sheets hindered the continuous production. To solve this problem, Sun et al. scrolled and drew the aligned MWCNT sheets through a Fe_3O_4 nanoparticle/water dispersion to continually produce the magnetic fiber electrodes (Figure 4.19c) [85]. The obtained Fe_3O_4/MWCNT hybrid fibers could be easily attracted by a magnet (Figure 4.19d). This method realized the continuous production of fibers electrodes even with lengths of hundreds of meters. To obtain higher electrocatalytic activity, Pt nanoparticles were electrodeposited onto hybrid fibers as the counter electrode. The PCE reached 8.03% outperformed 6.52% of the Pt wire counter electrode.

4.6 Summary and Perspectives

In this chapter, we introduce the mainstream fiber electrode materials of fiber-shaped solar cells, focusing on the preparation strategies and basic properties of different fiber electrodes and their applications in fiber-shaped solar cells. Despite of the impressive progress in recent years, some key aspects should be considered and improved to realize real-world applications.

First, the electrical conductivity needs further optimization, which is critical for producing fiber-shaped solar cells with lengths of meters or even kilometers. Up to date, the solution for fabricating long fiber-shaped devices lies in the use of metal electrodes with ultralow resistances (up to $2\,\Omega/m$) [86]. For instance, a long, high-performance fiber-shaped lithium-ion battery has been recently reported using Cu and Al as the anode and cathode, respectively, via industrial production equipment and process [87]. However, metal electrodes generally exhibit low yield strains (e.g. 0.3% for Cu), making them fragile after repeated bending. To this end, carbon nanomaterials with higher flexibility are promising candidates for the construction of long fiber devices. Still, their electrical resistance ($>100\,\Omega/m$) require significant reduction to fully compete with the metal wires.

Second, the continuous fabrication of low-cost fiber electrodes is of significance but remains challenging to date. Aligned MWCNT fibers with both superior electrical and mechanical properties can be readily realized via the dry-spinning approach using spinnable MWCNT arrays. However, the continuous and scale-up production is still plagued by the production complexity and cost. Floating-catalyst CVD can be a more promising alternative to this end, which allows for the direct spinning of kilometer-long MWCNT fibers with high throughput and low cost [51]. The reaction mechanism requires in-depth understanding, and the post-treatment process requires further optimizations. Ideally, the prices of MWCNT fibers can be potentially reduced (0.1–25 $/g) [88, 89], which can open up their large-scale applications in wearable solar cells.

Third, the morphologies of fiber electrodes require better regulation in order to obtain higher photovoltaic conversion performance. For example, the working electrodes with TiO_2 nanoparticles would result in high electrical resistances due to the numerous grain boundaries, which restrains the charge transport from semiconductors to electrodes. Vertically grown TiO_2 nanotube on Ti wire through an anodic oxidization approach can provide shortened charge transport pathways, and the increased porosity among the TiO_2 nanotube can greatly enhance the loading of photosensitive dyes, which increase the J_{SC} and PCEs of the resulted fiber-shaped solar cells, e.g. 15.42 mA/cm^2 and 6.93% [12]. Besides, the catalytic performance of Pt nanoparticles outperforms conventional Pt wire attributed to the much higher specific surface area [82]. As a result, Pt nanoparticles deposited on graphene fibers result in enhanced catalytic capability of I_3^- reduction, which contributes to a higher PCE of 8.45%.

Fourth, the unique interface between the one-dimensional fiber substrate and active components should be taken into serious consideration. The unstable interfacial interaction generally results in poor structural stability, particularly under various deformations. To this end, building a strong interfacial interaction such as covalent or hydrogen bonds represents a promising strategy. For instance, hydrogen-bond interaction between liquid metal and polymethacrylate has been proven to significantly enhance the stability of fiber-shaped electrodes under high-stress conditions [90, 91]. In addition, conductive polymers such as PEDOT:PSS have been fabricated into stretchable electrodes for organic electrochemical transistors with high conductivities [92, 93]. The introduction of these materials and interfaces to novel fiber electrodes is promising to further improve the photovoltaic conversion performance of wearable solar cells.

Last but not least, device functionalization represents an important step toward practical applications. For instance, fiber-shaped solar cells are required to adapt the frequent bending and stretching at specific positions such as elbows and knees, which depend on the stretchable capability. Besides, the delicate fiber device might be broken during use, and the self-healing capability will benefit the recovery of the mechanical and photovoltaic conversion properties. Functional fiber electrodes are the key building blocks to realize the functionalization of fiber devices. Therefore, more efforts should be devoted to the development of functional fiber electrodes and processing techniques for wearable solar cells and integration devices.

References

1 Sun, H., Zhang, Y., Zhang, J. et al. (2017). Energy harvesting and storage in 1D devices. *Nat. Rev. Mater.* 2 (6): 1–12.

2 Kang, M. and Kim, T.W. (2021). Recent advances in fiber-shaped electronic devices for wearable applications. *Appl. Sci.* 11 (13): 6131.

3 Wang, X., Jiang, K., and Shen, G. (2015). Flexible fiber energy storage and integrated devices: recent progress and perspectives. *Mater. Today* 18 (5): 265–272.

4 Hashemi, S.A., Ramakrishna, S., and Aberle, A.G. (2020). Recent progress in flexible-wearable solar cells for self-powered electronic devices. *Energy Environ. Sci.* 13 (3): 685–743.

5 Cheng, B. and Wu, P. (2021). Scalable fabrication of Kevlar/$Ti_3C_2T_x$ MXene intelligent wearable fabrics with multiple sensory capabilities. *ACS Nano* 15 (5): 8676–8685.

6 Zhang, X., Lin, H., Shang, H. et al. (2021). Recent advances in functional fiber electronics. *SusMat* 1 (1): 105–126.

7 Peng, M., Dong, B., and Zou, D. (2018). Three dimensional photovoltaic fibers for wearable energy harvesting and conversion. *J. Energy Chem.* 27 (3): 611–621.

8 Lee, S.-H., Park, J.H., and Kim, S.M. (2021). Synthesis, property, and application of carbon nanotube fiber. *J. Korean Ceram. Soc.* 58 (2): 148–159.

9 Li, L., Wang, L., Ye, T. et al. (2021). Stretchable energy storage devices based on carbon materials. *Small* 17 (48): 2005015.

10 Chen, M.X., Wang, Z., Li, K.W. et al. (2021). Elastic and stretchable functional fibers: a review of materials, fabrication methods, and applications. *Adv. Fiber Mater.* 3 (1): 1–13.

11 Peter, L.M. (2007). Dye-sensitized nanocrystalline solar cells. *Phys. Chem. Chem. Phys.* 9 (21): 2630–2642.

12 Lv, Z., Yu, J., Wu, H. et al. (2012). Highly efficient and completely flexible fiber-shaped dye-sensitized solar cell based on TiO_2 nanotube array. *Nanoscale* 4 (4): 1248–1253.

13 Liu, G., Peng, M., Song, W. et al. (2015). An 8.07% efficient fiber dye-sensitized solar cell based on a TiO_2 micron-core array and multilayer structure photoanode. *Nano Energy* 11: 341–347.

14 Lee, M.R., Eckert, R.D., Forberich, K. et al. (2009). Solar power wires based on organic photovoltaic materials. *Science* 324 (5924): 232–235.

15 Dong, B., Hu, J., Xiao, X. et al. (2019). High-efficiency fiber-shaped perovskite solar cell by vapor-assisted deposition with a record efficiency of 10.79%. *Adv. Mater. Technol.* 4 (7): 1900131.

16 Huang, S., Guo, X., Huang, X. et al. (2011). Highly efficient fibrous dye-sensitized solar cells based on TiO_2 nanotube arrays. *Nanotechnology* 22 (31): 315402.

17 Liu, D., Zhao, M., Li, Y. et al. (2012). Solid-state, polymer-based fiber solar cells with carbon nanotube electrodes. *ACS Nano* 6 (12): 11027–11034.

18 Fan, X., Chu, Z., Wang, F. et al. (2008). Wire-shaped flexible dye-sensitized solar cells. *Adv. Mater.* 20 (3): 592–595.

19 Gasana, E., Westbroek, P., Hakuzimana, J. et al. (2006). Electroconductive textile structures through electroless deposition of polypyrrole and copper at polyaramide surfaces. *Surf. Coat. Technol.* 201 (6): 3547–3551.

20 Kim, B., Koncar, V., Devaux, E. et al. (2004). Electrical and morphological properties of PP and PET conductive polymer fibers. *Synth. Met.* 146 (2): 167–174.

21 Hou, S.C., Lv, Z.B., Wu, H.W. et al. (2012). Flexible conductive threads for wearable dye-sensitized solar cells. *J. Mater. Chem.* 22 (14): 6549–6552.

22 Kaynak, A., Najar, S.S., and Foitzik, R.C. (2008). Conducting nylon, cotton and wool yarns by continuous vapor polymerization of pyrrole. *Synth. Met.* 158 (1–2): 1–5.

23 Abbasi, A.R., Akhbari, K., and Morsali, A. (2012). Dense coating of surface mounted CuBTC metal-organic framework nanostructures on silk fibers, prepared by layer-by-layer method under ultrasound irradiation with antibacterial activity. *Ultrason. Sonochem.* 19 (4): 846–852.

24 Marriam, I., Wang, X., Tebyetekerwa, M. et al. (2018). A bottom-up approach to design wearable and stretchable smart fibers with organic vapor sensing behaviors and energy storage properties. *J. Mater. Chem. A* 6 (28): 13633–13643.

25 Tebyetekerwa, M., Xu, Z., Li, W. et al. (2018). Surface self-assembly of functional electroactive nanofibers on textile yarns as a facile approach toward super flexible energy storage. *ACS Appl. Energy Mater.* 1 (2): 377–386.

26 Shao, W., Tebyetekerwa, M., Marriam, I. et al. (2018). Polyester@MXene nanofibers-based yarn electrodes. *J. Power Sources* 396: 683–690.

27 Qu, L., Dai, L., Stone, M. et al. (2008). Carbon nanotube arrays with strong shear binding-on and easy normal lifting-off. *Science* 322 (5899): 238–242.

28 Yang, Z., Deng, J., Sun, X. et al. (2014). Stretchable, wearable dye-sensitized solar cells. *Adv. Mater.* 26 (17): 2643–2647.

29 Yang, Z., Deng, J., Chen, X. et al. (2013). A highly stretchable, fiber-shaped supercapacitor. *Angew. Chem. Int. Ed.* 52 (50): 13453–13457.

30 Ma, R., Lee, J., Choi, D. et al. (2014). Knitted fabrics made from highly conductive stretchable fibers. *Nano Lett.* 14 (4): 1944–1951.

31 Hou, S., Cai, X., Fu, Y. et al. (2011). Transparent conductive oxide-less, flexible, and highly efficient dye-sensitized solar cells with commercialized carbon fiber as the counter electrode. *J. Mater. Chem.* 21 (36): 13776–13779.

32 Chen, L., Dai, H., Zhou, Y. et al. (2014). Porous, single crystalline titanium nitride nanoplates grown on carbon fibers: excellent counter electrodes for low-cost, high performance, fiber-shaped dye-sensitized solar cells. *Chem. Commun.* 50 (92): 14321–14324.

33 Chi, Z., Shen, J., Zhang, H. et al. (2017). $NiCo_2S_4$ nanosheets in situ grown on carbon fibers as an efficient counter electrode for fiber-shaped dye-sensitized solar cells. *J. Mater. Sci.: Mater. Electron.* 28 (14): 10640–10644.

34 Cai, X., Wu, H., Hou, S. et al. (2014). Dye-sensitized solar cells with vertically aligned TiO_2 nanowire arrays grown on carbon fibers. *ChemSusChem* 7 (2): 474–482.

35 Chen, H., Zhu, L., Liu, H. et al. (2012). Growth of ZnO nanowires on fibers for one-dimensional flexible quantum dot-sensitized solar cells. *Nanotechnology* 23 (7): 075402.

36 Cai, X., Hou, S., Wu, H. et al. (2012). All-carbon electrode-based fiber-shaped dye-sensitized solar cells. *Phys. Chem. Chem. Phys.* 14 (1): 125–130.

37 Wang, J.N., Luo, X.G., Wu, T. et al. (2014). High-strength carbon nanotube fibre-like ribbon with high ductility and high electrical conductivity. *Nat. Commun.* 5 (1): 3848.

38 Oh, E., Cho, H., Kim, J. et al. (2020). Super-strong carbon nanotube fibers achieved by engineering gas flow and postsynthesis treatment. *ACS Appl. Mater. Interfaces* 12 (11): 13107–13115.

39 Collins, P.G. and Avouris, P. (2000). Nanotubes for electronics. *Sci. Am.* 283 (6): 62–69.

40 Peng, B., Locascio, M., Zapol, P. et al. (2008). Measurements of near-ultimate strength for multiwalled carbon nanotubes and irradiation-induced crosslinking improvements. *Nat. Nanotechnol.* 3 (10): 626–631.

41 Iijima, S. (1991). Helical microtubules of graphitic carbon. *Nature* 354 (6348): 56–58.

42 Endo, M., Strano, M.S., and Ajayan, P.M. (2007). Potential applications of carbon nanotubes. In: *Carbon Nanotubes* (ed. A. Jorio, G. Dresselhaus, and M.S. Dresselhaus), 13–62. Springer.

43 Dalton, A.B., Collins, S., Munoz, E. et al. (2003). Super-tough carbon-nanotube fibres. *Nature* 423 (6941): 703–703.

44 Ericson, L.M., Fan, H., Peng, H. et al. (2004). Macroscopic, neat, single-walled carbon nanotube fibers. *Science* 305 (5689): 1447–1450.

45 Behabtu, N., Young, C.C., Tsentalovich, D.E. et al. (2013). Strong, light, multifunctional fibers of carbon nanotubes with ultrahigh conductivity. *Science* 339 (6116): 182–186.

46 Vigolo, B., Penicaud, A., Coulon, C. et al. (2000). Macroscopic fibers and ribbons of oriented carbon nanotubes. *Science* 290 (5495): 1331–1334.

47 Davis, V.A., Parra-Vasquez, A.N.G., Green, M.J. et al. (2009). True solutions of single-walled carbon nanotubes for assembly into macroscopic materials. *Nat. Nanotechnol.* 4 (12): 830–834.

48 Zhang, M., Atkinson, K.R., and Baughman, R.H. (2004). Multifunctional carbon nanotube yarns by downsizing an ancient technology. *Science* 306 (5700): 1358–1361.

49 Kuznetsov, A.A., Fonseca, A.F., Baughman, R.H. et al. (2011). Structural model for dry-drawing of sheets and yarns from carbon nanotube forests. *ACS Nano* 5 (2): 985–993.

50 Zhang, X., Jiang, K., Feng, C. et al. (2006). Spinning and processing continuous yarns from 4-inch wafer scale super-aligned carbon nanotube arrays. *Adv. Mater.* 18 (12): 1505–1510.

51 Li, Y.-L., Kinloch, I.A., and Windle, A.H. (2004). Direct spinning of carbon nanotube fibers from chemical vapor deposition synthesis. *Science* 304 (5668): 276–278.

52 Zhang, X., Li, Q., Holesinger, T.G. et al. (2007). Ultrastrong, stiff, and lightweight carbon-nanotube fibers. *Adv. Mater.* 19 (23): 4198–4201.

53 Wang, J., Luo, X., Wu, T. et al. (2014). High-strength carbon nanotube fibre-like ribbon with high ductility and high electrical conductivity. *Nat. Commun.* 5 (1): 1–8.

54 Meng, F., Zhao, J., Ye, Y. et al. (2012). Carbon nanotube fibers for electrochemical applications: effect of enhanced interfaces by an acid treatment. *Nanoscale* 4 (23): 7464–7468.

55 Wang, K., Li, M., Liu, Y.-N. et al. (2014). Effect of acidification conditions on the properties of carbon nanotube fibers. *Appl. Surf. Sci.* 292: 469–474.

56 Liu, K., Zhu, F., Liu, L. et al. (2012). Fabrication and processing of high-strength densely packed carbon nanotube yarns without solution processes. *Nanoscale* 4 (11): 3389–3393.

57 Motta, M., Moisala, A., Kinloch, I.A. et al. (2007). High performance fibres from 'dog bone' carbon nanotubes. *Adv. Mater.* 19 (21): 3721–3726.

58 Vilatela, J.J. and Windle, A.H. (2010). Yarn-like carbon nanotube fibers. *Adv. Mater.* 22 (44): 4959–4963.

59 Koziol, K., Vilatela, J., Moisala, A. et al. (2007). High-performance carbon nanotube fiber. *Science* 318 (5858): 1892–1895.

60 Chen, T., Wang, S., Yang, Z. et al. (2011). Flexible, light-weight, ultrastrong, and semiconductive carbon nanotube fibers for a highly efficient solar cell. *Angew. Chem. Int. Ed.* 123 (8): 1855–1859.

61 Chen, T., Qiu, L., Cai, Z. et al. (2012). Intertwined aligned carbon nanotube fiber based dye-sensitized solar cells. *Nano Lett.* 12 (5): 2568–2572.

62 Zhang, Z., Yang, Z., Wu, Z. et al. (2014). Weaving efficient polymer solar cell wires into flexible power textiles. *Adv. Energy Mater.* 4 (11): 1301750.

63 Li, R., Xiang, X., Tong, X. et al. (2015). Wearable double-twisted fibrous perovskite solar cell. *Adv. Mater.* 27 (25): 3831–3835.

64 Novoselov, K.S., Colombo, L., Gellert, P. et al. (2012). A roadmap for graphene. *Nature* 490 (7419): 192–200.

65 Mayorov, A.S., Gorbachev, R.V., Morozov, S.V. et al. (2011). Micrometer-scale ballistic transport in encapsulated graphene at room temperature. *Nano Lett.* 11 (6): 2396–2399.

66 Novoselov, K.S., Geim, A.K., Morozov, S.V. et al. (2004). Electric field effect in atomically thin carbon films. *Science* 306 (5696): 666–669.

67 Liu, Y., Xu, Z., Zhan, J. et al. (2016). Superb electrically conductive graphene fibers via doping strategy. *Adv. Mater.* 28 (36): 7941–7947.

68 Cruz-Silva, R., Morelos-Gomez, A., Kim, H.-i. et al. (2014). Super-stretchable graphene oxide macroscopic fibers with outstanding knotability fabricated by dry film scrolling. *ACS Nano* 8 (6): 5959–5967.

69 Fang, B., Xiao, Y., Xu, Z. et al. (2019). Handedness-controlled and solvent-driven actuators with twisted fibers. *Mater. Horiz.* 6 (6): 1207–1214.

70 Dong, Z., Jiang, C., Cheng, H. et al. (2012). Facile fabrication of light, flexible and multifunctional graphene fibers. *Adv. Mater.* 24 (14): 1856–1861.

71 Chen, T. and Dai, L. (2015). Macroscopic graphene fibers directly assembled from CVD-grown fiber-shaped hollow graphene tubes. *Angew. Chem. Int. Ed.* 127 (49): 15160–15163.

72 Li, J., Li, J., Li, L. et al. (2014). Flexible graphene fibers prepared by chemical reduction-induced self-assembly. *J. Mater. Chem. A* 2 (18): 6359–6362.

73 Xu, Z. and Gao, C. (2011). Graphene chiral liquid crystals and macroscopic assembled fibres. *Nat. Commun.* 2 (1): 571.

74 Xu, Z., Liu, Y., Zhao, X. et al. (2016). Ultrastiff and strong graphene fibers via full-scale synergetic defect engineering. *Adv. Mater.* 28 (30): 6449–6456.

75 Xu, Z., Sun, H., Zhao, X. et al. (2013). Ultrastrong fibers assembled from giant graphene oxide sheets. *Adv. Mater.* 25 (2): 188–193.

76 Jalili, R., Aboutalebi, S.H., Esrafilzadeh, D. et al. (2013). Scalable one-step wet-spinning of graphene fibers and yarns from liquid crystalline dispersions of graphene oxide: towards multifunctional textiles. *Adv. Funct. Mater.* 23 (43): 5345–5354.

77 Xu, Z., Liu, Z., Sun, H. et al. (2013). Highly electrically conductive Ag-doped graphene fibers as stretchable conductors. *Adv. Mater.* 25 (23): 3249–3253.

78 Xin, G., Yao, T., Sun, H. et al. (2015). Highly thermally conductive and mechanically strong graphene fibers. *Science* 349 (6252): 1083–1087.

79 Zhao, Y., Jiang, C., Hu, C. et al. (2013). Large-scale spinning assembly of neat, morphology-defined, graphene-based hollow fibers. *ACS Nano* 7 (3): 2406–2412.

80 Xu, Z., Zhang, Y., Li, P. et al. (2012). Strong, conductive, lightweight, neat graphene aerogel fibers with aligned pores. *ACS Nano* 6 (8): 7103–7113.

81 Sun, J., Li, Y., Peng, Q. et al. (2013). Macroscopic, flexible, high-performance graphene ribbons. *ACS Nano* 7 (11): 10225–10232.

82 Yang, Z., Sun, H., Chen, T. et al. (2013). Photovoltaic wire derived from a graphene composite fiber achieving an 8.45% energy conversion efficiency. *Angew. Chem.* 125 (29): 7693–7696.

83 Yao, Y., Lv, T., Li, N. et al. (2020). Selected functionalization of continuous graphene fibers for integrated energy conversion and storage. *Sci. Bull.* 65 (6): 486–495.

84 Lima, M.D., Fang, S.L., Lepro, X. et al. (2011). Biscrolling nanotube sheets and functional guests into yarns. *Science* 331 (6013): 51–55.

85 Sun, H., Yang, Z., Chen, X. et al. (2013). Photovoltaic wire with high efficiency attached onto and detached from a substrate using a magnetic field. *Angew. Chem. Int. Ed.* 52 (32): 8276–8280.

86 Goodno, B.J. and Gere, J.M. (2020). *Mechanics of Materials*. Cengage Learning.

87 He, J., Lu, C., Jiang, H. et al. (2021). Scalable production of high-performing woven lithium-ion fibre batteries. *Nature* 597 (7874): 57–63.

88 Zhang, Q., Huang, J.Q., Qian, W.Z. et al. (2013). The road for nanomaterials industry: a review of carbon nanotube production, post-treatment, and bulk applications for composites and energy storage. *Small* 9 (8): 1237–1265.

89 Lin, L., Peng, H., and Liu, Z. (2019). Synthesis challenges for graphene industry. *Nat. Mater.* 18 (6): 520–524.

90 Chen, G., Wang, H., Guo, R. et al. (2020). Superelastic EGaIn composite fibers sustaining 500% tensile strain with superior electrical conductivity for wearable electronics. *ACS Appl. Mater. Interfaces* 12 (5): 6112–6118.

91 Duan, M., Ren, Y., Sun, X. et al. (2021). EGaIn fiber enabled highly flexible supercapacitors. *ACS Omega* 6 (38): 24444–24449.

92 Zhang, Z., Wang, W., Jiang, Y. et al. (2022). High-brightness all-polymer stretchable LED with charge-trapping dilution. *Nature* 603 (7902): 624–630.

93 Su, X., Wu, X., Chen, S. et al. (2022). A highly conducting polymer for self-healable, printable and stretchable organic electrochemical transistor arrays and near hysteresis-free soft tactile sensors. *Adv. Mater.* 34 (19): 2200682.

5

Fiber-Shaped Dye-Sensitized Solar Cells

5.1 Introduction

Dye-sensitized solar cells (DSSCs) are promising alternatives to traditional silicon-based solar cells, attributed to their unique advantages such as low cost, facile fabrication, and high photovoltaic conversion performance, especially under low light intensity conditions. The first DSSC was invented by Grätzel et al. in 1991, based on transparent titanium dioxide (TiO_2) film for dye adsorption, which opened a new avenue for light harvesting [1]. A typical DSSC comprises a dye-adsorbed photoanode, a counter electrode with catalytic activity, and an electrolyte with redox species such as I^-/I_3^- to conduct photovoltaic conversion [2]. The photoanode generally comprises transparent conductive glass, i.e. fluorine-doped tin dioxide (FTO) and indium-doped tin oxide (ITO), with the incorporation of dye-sensitized semiconductor materials such as TiO_2 and ZnO [3, 4]. The counter electrode involves an electrocatalytic layer such as platinum (Pt) or carbon materials coating on FTO or ITO substrates [5]. After the injection of redox electrolyte, the obtained device is sealed to prevent electrolyte leakage and performance degradation [6]. Figure 5.1 illustrates the working mechanism of DSSC. Upon illumination, the ground-state electrons in the dye molecules get excited by photon energy, diffuse into the conduction band of semiconductor materials and accumulate at the conductive substrate. The oxidized dye molecules are regenerated by absorbing electrons from the redox electrolyte. The oxidized species in the electrolyte are further reduced on the counter electrode, accompanied by the electron flow in the external circuit [8, 9].

Tremendous efforts have been made to optimize DSSCs for high power conversion efficiency (PCE), long lifetime, and cost-effectiveness. For instance, various metal oxide semiconductors such as TiO_2, SnO_2, ZnO, Nb_2O_5, and Zn_2SnO_4 have been developed as photoanodes. In terms of photo-sensitized dyes, metal coordination complexes such as ruthenium-based dye molecules (N3, N719) are common benchmarking materials. Organic metal-free dyes have also been investigated via chromophores design toward desirable photophysical properties. To enhance the stability and safety of DSSCs, polymer-based solid/gel electrolytes have been actively explored in recent years toward safety enhancement. To overcome the high-cost issue of Pt-based counter electrodes, metal oxides, conductive polymers,

Wearable Solar Cells: Mechanisms, Materials, and Devices, First Edition.
Hao Sun, Zhibin Yang, Longbin Qiu, and Huisheng Peng.

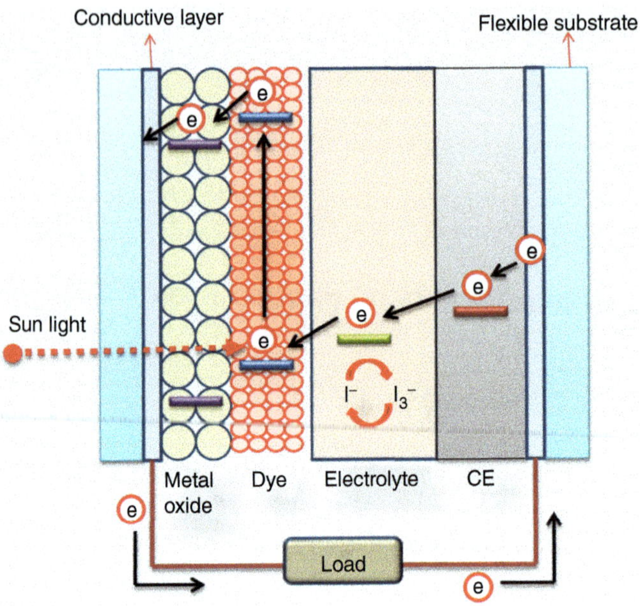

Figure 5.1 Schematic illustration of the working mechanism of a DSSC. Source: Devadiga et al. [7], © 2022/Elsevier.

and carbon materials have been investigated with promising performance. These endeavors have resulted in improved PCEs and stability of DSSCs [5, 10–12].

However, traditional bulky and rigid DSSCs cannot fully meet the demands for portable, flexible, and wearable electronics. Flexible DSSCs can be produced by using flexible conductive substrates. However, the resulted interfaces are not stable enough to fully accommodate the deformations in wearable applications, and the breathability fails to meet the requirements in wearable applications originating from the two-dimensional instincts. Therefore, wearable DSSCs based on fiber configuration have attracted broad interests owing to their flexibility, weavability, and wearability [13, 14]. According to the assembly mode of the photoanode and counter electrode, the structure of the fiber-shaped DSSCs can be divided into coaxial, twisting, and in-parallel structures [15]. An early trial on fiber-shaped DSSC made by Koyuncu [16] reported a stainless steel wire coated with nanosized TiO_2 powder, and a transparent conductive polymer and Cu-based dyes were incorporated inside a transparent polymer tube, coupling with a I^-/I_3^- liquid electrolyte. The obtained 10 cm-long fiber-shaped DSSCs exhibited a low working voltage of around 300–350 mV, which was far from practical requirements, and the flexibility and sealing methods were also unsatisfactory. The promising features, however, have inspired further investigations of fiber-shaped DSSCs. Since then, the optimizations of photoanode, electrolytes, and counter electrodes have greatly been accelerated for high-performance fiber-shaped DSSCs.

In this chapter, we first describe the mainstream configurations of fiber-shaped DSSCs, including coaxial and twisting structures, and analyze the corresponding materials, fabrication technique, and performance optimization. The main emphasis is then paid on functional fiber-shaped DSSCs, such as stretchability and magnetic response that promote compatibility in practical applications. Recent advancements in integrated fiber-shaped DSSCs with other electronic devices, e.g. fiber-shaped batteries and supercapacitors, have been also introduced. The remaining challenges and perspectives are further discussed to provide insights for future development.

5.2 Twisting Structure

The twisting structure represents a mainstream configuration of fiber-shaped DSSCs. To fabricate a fiber-shaped DSSC with a twisted structure, the counter electrode with better flexibility is generally twisted on the photoanode, forming the so-called "twisting structure" (Figure 5.2) [17]. Attributed to the efficient charge transport at the electrode/electrolyte interface, fiber-shaped DSSCs with different twisting angles and pitches can be obtained.

5.2.1 Photoanode

Metallic wires, such as stainless steel (SS) and Ti wires, were widely used as fiber substrates for the deposition of semiconductor materials. Compared with the SS wire, the Ti wire is more ideal for TiO_2 incorporation owing to the low interfacial resistance, which enables higher PCEs [18]. For instance, Fu et al. reported a photoanode using a layer of TiO_2 membrane coated on a Ti wire substrate, and the counter electrode was a Pt wire (Figure 5.3a,b) [19]. The obtained fiber-shaped DSSC was tested under a standard illumination intensity of AM 1.5 (100 mW/cm^2). The obtained fiber-shaped DSSC exhibited an average PCE of 6.7%, and the maximal PCE reached 7.02%. Furthermore, the maximum output power significantly

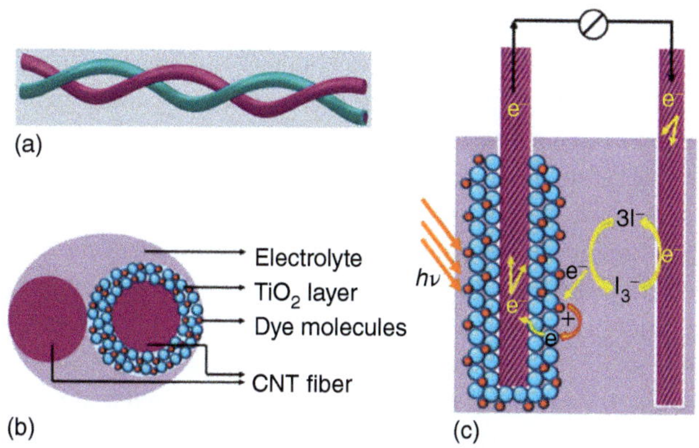

Figure 5.2 Schematic illustration of a fiber-shaped DSSC fabricated from two carbon nanotube fibers. (a) Two MWCNT fibers twined into a cell. (b) Top view of a cell. (c) Working mechanism. Source: Chen et al. [17], © 2012/American Chemical Society.

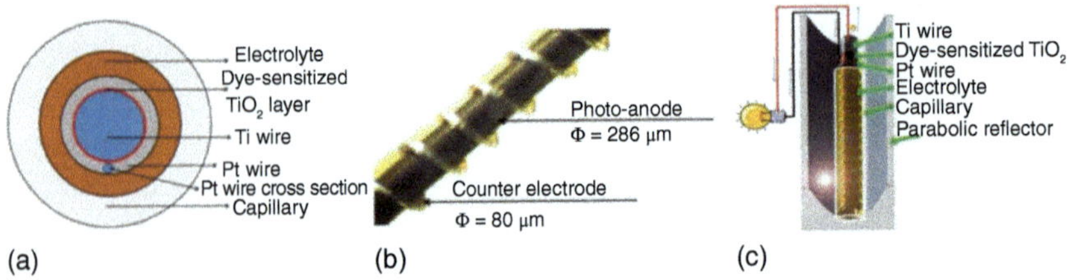

Figure 5.3 (a) Cross-sectional structure of the fiber-shaped DSSC. (b) The optical image of a fiber-shaped DSSC. (c) Schematic of a fiber-shaped DSSC housed in a parabolic condenser. Source: Fu et al. [19], © 2011/Royal Society of Chemistry.

increased from 0.72 to 4.78 mW through the employment of a high-efficiency light concentrator (Figure 5.3c).

It is expected that the construction of efficient heterojunctions is beneficial to improve the electron injection and collection and suppress the recombination of charge carriers, resulting in high efficiency of the photoanode [20]. For instance, SnO_2 (100–200 cm^2/VS) exhibits faster electron diffusion than TiO_2 (0.1–1.0 cm^2/VS) in terms of electron mobility [21, 22]. Thus, the photoanode based on SnO_2 may contribute to effective carrier collection. Peng et al. prepared SnO_2 nanoparticles on Ti wire via thermal sintering tin tetraisopropoxide and titanium tetraisopropoxide successively for the production of SnO_2-TiO_2 heterojunction [23]. The resulted fiber-shaped DSSC showed a higher short-circuit current density (J_{sc}) of 10.23 mA/cm^2 compared with 8.27 mA/cm^2 without the SnO_2 layer. The SnO_2-TiO_2 heterojunction is supposed to form a single-direction pathway for electron migration because the conduction band edge of SnO_2 is more favorable than TiO_2. The photovoltaic conversion performance was further improved by dip-coating TiO_2 nanoparticles outside the SnO_2-TiO_2 heterojunction, which showed an increased J_{sc} (11.73 mA/cm^2). As a result, the obtained fiber-shaped DSSC showed a PCE of 12.4% with the use of a diffuse light reflector (Figure 5.4).

In planar DSSCs, a light-scattering layer of large-sized TiO_2 particles was generally required to enhance the light-harvesting of the photoanode [24]. However, incorporating a light scattering layer on the semiconductor layer remains challenging for fiber-shaped DSSCs due to the large contact resistance and poor interface stability. Liu et al. developed a novel photoanode with a hierarchical structure based on a Ti/TiO_2 micron-cone array (TMCA) frame, which consisted of a compact TiO_2 semiconductor layer, a TiO_2 hollow spheres light-scattering layer, and a porous TiO_2 nanocrystal layer (Figure 5.5) [25]. The TMCA was grown inside the Ti wire, which effectively improved the binding between the TMCA and Ti substrate. Moreover, the TMCA frame could serve as the light-scattering center and provide electron transfer channels. As a result, the resulting fiber-shaped DSSC delivered a high J_{sc} and PCE of 16.04 mA/cm^2 and 8.07%, respectively. In addition, it exhibited decent flexibility that retained 92.01% of the original PCE after 100 bending cycles.

The performance of fiber-shaped DSSC is highly dependent on the interface properties of the photoanode because the electron exaction, injection, and transfer

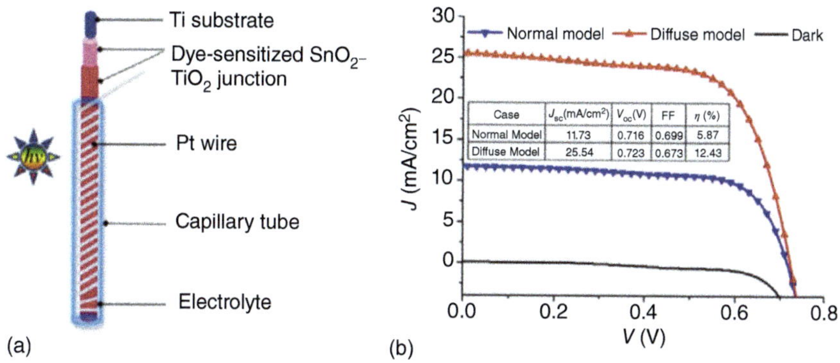

Figure 5.4 (a) Schematic illustration of a fiber-based DSSC. (b) Current–voltage curves of the fiber-shaped DSSC with SnO_2-TiO_2 (TTIP)-TiO_2 (NP) structured photoanode under AM 1.5 standard illumination (Normal Model) and the normal model with a diffuse light reflector underneath the cell (Diffuse Model). The reflector allowed the FDSC to collect stray light over a large area. Source: Peng et al. [23], © 2014/Elsevier.

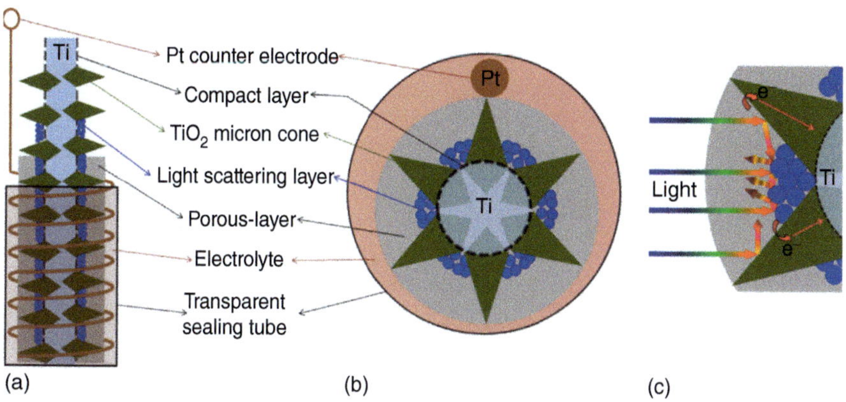

Figure 5.5 Schematic illustration of the novel fiber-shaped DSSC from the (a) side and (b) top views, and with (c) light scattering and electron transferring in the novel photoanode. Source: Liu et al. [25], © 2015/Elsevier.

mainly occur at these interfaces. The defects in TiO_2 might cause severe recombination of photoelectrons, oxidized dye molecules, and I_3^- acceptors [26]. Moreover, the esterification process between N719 dye and TiO_2 could result in the formation of H_2O byproduct [27, 28]. These additional defects further decreased the electron extraction/injection efficiencies. $TiCl_4$ has thus been used as an effective mediator to form an ultrathin TiO_2 layer via the hydrolysis reaction, which can effectively reduce photoelectron recombination [29, 30]. However, the high reactivity and irreversible degradation of $TiCl_4$ remain a grand challenge for large-scale applications.

To solve the above issues, Ardhi et al. optimized the TiO_2/dye interface by developing an environment-friendly, solution-processed, and low-cost poly(2-ethyl-2-oxazoline) (PEOx) interlayer (Figure 5.6) [26]. The permanent dipole moment of the PEOx interlayer could effectively improve the electron extraction/injection kinetics

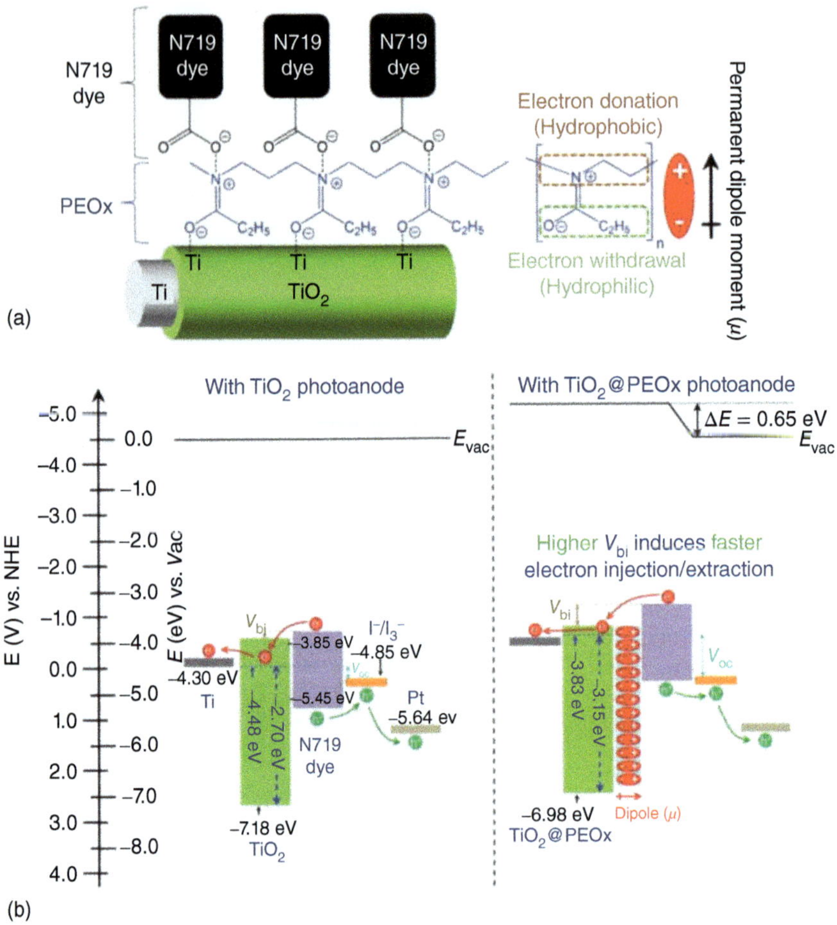

Figure 5.6 (a) PEOx-induced permanent dipole moment binding at TiO_2/N719 dye interface. (b) Schematic illustration of fiber-shaped DSSC energy diagram with (left panel) blanks TiO_2 and (right panel) TiO_2@PEOx photoanode. Source: Ardhi et al. [26], © 2020/Royal Society of Chemistry.

through the built-in electric field. Additionally, the generation of free electrons in the enolate group of the PEOx chain could decrease the defect sites, resulting in the reduction of both electronic recombination and interfacial ohmic interface at the TiO_2/dye interface. Moreover, the dye adsorption on TiO_2 could be improved, attributed to the dye anchoring sites provided by the iminium group with positive polarity in the PEOx chain. As a result, the PCE of the fiber-shaped DSSC reached 11.22%, which is among the highest values of fiber-shaped DSSCs using a Pt wire counter electrode.

It is well recognized that the interfacial contact between the semiconductor layer and substrate plays a critical role in efficient electron transport and suppressed electron recombination. For instance, Liu et al. reported a hierarchical structure with a Ti microridge/nanorod-modified wire as the substrate of a photoanode

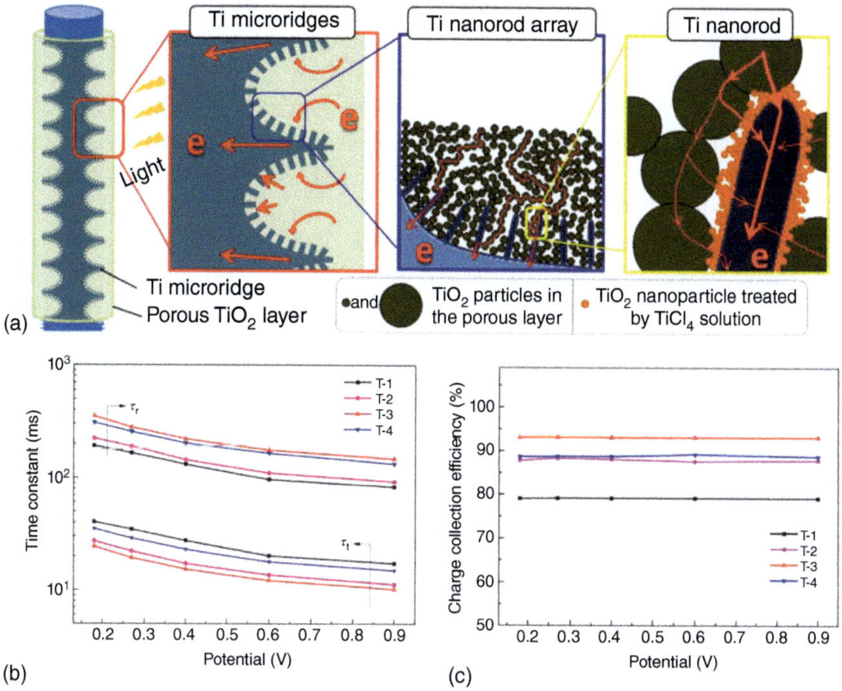

Figure 5.7 (a) Schematic illustration of the transmission path of electrons in the novel photoanode with the Ti microridge/nanorod-modified wire substrate. (b) Electron transport time (τ_t) and lifetime (τ_r) and (c) the charge collection efficiency, measured by intensity-modulated photocurrent/photovoltage spectroscopy (IMPS/IMVS). T-1-4 represent traditional devices, Ti microridges, Ti microridges/nanorods, and Ti nanorods, respectively. Source: Liu et al. [31], © 2018/Elsevier.

(Figure 5.7a) [31]. Specifically, the Ti microridges and micropit facilitated electron transport and incident light harnessing. Additionally, the Ti nanorod array was proved to improve the contact between TiO_2 and Ti wire substrate, which effectively suppressed the photoelectron recombination. The effective electron transport time (τt) and electron lifetime (τr) based on a variety of photoanodes, e.g. bare Ti wire, Ti microridges, Ti microridges/nanorods, and Ti nanorods, were evaluated (Figure 5.7b), in which short τt and long τr corresponded to rapid electron transfer and suppressed photoelectron recombination. Therefore, the fiber-shaped DSSCs using the optimized photoanode demonstrated a PCE and charge collection efficiency (CCE) up to 8.128% and 93.1%, respectively (Figure 5.7c).

5.2.2 Counter Electrode

An ideal counter electrode should simultaneously show high mechanical stability, electrical conductivity, and electrocatalytic activity. Pt metal represents a classic counter electrode for DSSC [5]. However, its high cost has hindered the scale-up application in fiber-shaped DSSCs. To this end, low-cost carbon materials are highly desirable to lower the material cost. For instance, aligned multi-wall carbon nanotube (MWCNT) fibers have been considered as promising alternatives to Pt electrodes owing to their high mechanical and electrical properties, as well as their decent electrocatalytic capability. A pioneer work by Chen et al. reported the use of aligned MWCNT fiber as a fiber-shaped counter electrode, and another MWCNT fiber incorporated with dye-adsorbed TiO_2 served as the photoanode. Based on a redox electrolyte comprised of I^-/I_3^-, the fiber-shaped DSSC based on aligned MWCNT fibers showed a high PCE of 2.94% [17]. Compared with fiber-shaped DSSCs using metal wires, the highly flexible MWCNT fibers enabled better weavability of the obtained fiber-shaped devices that could be woven into breathable textiles for wearable applications.

Bare MWCNT fibers generally show limited electrocatalytic properties that hinder the pursuit of high PCEs. Thus, the incorporation of a second phase with higher electrocatalytic performance is a promising strategy. Fu et al. proposed a core-sheath multi-wall carbon nanotube (CS-MWCNT) fiber by twisting stacked MWCNT sheets with hydrophobic and hydrophilic surfaces at two sides [32]. This asymmetric configuration enabled a new modification strategy, i.e. the hydrophilic sheath was beneficial for sufficient diffusion of K_2PtCl_6, and the oxygen-containing functional groups and defect sites created by O_2 plasma offered adequate nucleation sites for the deposition of Pt nanoparticles. Based on the hybrid fiber as the counter electrode, the resulting fiber-shaped DSSC showed a remarkable PCE of 10%. Besides, these fiber-shaped DSSCs were flexible and weavable (Figure 5.8) and could readily power wearable electronics such as a pedometer under the sunlight.

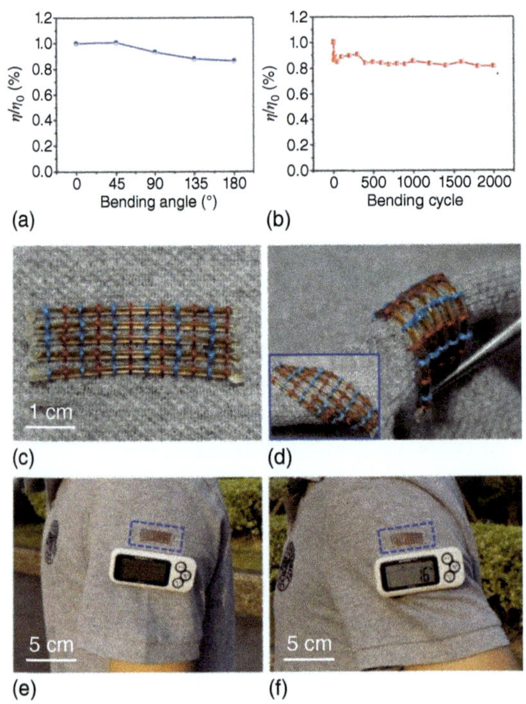

Figure 5.8 (a, b) PCEs of the fiber-shaped DSSC as a function of bending angle and cycle, respectively. η_0 and η correspond to the power conversion efficiencies before and after bending, respectively. The bending angle in (b) was 90°. (c) Five in-series connected fiber-shaped DSSCs with a length of 3 cm are woven onto a T-shirt. (d) Fiber-shaped DSSCs in (c) under bending and twisting states (the blue rectangle). (e) Fiber-shaped DSSCs in (c) and (d) are connected with a commercial pedometer. The inserted dashed rectangle highlights the fiber-shaped DSSCs. (f) Step number displayed on the pedometer after 16 arm swings under outdoor sunlight. Source: Fu et al. [32], © 2018/Royal Society of Chemistry.

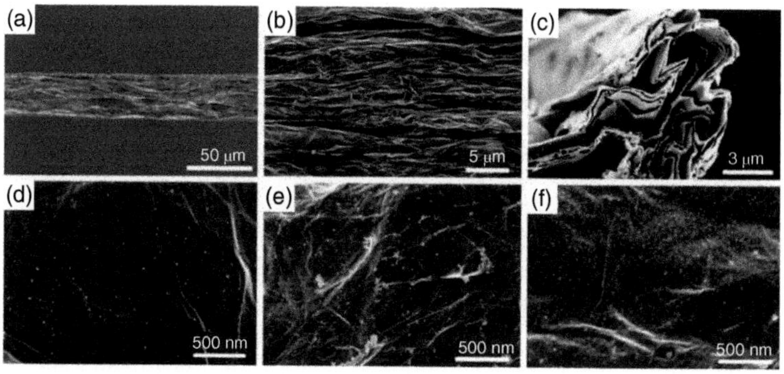

Figure 5.9 (a, b) A graphene fiber at different magnifications. (c) Cross-sectional view of a graphene fiber after failure during the mechanical test. (d–f) Side views of graphene/Pt composite fibers with increasing Pt contents of 4.6%, 7.1%, and 22.9%, respectively. Source: Yang et.al. [34], © 2013/John Wiley & Sons, Inc.

As another important branch of carbon-based fibers, graphene fibers have drawn broad attention owing to their attractive properties, such as high mechanical strength, electrical conductivity, and flexibility [33]. For instance, Yang et al. used a solution process to fabricate graphene fibers, which exhibited a strength of 10^2–10^3 MPa and an electrical conductivity of 10^2–10^3 S/cm. The excellent mechanical strength was derived from the closely stacked graphene sheets (Figure 5.9a–c). The modification of Pt nanoparticles further enhanced the electrocatalytic activity of graphene fiber (Figure 5.9d,e). When the Pt/graphene hybrid fiber was used as the counter electrode and paired with an optimized photoanode comprised of dye-absorbed TiO_2/Ti wire, the fiber-shaped DSSC demonstrated a high PCE of 8.45%, which outperformed the bare Pt wire counter electrode [34].

Transition metal nitrides, oxides, carbides, and chalcogenides have been widely recognized as excellent electrocatalysts as the counter electrodes for planar DSSCs [35–38]. However, they cannot form continuous fibers to serve as fiber counter electrode like Pt wire and carbon-based fibers. As a result, incorporating these transition metal compounds onto a fiber substrate represents a promising approach. Chen et al. reported a one-step strategy to prepare $CoNi_2S_4$ nanoribbons on carbon fibers (CF) as high-performing counter electrodes of fiber-shaped DSSCs, exhibiting a high PCE of 7.03% [39]. The impressive device performance was not only attributed to the high electrocatalytic activity of $CoNi_2S_4$ but also related to their unique polymorph structure. The exposed (110) crystal facet of the $CoNi_2S_4$ nanoribbon could expose more surface metal atoms than nanorod growing along (100), thus leading to enhanced reduction capability from I_3^- to I^-.

The various oxidation states of transition metal oxides afford suitable energy levels and efficient catalytic activities. For instance, molybdenum oxide (MoO_3) shows no contact energy barrier between electrolyte and counter electrode, suitable reduction/oxidation properties for counter electrodes, and multi-state energy bands. Utomo et al. reported a novel hierarchical MoO_3 catalyst and explored its application in all-solid-state fiber-shaped DSSCs [40]. The hierarchical structure and intrinsic catalytic activity of MoO_3 enabled facilitated reduction/oxidation process between I_3^- and I^- at the counter electrode/electrolyte interface. In this regard, the PCE had been improved by 60% compared with the use of Pt counter electrode, and the achieved PCE of 5.11% was the highest value among all-solid-state fiber-shaped DSSCs.

5.2.3 Electrolyte

The liquid electrolytes based on iodine-based species (I^-/I_3^-) suffer from poor temperature stability, electrolyte leakage, and volatilization that plague wearable applications. Attempts have thus been made to explore semi-solid-state and all-solid-state electrolytes for fiber-shaped DSSCs. An early trial by Wang et al. reported a cupric iodide (CuI) based solid-state electrolyte. By construction of mesoporous TiO_2 film with large pores using polymer additives as the template agent, the infiltration of solid electrolyte into the TiO_2 film could be effectively improved, and the obtained fiber-shaped DSSC achieved a PCE of 1.38% [41].

Another important step on polymer electrolyte involves the use of poly(vinylidene fluoride-*co*-hexafluoropropylene) (PVdF-HFP)/polyethylene oxide-*co*-polypropylene oxide-*co*-polyethylene oxide (P123) porous polymer electrolyte [42], which delivered a PCE of 1.029% through the optimization of the polymer composition. Although the achieved PCEs cannot fully compete with those based on conventional liquid electrolytes, the appealing features, such as high mechanical and thermal stability, are highly desirable in wearable applications. Other strategies have been developed for the optimization of ionic conductivity, such as the introduction of inorganic additives, copolymers, and ionic liquids. A representative work by Li et al. showed a novel gel electrolyte comprised of polymer and hydrophobic ionic liquid. In this case, the PVdF-HFP copolymer with high stability and gelling performance was used as the gelator, while a hydrophobic ionic liquid, 1-butyl-3-methylimidazolium bis(trifluoromethanesulfonyl) imide (BMITFSI) was used to improve ion transportation in the electrolyte [43]. Attributed to the synergistic effects of these components, the achieved fiber-shaped DSSCs could deliver a high PCE of 5.47% that outperformed other solid-state electrolytes and was even comparable to some liquid electrolytes.

The concentration of the gelator in gel electrolyte is relevant to the gelation and cross-linking extent, which determines the charge transfer and interfacial contact between the electrode and electrolytes. Xiao et al. constructed a fiber-shaped DSSC with a Pt wire counter electrode at center surrounded by the TiO_2 nanoarray/Ti wire photoanode. They examined the configuration of gel electrolytes on the curled photoanode based on different concentrations of PVdF-HFP [44]. With a proper amount of polymer, the gel electrolyte exhibited a porous structure on the surface of photoanode, which could accumulate more liquid electrolyte, and the outside gel electrolyte layer could inhibit the evaporation of the inner liquid electrolyte. The obtained fiber-shaped DSSC thus exhibited the highest PCE of 6.32%, attributed to the high charge collection efficiency, sufficient redox reaction, and optimal cross-linking structure.

5.3 Coaxial Structure

The coaxial structure of fiber-shaped DSSCs can be considered as wrapping a planar dye-sensitized solar cell into an ultrathin cylinder. The resulting "sandwich" configuration is composed of a fiber electrode as the core, a thin-film electrode as the shell, and the electrolyte in between. Fiber-shaped DSSCs with a coaxial structure show some unique advantages compared with the twisted structure. On one hand, the stability based on the coaxial structure is generally high. On the other hand, the contact between the electrode and electrolyte is basically more sufficient in a coaxial structure that benefits the fast reaction kinetics. However, the coaxial structure also faces limitations, such as the photovoltaic conversion performance and preparation complexity.

For the coaxial structure, the shell electrode (generally the counter electrode) plays a critical role in the device's performance. Ideally, it should exhibit comprehensively high optical transmittance, electrical conductivity, and mechanical strength. As a result, very few materials can simultaneously meet these requirements, which severely narrows the available material options. MWCNT film represents one of the promising candidates that has been investigated. For instance, Zhang et al. reported a coaxial fiber-shaped DSSC by wrapping a free-standing MWCNT film on a photoanode composed of dye-absorbed TiO_2 nanotube array on a Ti wire (Figure 5.10a) [45]. The obtained device exhibited high structure stability and flexibility with a well-reserved PCE at high bending angles up to 90°. It showed a PCE of 1.6% (Figure 5.10b), which could be further increased to 2.6% with the assistance of Ag or Cu metal wires as an additional electrode placed in parallel (Figure 5.10c,d).

The randomly dispersed MWCNT will decrease the electrical conductivity that deteriorates the PCEs. To this end, the employment of highly aligned MWCNT sheets in coaxial fiber-shaped DSSCs can be a promising approach for higher PCEs. Sun et al. used the highly aligned MWCNT sheets wrapped around the TiO_2 nanotube-modified Ti wire to obtain fiber-shaped DSSC, which exhibited a high PCE of 4.1% [46]. The obtained coaxial DSSC fibers also demonstrated remarkable flexibility, which showed slight PCE variations within 5% during 100 bending cycles.

Liang et al. designed a coaxial fiber-shaped DSSC with multiple working electrodes, which could facilitate photocurrent extraction from the Pt wire [47]. The fiber-shaped DSSC consisted of a Pt microwire counter electrode as the core and six Ti microwire photoanodes as the sheath. After the electrolyte injection, the fiber-shaped DSSC showed high PCEs of 6.6%, which could be further improved by modifying the TiO_2 nanotubes with the niobium isopropoxide solution, resulting in a remarkable PCE of 9.1%, and excellent flexibility under 180° bending.

The flexibility of coaxial fiber-shaped DSSC can be effectively improved by the employment of a quasi-solid-state electrolyte. Sun et al. fabricated a quasi-solid-state electrolyte for coaxial fiber-shaped DSSCs [48]. In specific, the eutectic melt comprised of 1-ethyl-3-methylimidazolium iodide, 1-propyl-3-methylimidazolium iodide ionic liquid, and iodine was used as the quasi-solid-state electrolyte. A transparent MWCNT sheet serving as the counter electrode was wrapped on a

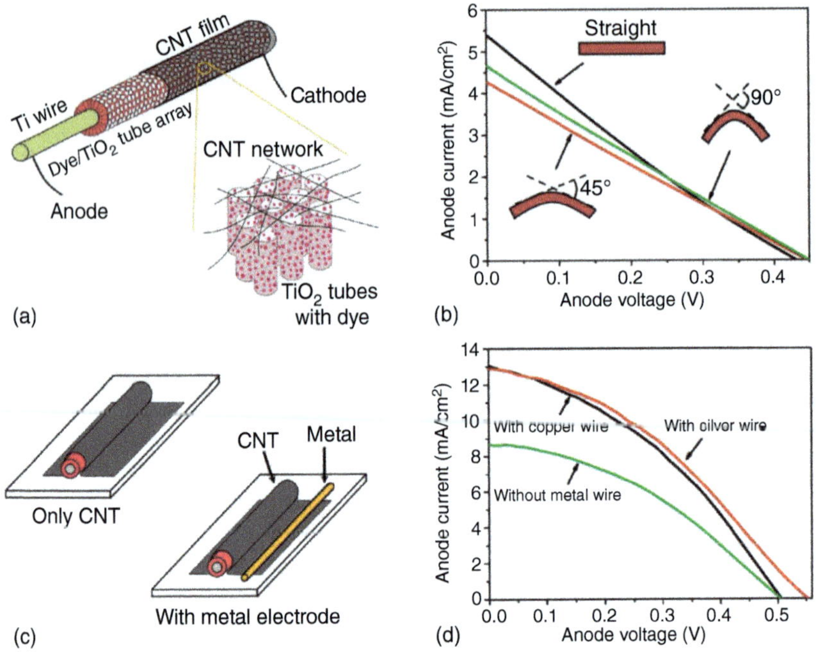

Figure 5.10 (a) Schematic illustration of a single-wire DSSC with a coaxial structure consisting of a Ti wire, dye-adsorbed TiO_2 nanotube arrays, and a transparent MWCNT film wrapping around the wire. (b) *J–V* curves of a flexible solar cell wire in a straight form and at bending angles of 45° and 90°, respectively. (c) Illustration of MWCNT-wrapped TiO_2–Ti solar cells without or with a conventional metal electrode (placed in parallel with the wire). (d) *J–V* characteristics of MWCNT-TiO_2-Ti solar cells without a metal electrode, with an additional Ag electrode, and with a Cu electrode, respectively. Source: Zhang et al. [45], © 2011/American Chemical Society.

modified Ti wire photoanode, coupling with the quasi-solid-state electrolyte to obtain the coaxial fiber-shaped DSSC (Figure 5.11a–e). It demonstrated stable photovoltaic conversion performance even at 120° (Figure 5.11f). In addition, the PCE could be maintained at 91% of the initial value after 50 heating/cooling cycles (Figure 5.11g). The high flexibility and thermal stability of these coaxial fiber-shaped DSSCs showed promising applications in wearable devices operated at a wide temperature range.

5.4 Functionalization and Integration

It is expected that the functionalization of fiber-shaped DSSCs can better accommodate the practical conditions in wearable applications, such as frequent bending, twisting, and stretching deformations. Meanwhile, an integrated device that realizes in-series or in-parallel connection of multiple device units can improve the output voltage and current, satisfying a variety of practical applications. The self-powered requirement for wearable devices also results in the integration of solar cells and

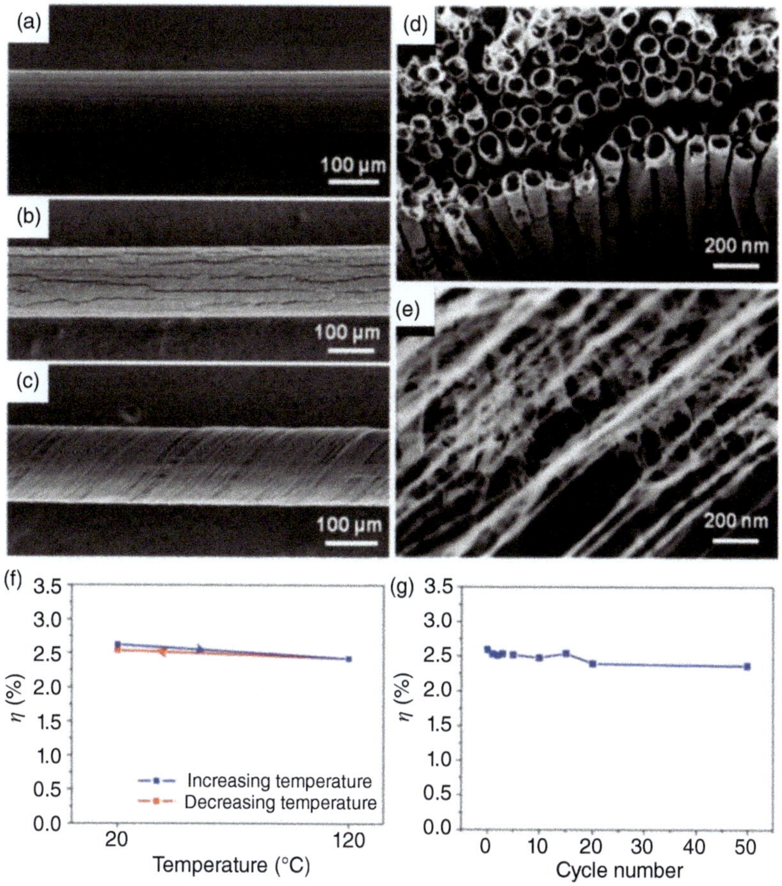

Figure 5.11 Scanning electron microscopy (SEM) images of the coaxial fiber-shaped DSSC. (a) A bare Ti wire. (b) Vertically aligned titania nanotubes grown on the Ti wire. (c) A single layer of highly aligned MWCNT sheet wrapped on (b). (d) Higher magnification of (b). (e) Higher magnification of (c). (f) PCE changes of a quasi-solid-state coaxial fiber-shaped DSSC after heating from 20° to 120° and then cooling down to 20°. (g) Dependence of the PCE on the cycle number for a typical quasi-solid-state coaxial fiber-shaped DSSC that was heated to 120° and cooled down to 20° for 50 cycles. Source: Sun et al. [48], © 2014/Royal Society of Chemistry.

energy storage devices in a single fiber, which overcomes the intermittence of solar light and outputs the stored energy when needed. The functionalization and integration of fiber-shaped DSSCs have become an important branch of flexible electronics with various impressive advancements.

5.4.1 Functional Fiber-Shaped DSSCs

Fiber-shaped DSSCs with sensitive responses to external stimuli (such as magnetic fields) can potentially regulate the fixing, movement, and assembly of the devices, leading to better processability and compatibility for real-world applications.

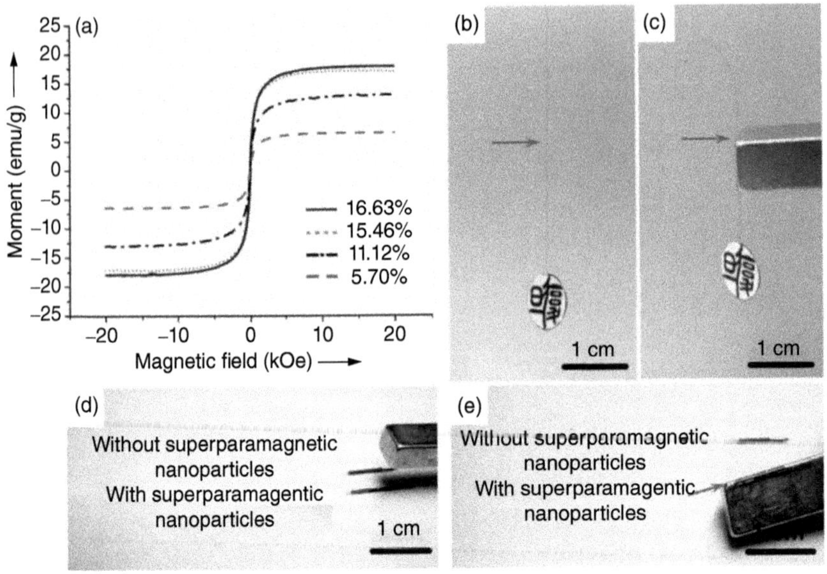

Figure 5.12 Magnetic responses of the superparamagnetic nanoparticle-based composite fiber and the resulting photovoltaic wire. (a) Hysteresis loop of the composite fibers with different weight percentages of superparamagnetic nanoparticles. (b, c) Photographs of a superparamagnetic nanoparticle/MWCNT composite fiber (at a diameter of about 25 mm) before and after being close to a magnet, respectively. (d, e) Photographs of two photovoltaic wires before and after a magnet was moved close to them at the same distance. The top device was fabricated by twisting a bare MWCNT fiber and a modified titanium wire. In contrast, the bottom was fabricated by twisting a superparamagnetic nanoparticle/MWCNT composite fiber and the same modified titanium wire. Source: Sun et.al. [49], © 2013/John Wiley & Sons, Inc.

A pioneer work by Sun et al. developed aligned MWCNT fibers spun from MWCNT arrays through the dry-spinning process, which was further incorporated with Fe_3O_4 nanoparticles to realize magnet response (Figure 5.12) [49]. The obtained MWCNT hybrid fiber exhibited a superparamagnetic behavior. It thus enabled sensitive magnetic responsiveness of the obtained fiber-shaped DSSC, which could be collected, fixed, and moved by a magnet. It was worth noting that the incorporation of Fe_3O_4 showed no negative effects on the device performance, e.g. J_{sc} and PCE, indicating the excellent compatibility of photovoltaic conversion performance and magnetic responsiveness. The PCE could be further improved to 8.03% with the incorporation of Pt nanoparticles into the Fe_3O_4/MWCNT fiber via electrochemical deposition, suggesting the high designability of the aligned MWCNT fiber.

Developing stretchable fiber-shaped DSSCs is crucial for practical wearable applications that involve various deformations. A pioneering work by Yang et al. prepared an elastic conductive fiber through a rotation-translation method, in which the MWCNT sheets were drawn from a spinnable MWCNT array and wrapped onto a rubber fiber with the two ends attached to two motors [50]. This setup enabled the fabrication of fiber-shaped devices in a continuous and well-controlled way. The elastic conducting fiber showed high stretchability, evidenced by the

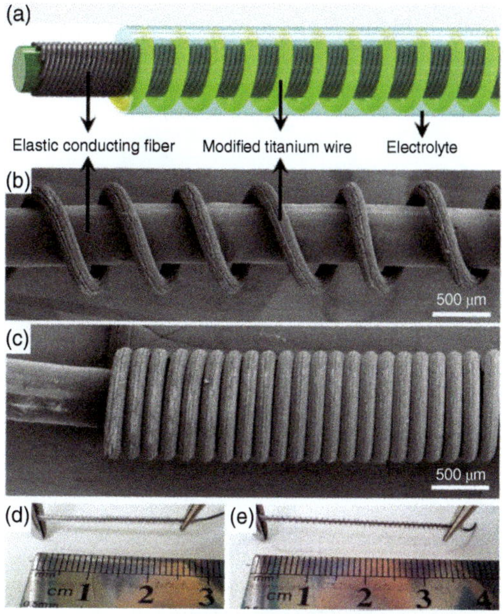

Figure 5.13 (a) Schematic illustration of a stretchable fiber-shaped DSSC. (b, c) SEM images of a stretchable fiber-shaped DSSC with pitch distances of 560 and 164 μm, respectively. (d, e) A stretchable fiber-shaped DSSC before and after stretch by 30%, respectively. Source: Yang et al. [50], © 2014/John Wiley & Sons, Inc.

well-maintained electrical resistances during repeated stretch by 100%. Stretchable fiber-shaped DSSC could thus be produced using the elastic conducting fiber as the counter electrode and a spring-shaped anodized Ti wire adsorbed with N719 dye as the photoanode (Figure 5.13). The stretchable fiber-shaped DSSC showed a high PCE of 7.13%, which could be maintained at a strain of 30%. This opens a new avenue for functional fiber-shaped DSSCs toward wearable applications.

Thermal stability is another key factor that requires serious evaluation for practical use. Conventional liquid electrolytes generally suffer from poor thermal stability. As a result, the combination of nonvolatile ionic liquid and thermally stable polymer is highly promising to unlock fiber-shaped DSSCs with remarkable thermal stability. Li et al. designed an ionic liquid-gel electrolyte comprised of polymer (PVdF-HFP) and ionic liquid BMITFSI [43]. Combined with the advantages of these components, the electrolyte showed excellent stability at high temperatures, e.g. without observable weight loss up to 300 °C (Figure 5.14). At a bending curvature of 5 mm^{-1}, the fiber-shaped DSSC using the optimal electrolyte showed no obvious change in PCE during 10 bending cycles. A PCE of 5.47% has thus been achieved and can remain by 90% after a 30-days reservation.

5.4.2 Integrated Fiber-Shaped DSSCs

As mentioned above, integrated systems such as solar/mechanical–electrical–chemical energy conversion have attracted tremendous interest for simultaneous power generation and electrochemical energy storage. Fiber-shaped energy harvesting and energy storage devices can be integrated into a coaxial or twisted structure, which generally shares at least one common electrode [51]. In 2012, Chen et al. developed an integrated fiber-shaped device comprised of a fiber-shaped

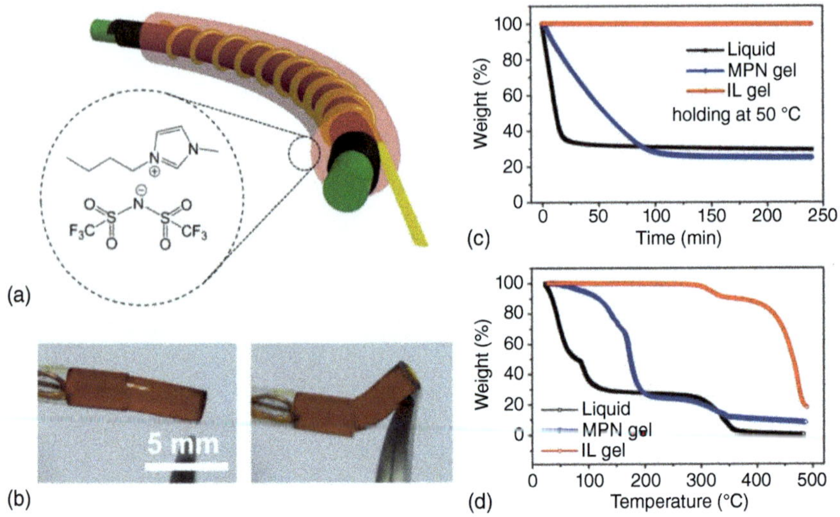

Figure 5.14 (a) Schematic illustration of the structure of a stretchable fiber-shaped DSSC and the chemical components of ionic liquid–gel electrolyte. (b) Photographs of ionic liquid-gel electrolyte under natural and bending states. (c) Weight losses of three different kinds of electrolytes were traced at the same temperature of 50 °C. (d) Weight losses of three different kinds of electrolytes from room temperature to 50 °C. Source: Li et al. [43], © 2015/John Wiley & Sons.

DSSC and a fiber-shaped supercapacitor (FSC) [52]. A Ti wire modified with TiO_2 nanotubes served as the common electrode, and the active materials for photoelectric conversion (PC) and energy storage (ES) were, respectively, coated on the corresponding parts (Figure 5.15). The integrated fiber-shaped device was prepared by twisting two aligned MWCNT fibers with the modified Ti wires. A switch was incorporated to control the energy conversion and storage processes. Upon exposure to light, the dye molecules of the PC part were excited with the injection of photoelectrons to the conduction band of TiO_2 on the photoanode, thus charging the ES part. The LS part could discharge when the switch was turned off. The integrated device showed an overall energy harvesting and storage efficiency of 1.5%, as well as impressive flexibility and mechanical stability under deformation.

An integrated fiber-shaped device with a coaxial structure was also developed based on a similar configuration. In 2011, an integrated coaxial device comprised of a fiber-shaped DSSC, nanogenerator, and FSC was developed [53]. It could simultaneously harvest the mechanical and optical energy and further store it by the FSC in the form of electric energy (Figure 5.16a), allowing for continuous power output regardless of the presence of sunlight. This integrated device used ZnO nanowires decorated with Kevlar fibers as the common electrode, and the transparent graphene on Cu mesh was wound around the common electrode as another electrode. However, the PCE of the fiber-shaped DSSC and the capacity of the supercapacitor were rather low (0.02% and 0.4 mF/cm^2, respectively), originating from the reduced incident light intensity shielded by the Cu mesh outside the device. To further improve the device performance, Chen et al. used a Ti wire modified with

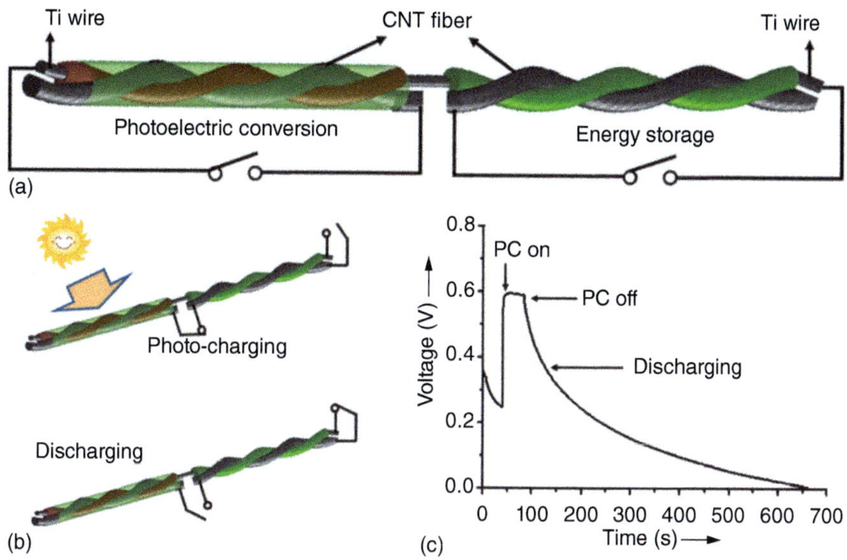

Figure 5.15 (a) Schematic illustration of the integrated wire-shaped device for photoelectric conversion (PC) and energy storage (ES). (b) Schematic illustration of the circuit connection during the charging and discharging processes, respectively. (c) Photocharging–discharging curve of a typical energy wire. The discharging current is 0.1 mA. Source: Chen et al. [52], © 2012/John Wiley & Sons.

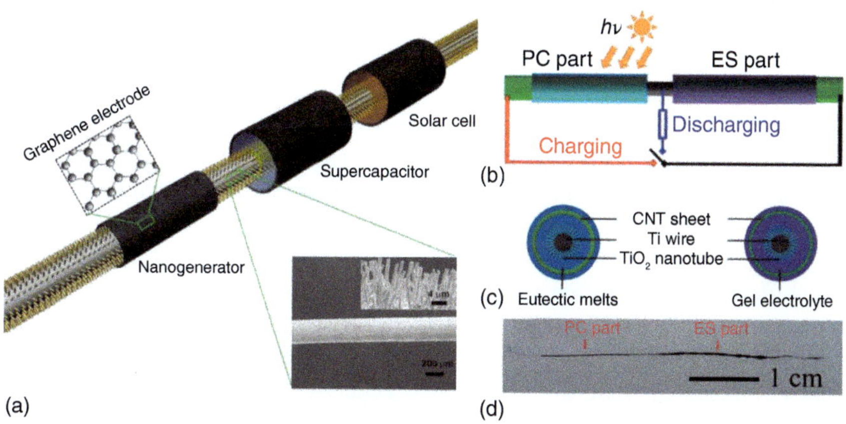

Figure 5.16 (a) Schematic of a fiber-based multi-energy device comprising a nanogenerator, solar cell, and supercapacitor. The inset is an SEM image of the plastic wire showing ZnO NW arrays grown along the radial direction [53], © 2011/John Wiley & Sons. (b) Schematic illustration of the structure of the coaxially integrated dye-sensitized solar cell and electrochemical capacitor into an energy fiber. (c) Cross-sectional views of the PC and ES parts of the energy fiber, respectively. (d) Photograph of an energy fiber. Source: Chen et al. [54], © 2014/Springer Nature.

TiO_2 nanotubes as a common electrode for energy conversion and energy storage devices. At the same time, the MWCNT sheet with high transparency was served as the other electrodes of the two devices (Figure 5.16b–d) [54]. The maximum PCE of the all-solid-state "energy fiber" was 2.73%, along with a remarkably high energy storage efficiency of 75.7%. This integrated energy fiber is highly promising in powering various wearable devices.

Conducting polymers have been widely investigated in integrated devices for performance improvement [55, 56]. Polyaniline (PANi) is promising for replacing the highly expensive Pt counter electrode [57]. Fu et al. reported an integrated fiber-shaped device using PANi-SS wire as both the counter electrode of the fiber-shaped DSSC and the electrode of the FSC [58]. The integrated fiber could deliver an overall PCE of 2.1%. In addition, poly(3,4-ethylenedioxythiophene) polystyrene sulfonate (PEDOT:PSS) has been widely used for solar cells and supercapacitors due to its high catalytic performance and good capacitive characteristics. Wang et al. developed a flexible PEDOT:PSS fiber with a high electrical conductivity and mechanical strength of 1055.6 S/cm and 179.9 MPa, respectively. The PEDOT:PSS fiber could serve as the common electrode of the integrated fiber-shaped device, which achieved a high overall PCE of up to 5.1% and outstanding energy storage stability of 1700 cycles for 63 days, representing an important step toward practical applications [59].

Efforts have been also made to enhance the integration of fiber-shaped DSSCs. Sun et al. designed a novel fiber-shaped energy device that enabled energy harvesting and storage in a two-electrode-based device [60]. This integrated fiber-shaped device showed simple components, i.e. a Ti wire modified with dye-sensitized TiO_2 nanotubes twisted by an aligned MWCNT fiber (Figure 5.17). The device could be employed as a DSSC and supercapacitor without relatively independent function and performance. The resulting "energy wire" exhibited a high PCE and specific capacitance of 6.58% and 85.03 $\mu F/cm^2$, respectively. Both the energy storage and photovoltaic conversion performance could be well maintained during alternate operations. As a proof of concept, it could be readily woven into a lightweight and flexible energy textile with integrated functions.

A key challenge of the above-integrated devices lies in the power mismatch between the energy harvesting and storage parts. For instance, the photo-charging process generally completes within a few seconds, indicating that the limited energy storage capability fails to match the energy harvesting capability. Sun et al. developed integrated DSSC and lithium-ion battery (LIB) fiber devices based on a core-sheath structure with multiple in-series connected DSSC at the sheath and a fiber-shaped LIB at core [61]. The photovoltaic conversion and energy storage performances were enhanced with the increasing amounts of PC parts, which exhibited a decrease in charging time, i.e. from over 600 to 86 seconds for 5 and 8 units, respectively (Figure 5.18a). Variation of the incident light angle from 0° to 180° showed no significant changes in the photovoltaic conversion performance since the fiber-shaped DSSCs were capable of harvesting light from all directions (Figure 5.18b). The integrated device exhibited an output voltage of up to 2.2 V and could be further raised to 2.6 V with the incorporation of amorphous carbon

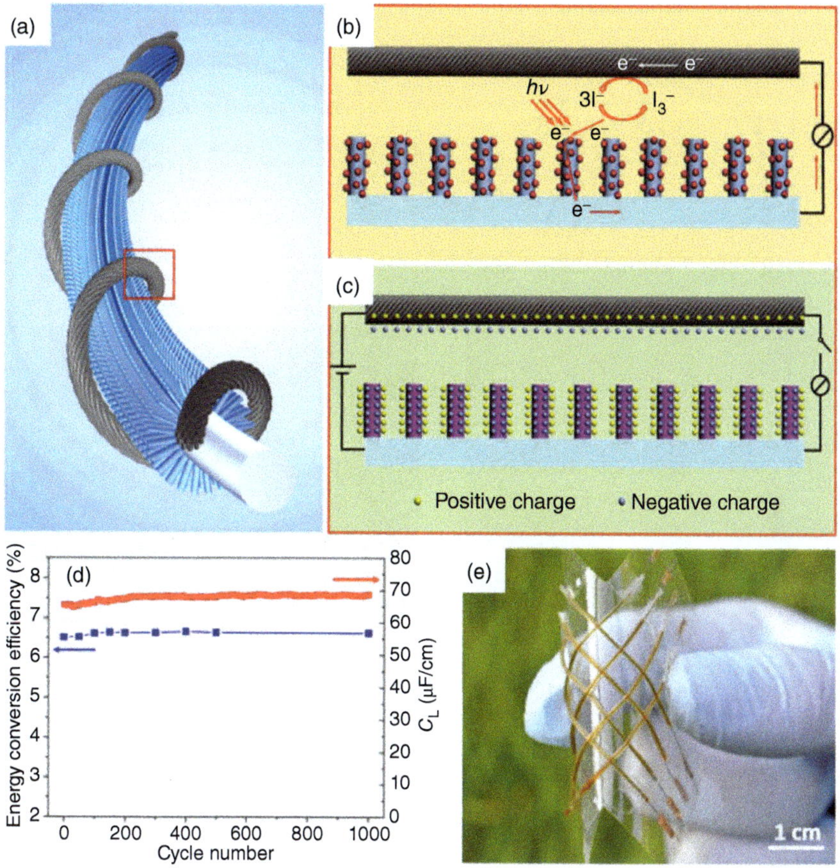

Figure 5.17 (a) Structure of the wire-shaped energy device. (b) Working mechanism as a DSSC. (c) Working mechanism as an electrochemical capacitor. (b) and (c) are cross-sections of the rectangle in (a). (d) A wire-shaped energy device is alternately operated as a solar cell and electrochemical capacitor for continuous 1000 cycles. The galvanostatic charge–discharge performance was conducted at a current density of 0.64 mA/cm^2. (e) Photograph of the wire-shaped device woven into a flexible textile. Source: Sun et al. [60], © 2014/John Wiley & Sons.

into the hybrid electrode, which could power wearable electronic devices such as light-emitting diode (LED) lamps (Figure 5.18c,d).

The integration of multiple devices that can scavenge different types of energy from the surrounding environment provides intriguing possibilities for wearable devices. For example, Wang and coworkers proposed a highly elastic and wearable self-charging power bracelet that could scavenge ambient sunshine and mechanical energy simultaneously [62]. Specifically, three in-series fiber-shaped DSSCs were connected with the rectified in-parallel triboelectric nanogenerators (TENGs), and these energy harvesting devices were further integrated into a flexible silicone rubber belt. The harvested energy from the sunlight and arm motions could be stored in in-series connected SCs. The charging time of the SC could be greatly shortened

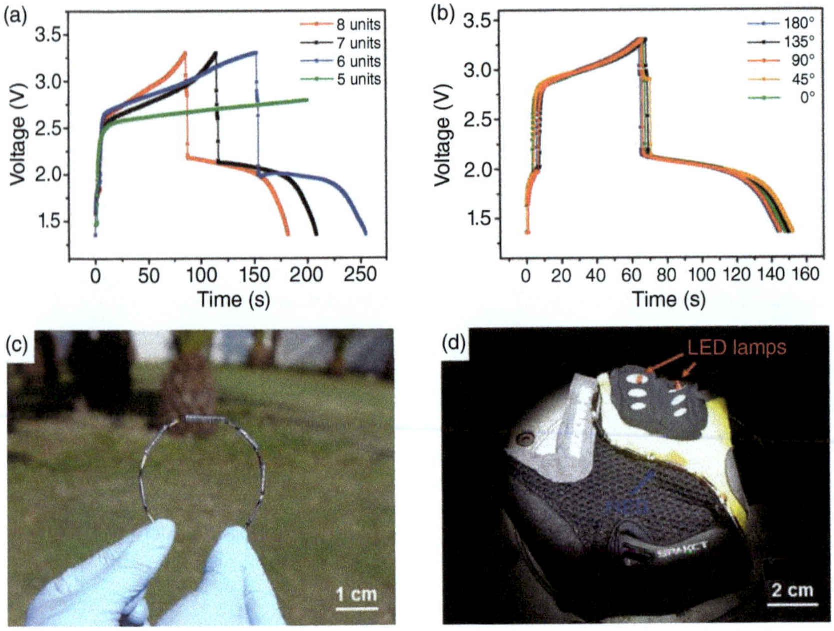

Figure 5.18 (a) Photocharging and discharging curves with increasing PC units. The photocharging process was performed under the illumination of simulated AM 1.5 solar light (100 mW/cm^2). Discharging current, 0.05 mA. (b) Photocharging and discharging curves at different angles of incident light. Discharging current, 0.05 mA. (c) Photograph of an integrated fiber-shaped device under bending. (d) Photograph of an integrated fiber-shaped device being woven into a sports glove for powering two light-emitting diodes in the dark. Source: Sun et al. [61]/Royal Society of Chemistry.

via the integration of TENGs. In practical wearable applications, the charged SC could stably power the electronic watch for 85 seconds. The output voltage of SCs can be readily increased by connecting more SCs for high-voltage applications.

5.5 Summary and Prospects

Over the past decade, the device performances of fiber-shaped DSSCs have been significantly promoted, along with the inspiring advancements in preparation, functionalization, and integration. However, it should be noted that the state-of-the-art PCEs of fiber-shaped DSSCs are still much lower compared with the conventional planar counterparts. In addition, the integration of fiber-shaped DSSCs-based devices into textiles with comfortable human/textile interfaces requires more efficient processing technologies. Thus, great efforts should be made to further improve the flexibility, weavability, and wearability of the fiber-shaped DSSCs.

From the aspect of materials, it is still demanded to develop high-performing materials for performance improvement. For instance, high-performing photoanodes are highly demanded for achievement of sufficient electron collection/diffusion, effective dye adsorption, and high corrosion resistance. Conventional Pt counter electrodes suffer from the high-cost issue, which requires the development of low-cost catalysts such as carbon, conducting polymer, and sustainable transition metal compounds. A rational combination of multiple components to prepare the fiber electrodes can make full use of their advantages. In addition, the widely used liquid electrolytes with volatility and instability have hindered the practical applications of fiber-shaped DSSCs. To this end, quasi-solid-state and all-solid-state electrolytes are promising candidates to greatly enhance structure integrity and electrochemical stability during use. However, their development has been impeded by the low ionic conductivity and insufficient electrode wetting. Polymer-based gel electrolytes can be also used for fiber-shaped DSSCs through rational optimization of the preparation techniques and interfaces.

For the functionalization of fiber-shaped DSSCs, it requires further improvement for real-world applications. For instance, the compromise between the stretchability and electrical conductivity of fiber electrodes has plagued the achievement of high-performance fiber-shaped DSSCs. Therefore, rational designing and screening of conductive and stretchable components should be taken into consideration toward balanced electrical and mechanical properties. In addition, the integration of energy harvesting and storage devices can greatly extend the applications of fiber-shaped DSSCs. However, the duration, working voltage/current, and total energy efficiency require further improvement. In regard to the weaving process for wearable photovoltaic textiles, the sufficient and reliable interconnection of fiber-shaped DSSCs remains challenging. Therefore, we should develop new weaving techniques for better integration of multiple devices into a breathable textile, and the structural stability and washability also require further evaluation and optimization.

Last but not least, the state-of-the-art production techniques of fiber-shaped DSSCs are still not far from the requirements of practical production. The performance of fiber-shaped DSSCs usually deteriorates with increased length due to the significantly increased resistance. As a result, the production of large-area photovoltaic textiles that can output practically useful power is still rather challenging. To this end, the development of highly conductive fiber electrodes is highly required to achieve high device performance for long devices at the meter or even kilometer scale. Meanwhile, continuous preparation methods should be developed toward high production efficiency. Fortunately, the recent breakthrough in the continuous production of fiber-shaped lithium-ion batteries with commercially available equipment and materials can greatly inspire the innovation of fiber-shaped DSSCs.

References

1 O'Regan, B. and Grätzel, M. (1991). A low-cost, high-efficiency solar cell based on dye-sensitized colloidal TiO_2 films. *Nature* 353 (6346): 737–740.

2 Muñoz-García, A.B., Benesperi, I., Boschloo, G. et al. (2021). Dye-sensitized solar cells strike back. *Chem. Soc. Rev.* 50 (22): 12450–12550.

3 Chu, L., Su, J., Ahmad, W. et al. (2015). Facile, rapid and in-situ synthesis of ZnO nanoparticle films on Zn wires for fiber dye-sensitized solar cells. *Mater. Res. Bull.* 66: 244–248.

4 Shakeel Ahmad, M., Pandey, A.K., and Abd Rahim, N. (2017). Advancements in the development of TiO_2 photoanodes and its fabrication methods for dye sensitized solar cell (DSSC) applications. A review. *Renewable Sustainable Energy Rev.* 77: 89–108.

5 Wu, J., Lan, Z., Lin, J. et al. (2017). Counter electrodes in dye-sensitized solar cells. *Chem. Soc. Rev.* 46 (19): 5975–6023.

6 Ling, C.K., Aung, M.M., Abdullah, L.C. et al. (2020). A short review of iodide salt usage and properties in dye sensitized solar cell application: single vs binary salt system. *Sol. Energy* 206: 1033–1038.

7 Devadiga, D., Selvakumar, M., Shetty, P. et al. (2022). The integration of flexible dye-sensitized solar cells and storage devices towards wearable self-charging power systems: a review. *Renewable Sustainable Energy Rev.* 159: 112252.

8 Barnes, P.R., Miettunen, K., Li, X. et al. (2013). Interpretation of optoelectronic transient and charge extraction measurements in dye-sensitized solar cells. *Adv. Mater.* 25 (13): 1881–1922.

9 Katoh, R. and Furube, A. (2014). Electron injection efficiency in dye-sensitized solar cells. *J. Photochem. Photobiol., C* 20: 1–16.

10 Hashmi, S.G., Özkan, M., Halme, J. et al. (2016). Dye-sensitized solar cells with inkjet-printed dyes. *Energy Environ. Sci.* 9 (7): 2453–2462.

11 Lee, C.-P., Li, C.-T., and Ho, K.-C. (2017). Use of organic materials in dye-sensitized solar cells. *Mater. Today* 20 (5): 267–283.

12 Chen, C.C., Nguyen, V.S., Chiu, H.C. et al. (2022). Anthracene-bridged sensitizers for dye-sensitized solar cells with 37% efficiency under dim light. *Adv. Energy Mater.* 12: 2104051.

13 Zhang, Y., Wang, H., Lu, H. et al. (2021). Electronic fibers and textiles: recent progress and perspective. *iScience* 24 (7): 102716.

14 Fakharuddin, A., Li, H., Di Giacomo, F. et al. (2021). Fiber-shaped electronic devices. *Adv. Energy Mater.* 11 (34): 2101443.

15 Sun, H., Zhang, Y., Zhang, J. et al. (2017). Energy harvesting and storage in 1D devices. *Nat. Rev. Mater.* 2 (6): 17023.

16 Baps, B., Eber-Koyuncu, M., and Koyuncu, M. (2001). Ceramic based solar cells in fiber form. *Key Eng. Mater.* 206–213: 937–940.

17 Chen, T., Qiu, L., Cai, Z. et al. (2012). Intertwined aligned carbon nanotube fiber based dye-sensitized solar cells. *Nano Lett.* 12 (5): 2568–2572.

18 Fan, X., Chu, Z.Z., Wang, F.Z. et al. (2008). Wire-shaped flexible dye-sensitized solar cells. *Adv. Mater.* 20 (3): 592–595.

19 Fu, Y., Lv, Z., Hou, S. et al. (2011). Conjunction of fiber solar cells with groovy micro-reflectors as highly efficient energy harvesters. *Energy Environ. Sci.* 4 (9): 3379.

20 Kim, C.W., Suh, S.P., Choi, M.J. et al. (2013). Fabrication of $SrTiO_3$–TiO_2 heterojunction photoanode with enlarged pore diameter for dye-sensitized solar cells. *J. Mater. Chem. A* 1 (38): 11820.

21 Fonstad, C.G. and Rediker, R.H. (1971). Electrical properties of high-quality stannic oxide crystals. *J. Appl. Phys.* 42 (7): 2911–2918.

22 Breckenridge, R.G. and Hosler, W.R. (1953). Electrical properties of titanium dioxide semiconductors. *Phys. Rev.* 91 (4): 793–802.

23 Peng, M., Cai, X., Fu, Y. et al. (2014). Facial synthesis of SnO_2 nanoparticle film for efficient fiber-shaped dye-sensitized solar cells. *J. Power Sources* 247: 249–255.

24 Kim, J., Lee, H., Kim, D.Y. et al. (2014). Resonant multiple light scattering for enhanced photon harvesting in dye-sensitized solar cells. *Adv. Mater.* 26 (30): 5192–5197.

25 Liu, G., Peng, M., Song, W. et al. (2015). An 8.07% efficient fiber dye-sensitized solar cell based on a TiO_2 micron-core array and multilayer structure photoanode. *Nano Energy* 11: 341–347.

26 Ardhi, R.E.A., Tran, M.X., Wang, M. et al. (2020). Chemically tuned, bi-functional polar interlayer for TiO_2 photoanodes in fibre-shaped dye-sensitized solar cells. *J. Mater. Chem. A* 8 (5): 2549–2562.

27 Finnie, K.S., Bartlett, J.R., and Woolfrey, J.L. (1998). Vibrational spectroscopic study of the coordination of (2,2′-ipyridyl-4,4′-dicarboxylic acid)ruthenium(II) complexes to the surface of nanocrystalline titania. *Langmuir* 14 (10): 2744–2749.

28 Zakeeruddin, S.M., Nazeeruddin, M.K., Humphry-Baker, R. et al. (2002). Design, synthesis, and application of amphiphilic ruthenium polypyridyl photosensitizers in solar cells based on nanocrystalline TiO_2 films. *Langmuir* 18 (3): 952–954.

29 Zhang, S., Ji, C., Bian, Z. et al. (2012). Porous, platinum nanoparticle-adsorbed carbon nanotube yarns for efficient fiber solar cells. *ACS Nano* 6 (8): 7191–7198.

30 Ali, A., Shehzad, K., Ur-Rahman, F. et al. (2016). Flexible, low cost, and platinum-free counter electrode for efficient dye-sensitized solar cells. *ACS Appl. Mater. Interfaces* 8 (38): 25353–25360.

31 Liu, G., Wang, M., Wang, H. et al. (2018). Hierarchically structured photoanode with enhanced charge collection and light harvesting abilities for fiber-shaped dye-sensitized solar cells. *Nano Energy* 49: 95–102.

32 Fu, X., Sun, H., Xie, S. et al. (2018). A fiber-shaped solar cell showing a record power conversion efficiency of 10%. *J. Mater. Chem. A* 6 (1): 45–51.

33 Xu, Z. and Gao, C. (2015). Graphene fiber: a new trend in carbon fibers. *Mater. Today* 18 (9): 480–492.

34 Yang, Z., Sun, H., Chen, T. et al. (2013). Photovoltaic wire derived from a graphene composite fiber achieving an 8.45% energy conversion efficiency. *Angew. Chem. Int. Ed.* 52 (29): 7545–7548.

35 Wang, M., Anghel, A.M., Marsan, B. et al. (2009). CoS supersedes Pt as efficient electrocatalyst for triiodide reduction in dye-sensitized solar cells. *JACS* 131 (44): 15976–15977.

36 Wu, M., Lin, X., Hagfeldt, A. et al. (2011). A novel catalyst of WO_2 nanorod for the counter electrode of dye-sensitized solar cells. *Chem. Commun.* 47 (15): 4535–4537.

37 Jiang, Q.W., Li, G.R., and Gao, X.P. (2009). Highly ordered TiN nanotube arrays as counter electrodes for dye-sensitized solar cells. *Chem. Commun.* 44: 6720–6722.

38 Wu, M., Lin, X., Wang, Y. et al. (2012). Economical Pt-free catalysts for counter electrodes of dye-sensitized solar cells. *JACS* 134 (7): 3419–3428.

39 Chen, L., Zhou, Y., Dai, H. et al. (2015). One-step growth of $CoNi_2S_4$ nanoribbons on carbon fibers as platinum-free counter electrodes for fiber-shaped dye-sensitized solar cells with high performance: polymorph-dependent conversion efficiency. *Nano Energy* 11: 697–703.

40 Utomo, D.S., Kim, J.H., Lee, D. et al. (2021). Fractional structured molybdenum oxide catalyst as counter electrodes of all-solid-state fiber dye-sensitized solar cells. *J. Colloid Interface Sci.* 584: 520–527.

41 Wang, D., Hou, S., Wu, H. et al. (2011). Fiber-shaped all-solid state dye sensitized solar cell with remarkably enhanced performance via substrate surface engineering and TiO_2 film modification. *J. Mater. Chem.* 21 (17): 6383–6388.

42 Kim, J.H., Chi, Y.S., and Kang, T.J. (2013). Optimization of quasi-solid-state dye-sensitized photovoltaic fibers using porous polymer electrolyte membranes. *J. Power Sources* 229: 84–94.

43 Li, H., Guo, J., Sun, H. et al. (2015). Stable hydrophobic ionic liquid gel electrolyte for stretchable fiber-shaped dye-sensitized solar cell. *ChemNanoMat* 1 (6): 399–402.

44 Xiao, B.C. and Lin, L.Y. (2020). Tuning electrolyte configuration and composition for fiber-shaped dye-sensitized solar cell with poly(vinylidene fluoride-*co*-hexafluoropropylene) gel electrolyte. *J. Colloid Interface Sci.* 571: 126–133.

45 Zhang, S., Ji, C., Bian, Z. et al. (2011). Single-wire dye-sensitized solar cells wrapped by carbon nanotube film electrodes. *Nano Lett.* 11 (8): 3383–3387.

46 Sun, H., You, X., Yang, Z. et al. (2013). Winding ultrathin, transparent, and electrically conductive carbon nanotube sheets into high-performance fiber-shaped dye-sensitized solar cells. *J. Mater. Chem. A* 1 (40): 12422–12425.

47 Liang, J., Zhang, G., Sun, W. et al. (2015). High efficiency flexible fiber-type dye-sensitized solar cells with multi-working electrodes. *Nano Energy* 12: 501–509.

48 Sun, H., Li, H., You, X. et al. (2014). Quasi-solid-state, coaxial, fiber-shaped dye-sensitized solar cells. *J. Mater. Chem. A* 2 (2): 345–349.

49 Sun, H., Yang, Z., Chen, X. et al. (2013). Photovoltaic wire with high efficiency attached onto and detached from a substrate using a magnetic field. *Angew. Chem. Int. Ed.* 52 (32): 8276–8280.

50 Yang, Z., Deng, J., Sun, X. et al. (2014). Stretchable, wearable dye-sensitized solar cells. *Adv. Mater.* 26 (17): 2643–2647.

51 Ma, W., Zhang, Y., Pan, S. et al. (2021). Smart fibers for energy conversion and storage. *Chem. Soc. Rev.* 50 (12): 7009–7061.

52 Chen, T., Qiu, L., Yang, Z. et al. (2012). An integrated "energy wire" for both photoelectric conversion and energy storage. *Angew. Chem. Int. Ed.* 51 (48): 11977–11980.

53 Bae, J., Park, Y.J., Lee, M. et al. (2011). Single-fiber-based hybridization of energy converters and storage units using graphene as electrodes. *Adv. Mater.* 23 (30): 3446–3449.

54 Chen, X., Sun, H., Yang, Z. et al. (2014). A novel "energy fiber" by coaxially integrating dye-sensitized solar cell and electrochemical capacitor. *J. Mater. Chem. A* 2 (6): 1897–1902.

55 Kim, T., Kim, G., Kim, H. et al. (2019). Megahertz-wave-transmitting conducting polymer electrode for device-to-device integration. *Nat. Commun.* 10 (1): 653.

56 Luo, H., Kaneti, Y.V., Ai, Y. et al. (2021). Nanoarchitectured porous conducting polymers: from controlled synthesis to advanced applications. *Adv. Mater.* 33 (29): e2007318.

57 Tai, Q., Chen, B., Guo, F. et al. (2011). In situ prepared transparent polyaniline electrode and its application in bifacial dye-sensitized solar cells. *ACS Nano* 5 (5): 3795–3799.

58 Fu, Y., Wu, H., Ye, S. et al. (2013). Integrated power fiber for energy conversion and storage. *Energy Environ. Sci.* 6 (3): 805–812.

59 Wang, Z., Cheng, J., Huang, H. et al. (2020). Flexible self-powered fiber-shaped photocapacitors with ultralong cyclelife and total energy efficiency of 5.1%. *Energy Storage Mater.* 24: 255–264.

60 Sun, H., You, X., Deng, J. et al. (2014). A twisted wire-shaped dual-function energy device for photoelectric conversion and electrochemical storage. *Angew. Chem. Int. Ed.* 53 (26): 6664–6668.

61 Sun, H., Jiang, Y., Xie, S. et al. (2016). Integrating photovoltaic conversion and lithium ion storage into a flexible fiber. *J. Mater. Chem. A* 4 (20): 7601–7605.

62 Song, W., Yin, X., Liu, D. et al. (2019). A highly elastic self-charging power system for simultaneously harvesting solar and mechanical energy. *Nano Energy* 65: 103997.

6

Fiber Polymer Solar Cells

Polymer solar cells (PSCs) have attracted broad interest due to their unique properties, including flexibility, solution processability, and lightweight. These merits make them ideal candidates for fabrication into the fiber format. In this chapter, we will first introduce the history and mechanism of PSCs and fiber-shaped PSCs. Then, the advancement of flexible fiber-shaped PSCs will be mainly introduced, focusing on device structure, fiber electrode, fabrication method, and their influence on power conversion efficiency (PCE), flexibility, and stability of fiber-shaped PSCs. Finally, the perspective and applications of fiber-shaped PSCs will be briefly introduced from the viewpoint of large-scale production.

6.1 Overview of Polymer Solar Cells

The PSC is a type of photovoltaic device based on organic semiconductors. Tang et al. introduced a donor-acceptor bilayer planar heterojunction into organic photovoltaics (OPV) in 1979 and achieved a PCE of around 1% [1]. Later, fullerene and its derivatives with high electron mobility and strong electronegativity were adopted as acceptors in OPV since the electron transfer was discovered between a conjugated polymer and fullerene derivatives [2, 3]. Based on these materials, the first PSC with a planar heterojunction was demonstrated in 1993 [4].

However, the PSCs with planar heterojunction suffer from severe recombination owing to the limited contact area between donor and acceptor and the short carrier lifetime of both electrons and holes. Therefore, a bulk heterojunction is proposed for PSCs by mixing the donor and acceptor material in the active layer, as shown in Figure 6.1a [5]. In this bulk heterojunction structure, the donor and acceptor can contact sufficiently for exciton dissociation and extraction. These classic polymer-fullerene systems promoted the development in the earlier period and broke the PCE to over 10% [5–7]. Recently, nonfullerene small molecule acceptors were used in PSCs to construct polymer-nonfullerene systems for better power conversion efficiency [8–10]. Compared to the polymer-fullerene systems, the polymer-nonfullerene systems exhibited broad absorption spectra for a higher short-circuit current density (J_{SC}), a low energy loss in charge generation and dissociation for higher open-circuit voltages (V_{OC}), as well as the enhanced thermal

Wearable Solar Cells: Mechanisms, Materials, and Devices, First Edition.
Hao Sun, Zhibin Yang, Longbin Qiu, and Huisheng Peng.

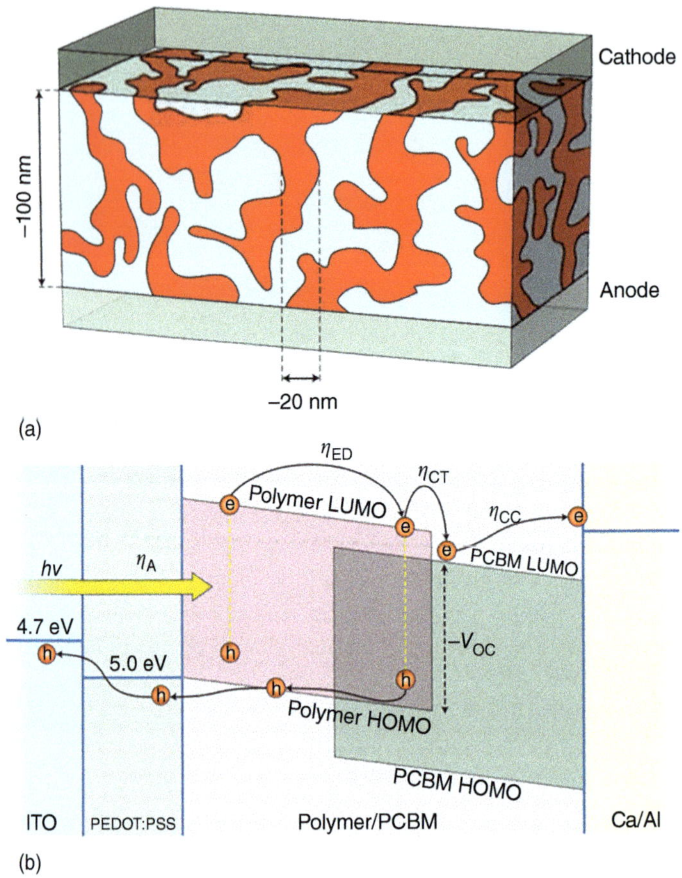

Figure 6.1 Schematic illustration (a) and working principle (b) of a typical film PSC. Source: Li et al. [5]/Springer Nature.

and photostability. With the appearance of a nonfullerene acceptor, the PCEs of single-junction organic solar cells have been promoted to over 19% in a short time [11].

A typical PSC shows a multilayered structure composed of transparent anode/hole transporting layer/intercalated donor-acceptor materials/electron transporting layer/cathode (Figure 6.1b). Normally, a transparent conducting substrate such as indium tin oxide (ITO) glass is used as the anode, and a lower-work-function metal like aluminum or silver is used as the cathode. A hole transport layer was coated onto the ITO glass to improve hole extraction, such as poly(ethylene-dioxythiophene)/polystyrene sulfonic acid (PEDOT: PSS). Similarly, an electron transporting layer was coated before depositing the metal cathode for better electron extraction, such as lithium fluoride (LiF). In the middle, the photo-active layer is prepared by coating a donor-acceptor mixture to form a bulk heterojunction.

Figure 6.1b shows the working principle of a typical bulk heterojunction PSC with polymer and [6,6]-phenyl-C61-butyric acid methyl ester (PCBM) as donor and acceptor materials [5]. Once a PSC is illuminated by light, Frenkel excitons form in the polymer and PCBM bulk. The energy level difference between the lowest unoccupied molecular orbital (LUMO) of the donor polymer and the highest occupied molecular orbital (HOMO) of the acceptor PCBM forms an internal heterojunction and a driving force to dissociate the excitons. Then, the dissociated holes and electron carriers are diffused and collected at the anode and cathode, respectively.

6.2 Overview of Fiber-Shaped Polymer Solar Cells

With the development of wearable devices, flexible and wearable solar cells are highly required for power supplies. The lightweight fiber solar cells show great potential as the power sources for these electronics. Compared to the fiber-shaped dye-sensitized solar cell (DSSC) using liquid electrolytes, fiber-shaped PSCs are all-solid-state, which is more desirable for practical applications. In addition, fiber-shaped PSCs can be fabricated via solution processes with low cost, and the essential properties of polymers make them more favorable in a flexible format.

6.2.1 Fiber-Shaped Polymer Solar Cells with Twisting Structure

The first fiber-shaped PSC was reported by Russell et al. in 2009 with a twisting structure [12]. As shown in Figure 6.2, a stainless steel wire with a diameter

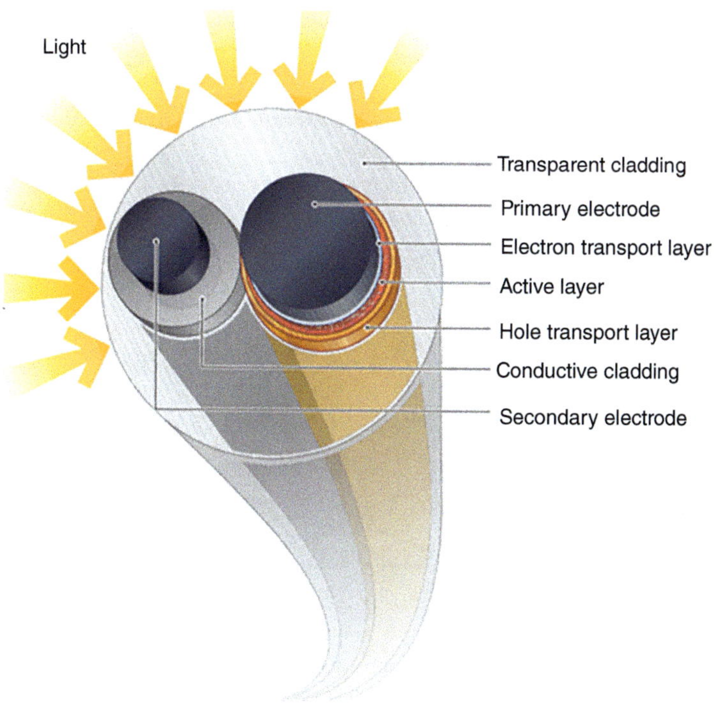

Figure 6.2 Schematic illustration of the first PSC based on a twisting structure. Source: Lee et al. [12]/American Association for the Advancement of Science.

of 100 μm was used as a primary electrode. It was sequentially dip-coated with electron-transporting material (TiO_x), photoactive material (P3HT: PCBM), and hole-transporting material (PEDOT: PSS). The resulting fiber electrode was twisted with another metal electrode with conductive cladding to obtain the fiber-shaped PSC. Under illumination, the electron carriers are extracted to the inner primary fiber electrode, and the hole carriers transfer to the second electrode by crossing the interface of these contacted electrodes. These fiber-shaped PSCs give an average PCE of 2.99% and a champion efficiency of 3.81% under AM1.5G illumination.

The above fiber-shaped PSCs suffered from severe internal stress when two stainless-steel wire electrodes were twisted together, which resulted in the exfoliation of photoactive material and even a short circuit. Therefore, a fiber electrode with higher flexibility and lower hardness is preferred for constructing a mechanically stable fiber solar cell. To this end, carbon nanotube (CNT) fibers with excellent flexibility and conductivity are introduced to fiber-shaped PSCs by Peng et al. [13]. As shown in Figure 6.3a, a Ti wire with TiO_2 nanotube arrays was used as the cathode for electron extraction. TiO_2 nanotube arrays with tunable lengths were prepared by an electrochemical anodic oxidation method. This made the electron transporting layer more controllable, where the length of the TiO_2 nanotube could be tuned from nanometers to micrometers [13]. In addition, the nanotube structure could improve the electron transfer efficiency from the active material to the inner Ti electrode. Such Ti–TiO_2 electrode was sequentially dip-coated with active material (P3HT: PCBM) and hole transporting material (PEDOT: PSS) as a primary electrode. At last, a fiber-shaped PSC was obtained by twining a flexible CNT fiber around the primary electrode. A fiber-shaped PSC based on such a device structure should exhibit high photovoltaic conversion performance in theory but only deliver a PCE of 0.15%.

The inferior PCE of the above fiber solar cells was mainly due to the hydrophobic surface of TiO_2 nanotube arrays that caused a barrier for electron transfer. Thus, the TiO_2 nanotube arrays were further treated in a $TiCl_4$ solution to form thin

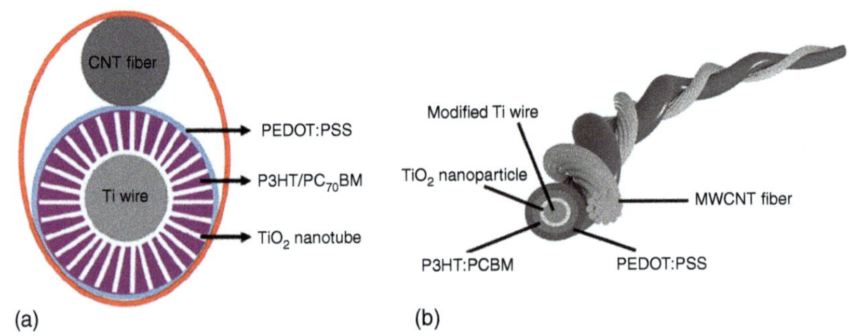

Figure 6.3 Schematic illustration of a twisting structural fiber-shaped PSC with CNT electrodes. (a) A PSC with TiO_2 nanotube arrays as cathode. Source: Chen et al. [13]/Royal Society of Chemistry. (b) A PSC with TiO_2 nanoparticles as cathode. Source: Zhang et al. [14]/John Wiley & Sons.

TiO_2 nanoparticles for better hydrophily (Figure 6.3b) [14]. With the modification of TiO_2 nanoparticles, the optimized fiber-shaped PSCs exhibited a V_{OC} of 0.52 V, J_{SC} of 9.06 mA/cm^2, and FF of 38%, which resulted in an improved PCE of 1.78%. Beneficial from the flexible CNT electrodes and enhanced connection between Ti wire and TiO_2, these fiber-shaped PSCs showed excellent bending stability, which maintained 85% of the PCE after 1000 bending cycles. Besides, it also demonstrated remarkable ambient stability that retained 70% of the initial PCE after storing it in air for 16 days.

For wearable applications, fiber-shaped PSCs will inevitably encounter bending and stretching during use. Therefore, they might be damaged by the internal stress under deformation. Zhang et al. developed a stretchable fiber-shaped PSC based on a spring-like structure [15]. As shown in Figure 6.4a, a spring-shaped Ti wire was prepared by winding the Ti wire onto a steel wire and was then removed from the

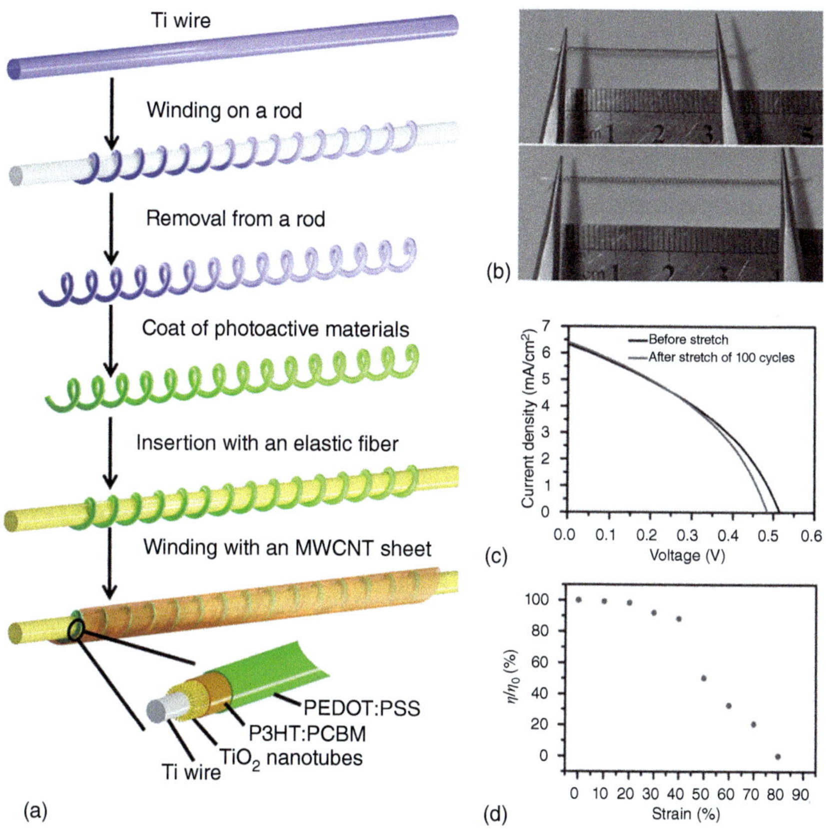

Figure 6.4 (a) Schematic illustration of the fabrication process of a stretchable fiber-shaped PSC. (b) Photograph of a fiber-shaped PSC before and after stretching. (c) J–V curve of a fiber-shaped PSC before and after being stretched by 30% for 100 times. (d) The PCE evolution with the strain of a fiber-shaped PSC. η and η_0 correspond to the PCEs before and after being stretched with different strains, respectively. Source: Zhang et al. [15]/John Wiley & Sons.

steel wire. Aligned TiO_2 nanotube arrays were grown on the spring-type Ti wire by an electrochemical anodic oxidation method. The modified Ti spring was then sequentially coated with photoactive materials (P3HT: PCBM) and the hole-transfer layer (PEDOT: PSS) as the primary electrode.

At last, an elastic fiber was inserted into the obtained Ti spring, followed by winding an ultrathin CNT sheet onto it. This stretchable PSC exhibited a PCE of 1.23% with a V_{OC} of 0.51 V, a J_{SC} of 6.33 mA/cm^2, and an FF of 38%. Figure 6.4b showed a typical PSC before and after stretching by 30%, which exhibited a well-maintained structure. The photovoltaic performance of the stretchable fiber-shaped PSCs was almost the same before and after stretching for 100 times, and the PCEs slightly decreased from 1.23% to 1.19% (Figure 6.4c). Figure 6.4d shows the PCEs at increasing strains beyond 30%. The PCE maintained above 88% when the strain was below 40% rapidly decreasing to less than 50% if the strain was further increased to 50%.

With the emerging donor and nonfullerene acceptor materials, the PCEs of planar PSCs have been increased to above 18%. These active materials can also benefit the performance of fiber-shaped PSCs. For instance, Liu et al. introduced these polymers into fiber-shaped PSCs and achieved a high PCE of 9.4% (Figure 6.5a) [16]. In their study, a nonfullerene acceptor-based ternary active material of PM6:Y6:PC_{71}BM was used as the light absorption layer, which was proved to deliver a high PCE for planar PSCs [16]. In addition, an electron-transporting material (N,N-dimethyl-ammonium N-oxide) propyl perylene diimide (PDINO) was used to

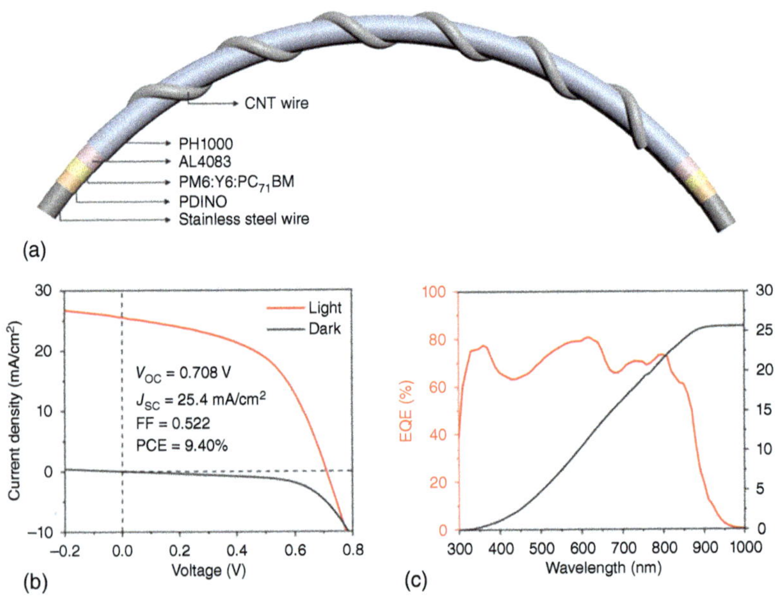

Figure 6.5 (a) Schematic illustration of a fiber-shaped PSC with active materials of PM6:Y6:PCBM. (b) J–V curves of the best fiber-shaped PSC fabricated under relative humidity of 10–15%. (c) The external quantum efficiency of the fiber-shaped PSC and the integrated J_{SC}. Source: Lv et al. [16]/Springer Nature/CC BY 4.0.

replace the conventional TiO_2 layer. These optimizations offered fiber-shaped PSC a high PCE of 9.40% with a V_{OC} of 0.708 V, a J_{SC} of 25.4 mA/cm^2, and an FF of 0.522 (Figure 6.5b). The fiber device exhibited a broad absorption from 300 to 950 nm in the EQE spectrum with high values from 60% to 80% (Figure 6.5c).

6.2.2 Fiber Polymer Solar Cells with a Coaxial Structure

Compared to fiber-shaped PSCs based on a twisting structure, the devices with a coaxial structure are more similar to the planar ones from a structure point of view, which is fabricated layer by layer from anode to cathode. The charge carriers transfer in a vertical direction in such coaxial structure rather than transferring from one fiber to another fiber in the twisted structure. Therefore, the fiber solar cells with a coaxial structure should possess a higher PCE in theory.

In 2007, Liu et al. developed a fiber-shaped PSC based on a standard multimode optical fiber (Figure 6.6) [17]. They first stripped off the hard polymer cladding from the optical fiber and coated ITO onto the fiber with a dip-coating process. The hole transporting material PEDOT: PSS and active layer P3HT: PCBM were sequentially dip-coated onto the optical fiber, followed by depositing LiF and Al electrodes via thermal evaporation (Figure 6.6). Thus, a novel fiber solar cell was produced, which could absorb light from the inner optical fiber. Under an average angle of incident

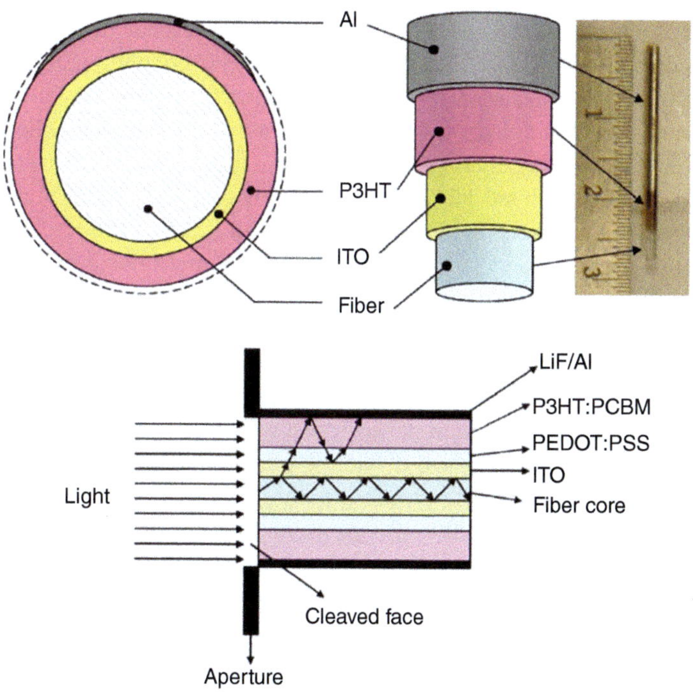

Figure 6.6 Schematic illustration of a fiber-shaped PSC's structure and working principle. Source: Liu et al. [17]/AIP Publishing.

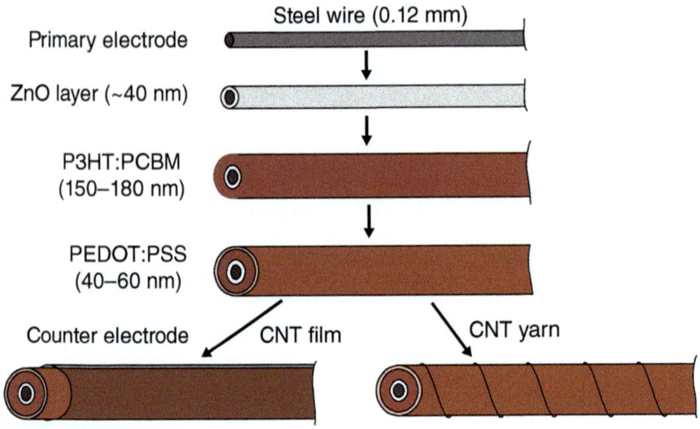

Figure 6.7 Schematic illustration of fabrication process of fiber-shaped PSCs with a coaxial and twisted structure. Source: Liu et al. [18]/American Chemical Society.

light, the J_{SC} of fiber-shaped PSCs increases significantly as the diameter of the fiber decreases. That is because most of the light transmits straight through the back in the large multimode fiber, and the numerical aperture of the fiber changes, and refractive losses into the cladding increase as the diameter of the fiber decreases. A maximum PCE of ~1.1% was achieved when an optical fiber with a diameter of 200 μm was used [17].

In 2013, Cao et al. developed fiber-shaped PSCs based on a coaxial structure, which was fabricated by sequentially dip-casting ZnO, P3HT: PCBM, and PEDOT: PSS layers onto the surface of stainless-steel wires as shown in Figure 6.7 [18]. At last, an ultrathin and semitransparent CNT sheet was wrapped around the resulting fiber as a transparent conducting electrode for the fiber device. The most efficient fiber-shaped PSCs exhibited PCEs of 2.26–2.31% with a V_{OC} of 0.55 V, a J_{SC} of 8.1 mA/cm^2, and a FF above 0.50. The CNT sheets with small tensile strains can effectively wrap the inner fibers, which makes the device more stable under bending. The J–V cure maintained almost the same when the fiber-shaped PSC was bent from 0° to 45° or 90°.

Single-layer graphene film possesses excellent properties of high light transmittance (>95%) and low sheet resistance(<100 Ω/□), which makes it an excellent material for optoelectronic devices. Liu et al. applied single-layer graphene films as the transparent electrodes in fiber-shaped PSCs (Figure 6.8a) [19]. The main challenge lies in transferring the graphene electrode to the surface of a fiber-shaped device. Thus, the single-layer graphene film was grown on a Cu foil via a chemical vapor deposition method. After that, a thin poly(methyl methacrylate) (PMMA) layer was spin-coated onto the graphene/Cu foil, followed by etching with a $FeCl_3$ solution. The graphene films were doped by Au nanoparticles via a $HAuCl_4$ solution. The doped graphene/PMMA sheets were then wrapped around the fiber-shaped PSCs. At last, the PMMA was washed away by acetone to obtain the fiber solar cell (Figure 6.8b). It exhibited a PCE of 2.53% with a V_{OC} of 0.57 V, a J_{SC} of 8.14 mA/cm^2, and an FF of above 0.55 [19].

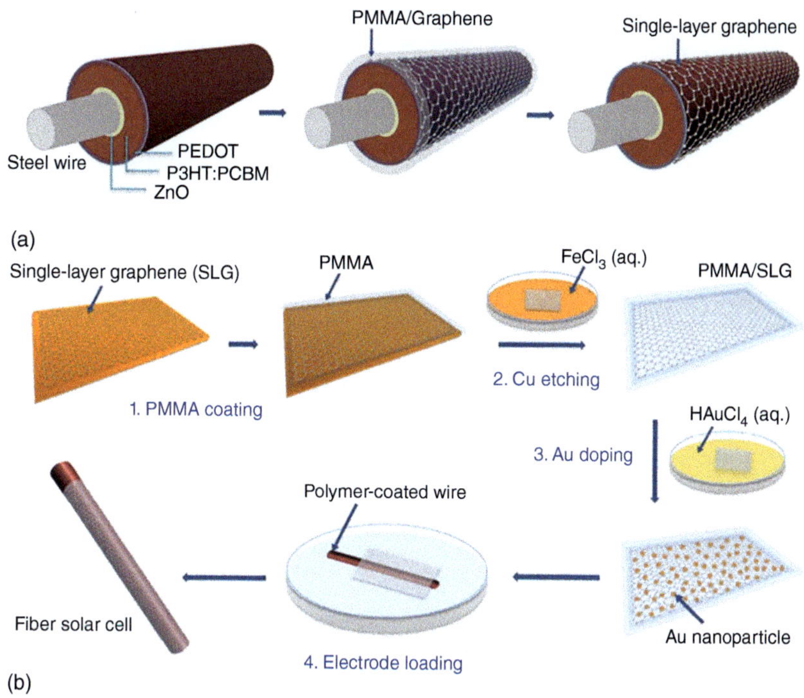

Figure 6.8 Schematic illustration of the fabrication process of a fiber-shaped PSC with a single graphene electrode. (a) Schematic illustration of the structure of a fiber-shaped PSC with a graphene electrode. (b) Procedure of transferring the graphene layer onto the fiber solar cells as the outer electrode. Source: Liu et al. [19]/Royal Society of Chemistry.

6.3 Polymer Solar Cell Textiles

For practical wearable applications, the obtained fiber PSCs should be woven into textiles in series or parallel connection for desirable output voltage and current. Stretchable polymer fiber solar cells could be integrated into clothes for demonstration (Figure 6.9b). Their PCE retained 90% of the initial value after 1000 bending cycles (Figure 6.9c). Besides, the PCE loss was less than 10% when the textile was 130% stretched for 50 cycles (Figure 6.9d) [15].

The weaving process might damage the surface of fiber solar cells and cause short circuits. To solve this problem, Zhang et al. developed a new approach for textile PSCs [20]. As shown in Figure 6.10a, a Ti textile was firstly prepared by weaving Ti wires together, followed by growing with vertically aligned TiO_2 nanotube arrays on the surfaces by anodization method and modifying with a layer of TiO_2 nanoparticles. Then, the textile was sequentially dip-coated with the P3HT: PCBM active layer and PEDOT: PSS hole transporting layer. Two flexible, aligned CNT sheets were then coated onto both sides of the resulting Ti textile to complete the PSC textile. The optimized PSC textile exhibited a maximum PCE of 1.08%. Based on this new strategy, the structure of the PSC textile can be well maintained after

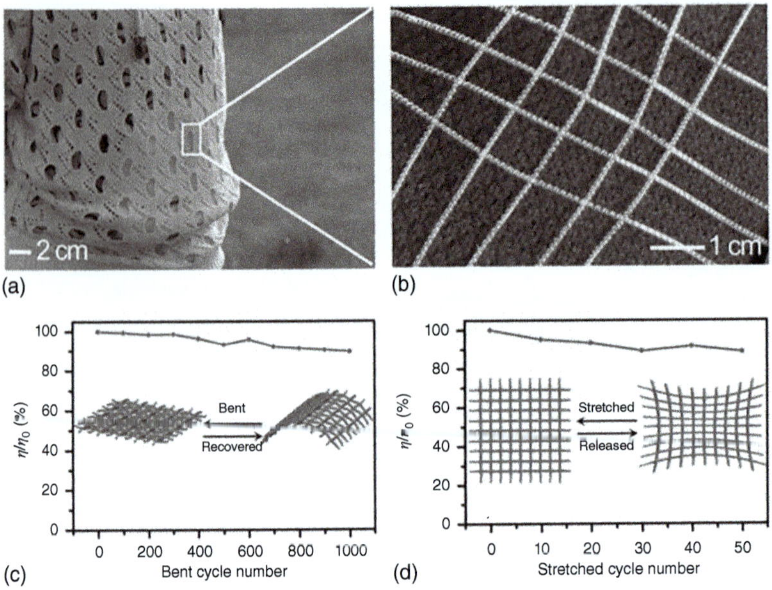

Figure 6.9 (a) and (b), A stretchable PSC textile being integrated into clothes. (c) PCE evolution with bending cycle number of PSC textiles. η_0 and η correspond to the PCE before and after bending, respectively. (d) PCE evolution with stretching cycle number of PSC textiles. η_0 and η correspond to the PCE before and after bending, respectively. Source: Zhang et al. [15]/John Wiley & Sons.

bending (Figure 6.10b,c). The PCE loss was verified to be less than 3% after 200 bending cycles [20].

It remains difficult to weave fiber-shaped PSCs into a photovoltaic textile with a large size. The main challenge lies in the extraction and connection of electrodes of the photovoltaic textiles. Liu et al. developed an interlaced structure to fabricate PSC textiles with a simplified connection configuration. To achieve that, they fabricated PSC textiles based on module design, which can be connected in series or parallel. As shown in Figure 6.11a, there is an interlaced structure in the warp and weft yarns of each module. Both warp and weft yarns can be easily connected as anode and cathode in the textiles (Figure 6.11b). In such a structure, the current and voltage can be increased by connecting the electrodes in parallel and series, respectively (Figure 6.11c,d).

6.4 Summary and Perspective

In the past two decades, fiber-shaped PSCs have been actively pursued with significant advancement in device performance, with the current record PCE of 9.4%. Meanwhile, various weaving methods have been developed to realize fiber-shaped PSC textiles. Compared to the planar PSCs (>18%), however, the PCEs of fiber-shaped PSCs are still far from satisfactory for practical application. Several challenges need to be solved to enable real-world applications.

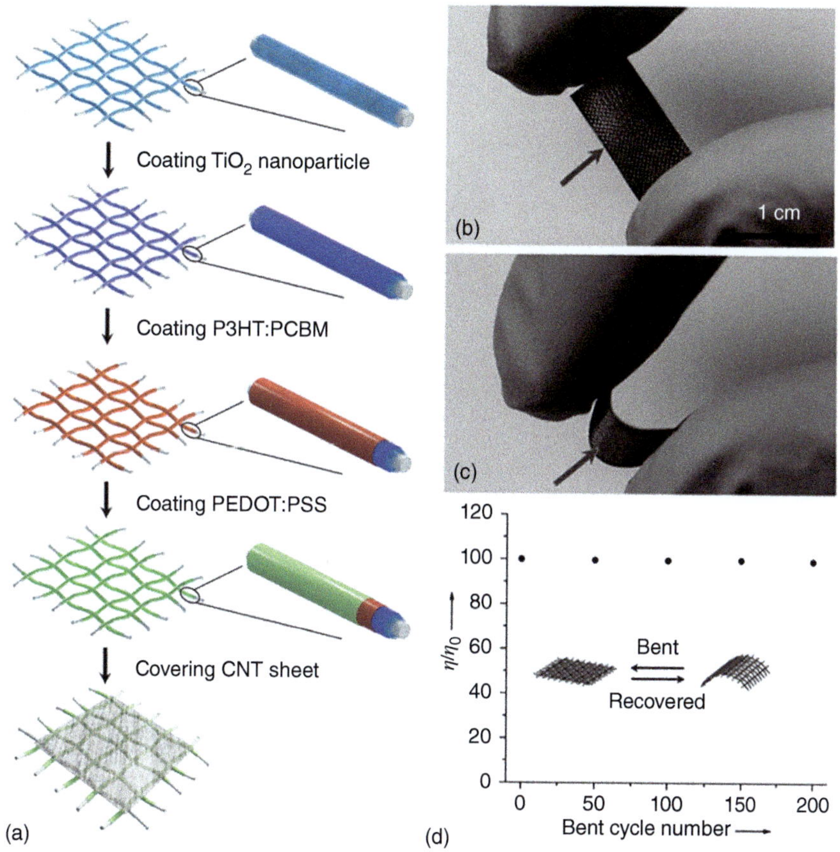

Figure 6.10 (a) Schematic illustration of fabrication process of PSC textiles. (b) and (c) Photographs of the PSC textile before (a) and after (b) being bent. (d) PCE evolution with bending cycle number of PSC textiles. η_0 and η correspond to the PCE before and after bending, respectively. Source: Zhang et al. [20]/John Wiley & Sons.

1. Effective coating techniques on fibers. For a fiber photovoltaic device, the active layers and carrier transporting layers need to be uniformly coated onto fibers with high quality. However, a manual dip-coating method is still used, which cannot guarantee uniformity towards high PCEs.
2. Fabrication of long fiber-shaped PSCs in an applicable level. The lengths of reported fiber-shaped PSCs mainly ranged from 1 to 10 cm, and the PCE might rapidly decay when the length was further increased, owing to the enlarged electrical resistance of the fiber electrodes. This requires further design and optimization to unlock long and continuous fiber-shaped PSCs with high PCEs, which will be the key for the subsequent weaving process and wearing application.
3. Electrode extraction and connection for fiber and textile PSCs. Despite many attempts that have been made, the extraction and connection of fiber and textile

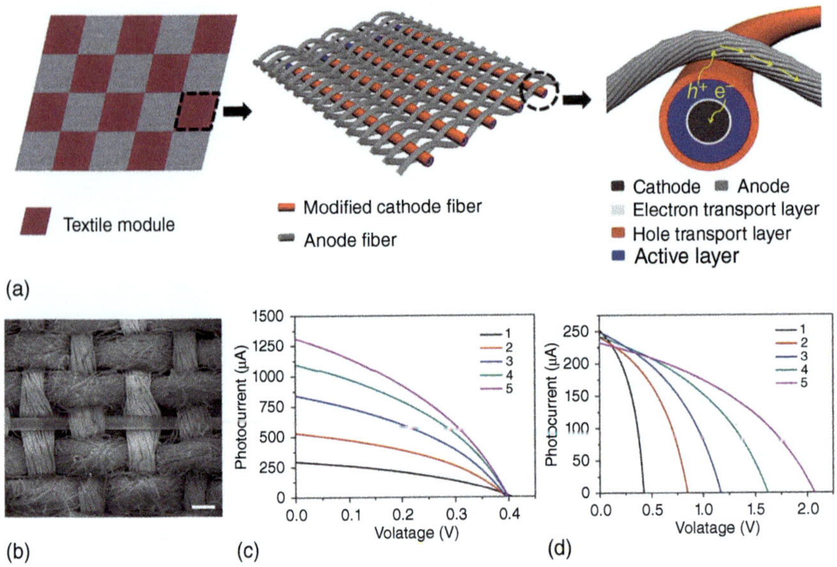

Figure 6.11 (a) Schematic illustration of the PSC textile in different degrees of amplification. (b) Top view SEM image of the PSC textiles. Scale bar, 200 mm. (c) and (d) JV curves of the PSC textile by connecting electrodes in parallel (c) and series (d), respectively. Source: Liu et al. [21]/Royal Society of Chemistry.

PSCs remains an unsolved issue. To achieve this goal, there is an urgent need for easy and reliable electrode extraction and connection.

4. Material innovation toward replacement of metal wire electrodes. Metal wire electrodes such as Ti and stainless steel are widely used for fiber-shaped PSCs due to their high conductivity and mechanical robustness. However, the high mass density, non-stretchability, and limited flexibility have plagued their use for wearable electronics. Therefore, developing new electrode materials with comprehensively high properties will be important for the wearable applications of fiber PSCs.

References

1 Tang, C.W. (1979). United States, pp. Medium: X; Size: Pages: v 2008-2002-2007.

2 Morita, S., Zakhidov, A.A., and Yoshino, K. (1992). Doping effect of buckminsterfullerene in conducting polymer: change of absorption spectrum and quenching of luminescene. *Solid State Commun.* 82 (4): 249–252.

3 Sariciftci, N.S., Smilowitz, L., Heeger, A.J. et al. (1992). Photoinduced electron transfer from a conducting polymer to buckminsterfullerene. *Science* 258 (5087): 1474–1476.

4 Sariciftci, N.S., Smilowitz, L., Heeger, A.J. et al. (1993). Semiconducting polymers (as donors) and buckminsterfullerene (as acceptor): photoinduced electron transfer and heterojunction devices. *Synth. Met.* 59 (3): 333–352.

5 Li, G., Zhu, R., and Yang, Y. (2012). Polymer solar cells. *Nat. Photonics* 6 (3): 153–161.

6 Wang, G., Melkonyan, F.S., Facchetti, A. et al. (2019). All-polymer solar cells: recent progress, challenges, and prospects. *Chem. Rev.* 58 (13): 4129–4142.

7 Lee, C., Lee, S., Kim, G.-U. et al. (2019). Recent advances, design guidelines, and prospects of all-polymer solar cells. *Chem. Rev.* 119 (13): 8028–8086.

8 Karki, A., Gillett, A.J., Friend, R.H. et al. (2021). The path to 20% power conversion efficiencies in nonfullerene acceptor organic solar cells. *Adv. Energy Mater.* 11 (15): 2003441.

9 Yan, C., Barlow, S., Wang, Z. et al. (2018). Non-fullerene acceptors for organic solar cells. *Nat. Rev. Mater.* 3 (3): 18003.

10 Cheng, P., Li, G., Zhan, X. et al. (2018). Next-generation organic photovoltaics based on non-fullerene acceptors. *Nat. Photonics* 12 (3): 131–142.

11 Best Research-Cell Efficiencies Chart-NREL. July 2022. *Best Research-Cell Efficiencies Chart-NREL. June 2023.*

12 Lee, M.R., Eckert, R.D., Forberich, K. et al. (2009). Solar power wires based on organic photovoltaic materials. *Science* 324 (5924): 232–235.

13 Chen, T., Qiu, L., Li, H. et al. (2012). Polymer photovoltaic wires based on aligned carbon nanotube fibers. *J. Mater. Chem.* 22 (44): 23655–23658.

14 Zhang, Z., Yang, Z., Wu, Z. et al. (2014). Weaving efficient polymer solar cell wires into flexible power textiles. *Adv. Energy Mater.* 4 (11): 1301750.

15 Zhang, Z., Yang, Z., Deng, J. et al. (2015). Stretchable polymer solar cell fibers. *Small* 11 (6): 675–680.

16 Lv, D., Jiang, Q., Shang, Y. et al. (2022). Highly efficient fiber-shaped organic solar cells toward wearable flexible electronics. *npj Flexible Electron.* 6 (1): 38.

17 Liu, J., Namboothiry, M.A., and Carroll, D.L.J.A.P.L. (2007). Optical geometries for fiber-based organic photovoltaics. *Appl. Phys. Lett.* 90 (13): 133515.

18 Liu, D., Zhao, M., Li, Y. et al. (2012). Solid-state, polymer-based fiber solar cells with carbon nanotube electrodes. *ACS Nano* 6 (12): 11027–11034.

19 Liu, D., Li, Y., Zhao, S. et al. (2013). Single-layer graphene sheets as counter electrodes for fiber-shaped polymer solar cells. *RSC Adv.* 3 (33): 13720–13727.

20 Zhang, Z., Li, X., Guan, G. et al. (2014). A lightweight polymer solar cell textile that functions when illuminated from either side. *Angew. Chem. Int. Ed.* 53 (43): 11571–11574.

21 Liu, P., Gao, Z., Xu, L. et al. (2018). Polymer solar cell textiles with interlaced cathode and anode fibers. *J. Mater. Chem. A* 6 (41): 19947–19953.

7

Fiber Perovskite Solar Cells

7.1 Introduction

Aiming at wearable applications, fiber solar cells must be lightweight and flexible. To this end, perovskite materials are most intriguing for application in a fiber structure, considering their ultra-high light absorption coefficient [1], long charge carrier lifetime (diffusion length) [2], small exciton binding energy [3], and tunable bandgap [4] (Figure 7.1). With an ultrahigh light absorption coefficient, an ultra-thin perovskite layer of ~300 nm is thick enough to absorb most of the incident light [5], which can meet the requirement of lightweight. Combining an ultra-thin flexible substrate, perovskite solar cells with an ultrahigh power-to-weight ratio of 23 W/g have been achieved [6], and the thin layer structure diminishes the tensile strain with a neutral-plan design [7]. On the other hand, perovskite solar cells have shown a high power conversion efficiency (PCE) of certified 25.7% comparable to crystalline silicon solar cells and higher than the other thin film solar cells after around 14 years of development (Figure 7.2) [19]. The strategies developed for planar structure solar cells are also promising for the development of highly efficient fiber perovskite solar cells with a rapidly growing performance recently (Figure 7.2) [20]. The fiber-shaped perovskite solar cells also provide various multifunctional applications by integrating with other electronics, like light emitting devices [21], energy storage fibers [22], and fiber actuators [23]. This chapter refers to the recent breakthroughs of fiber perovskite solar cells as the pioneer in the next-generation photovoltaics. The materials and structures of fiber-shaped perovskite solar cells are first introduced, focusing on the innovative deposition of high-quality perovskite photoactive layer on the curve surface of fiber/wire electrode and efficient charge transport and separation process by designing aligned one-dimensional nanostructures charge extraction and transport layers. The realization of stretchable fiber perovskite solar cells is then presented through a delicate spring structure design [12]. The pros and cons of fiber perovskite solar cells are compared with fiber dye-sensitized and polymer solar cells. Finally, the challenges that inhibit the further application of fiber perovskite solar cells have been summarized.

Wearable Solar Cells: Mechanisms, Materials, and Devices, First Edition.
Hao Sun, Zhibin Yang, Longbin Qiu, and Huisheng Peng.

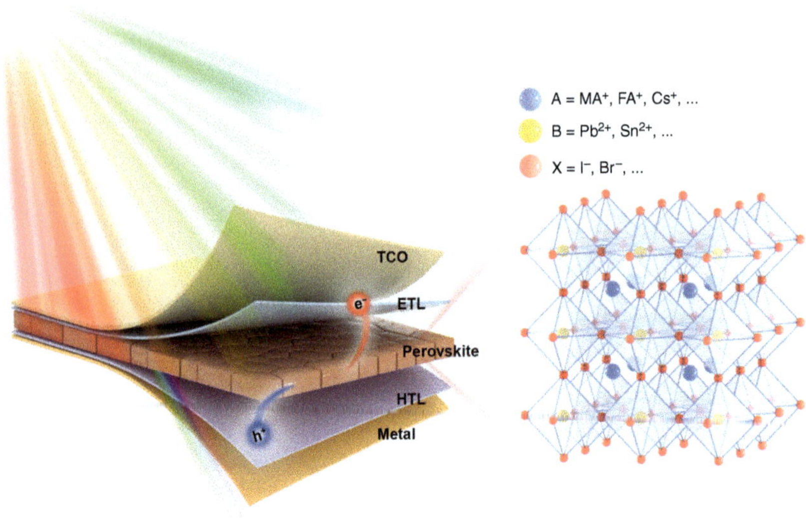

Figure 7.1 Schematic drawing shows the typical structure of perovskite materials and perovskite solar cells. The ultra-high absorbance could absorb most of the incident light. A small binding energy induces free charge carriers. And the long lifetime and diffusion length contribute to the charge extraction efficiency.

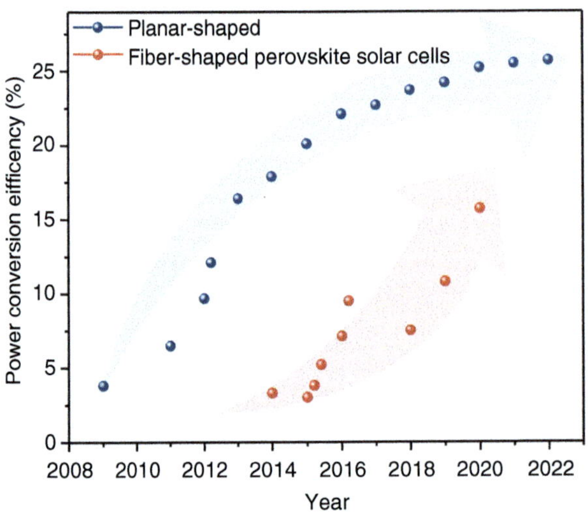

Figure 7.2 The performance of planar-shaped and fiber-shaped perovskite solar cells. The data for planar-shaped perovskite solar cells were collected from the certified efficiency table. The PCE for fiber-shaped perovskite solar cells is listed in Table 7.1.

7.2 Perovskite Photoactive Materials and Solar Cells

Perovskite is a kind of crystal structure that has a general form of ABX_3, where A is an organic or inorganic cation (MA^+, FA^+, or Cs^+), B is a divalent metal cation such as Pb^{2+} and Sn^{2+}, and X is halide elements like Br^- and I^- (Figure 7.1) [24].

Table 7.1 Fiber-shaped perovskite solar cell structure and corresponding photovoltaic parameters.

Configuration	V_{OC} (V)	J_{SC} (mA/cm²)	FF	PCE (%)	References
Coaxial	0.667	10.2	0.487	3.3	[8]
Coaxial	0.85	14.20	0.56	7.1	[9]
Coaxial	0.731	11.97	0.44	3.85	[10]
Coaxial	0.91	15.9	0.656	9.49	[11]
Coaxial/spring	NA	NA	NA	5.2	[12]
Double twisted	0.615	8.75	0.564	3.03	[13]
Double twisted	0.714	12.32	0.609	5.35	[14]
Double twisted	0.96	14.18	0.66	7.53	[15]
Double twisted	0.95	15.14	0.75	10.79	[16]
Double twisted	1.52	4.75	0.725	7.49	[17]
Double twisted	1.92	11.94	0.542	15.7	[18]

The specific composition of recently applied perovskite materials was introduced by Mitzi et al. in the 1990s, while at that time was aimed for transistors [25]. The developed perovskite materials used for solar cells are $CH_3NH_3PbBr_3$ ($MAPbBr_3$) and $CH_3NH_3PbI_3$ ($MAPbI_3$), introduced by Miyasaka et al. in 2009 [26]. It was proposed as an organic–inorganic hybrid dye to replace the most popular organic dye, N719, in dye-sensitized solar cells (see Chapter 5). The first solar cell based on the $MAPbBr_3$ showed a PCE of 3.8%. However, due to the dissolvability of this organic–inorganic hybrid perovskite in the liquid electrolyte, the lifetime is only several tens of minutes [26]. With the introduction of solid-state hole transport material spiro-MeOTAD into the perovskite-sensitized structure, the solar cell PCE rapidly improved to over 10% [5, 27]. Furthermore, with the discovery of the bi-polar transportation of charge carriers [28, 29] in the perovskite layer and the ultra-high lifetime and transport and diffusion length of electron-hole [2, 30], the design of bulk perovskite layer function as photoactive layer has been developed [31]. Then, a new kind of thin film perovskite solar cells that differ from dye-sensitized solar cells has been developed [31]. The bandgap of perovskite materials is readily tuned in the range of 1.2–2.3 eV by changing the A-site cation, B-site metal, and X-site halide, which is promising for various indoor and outdoor applications that requested to absorb different ranges of light [4].

A typical structure of perovskite solar cell is shown in Figure 7.1. The perovskite photoactive layer is sandwiched between the electron transport layer (ETL) and hole transport layer (HTL). A transparent conductive electrode (TCE) and metal electrode are on each side for the collection of electrons and holes, respectively. Depending on the sequence of the functional layers, n–i–p (TCE/ETL/PVSK/HTL/metal electrode) and p–i–n (TCE/HTL/PVSK/ETL/metal electrode) cell structures have been developed, with p–i–n structure showed slightly lower performance but higher stability at current stage [32–34]. As depicted in Chapter 2, the working mechanism of solar cells, with the incident sunlight, the electrons are excited to higher energy levels above the bandgap. Due to small binding energy, the excitons will separate into free electrons and holes under thermal energy [3]. And the free carriers of electrons and holes are easily transported and extracted in the corresponding charge carrier selective layer and collected by electrodes. The development of high-performance solar cells went through several stages, including the realization of solid-state structure [5, 27], high-quality perovskite layer deposition [35–41], interface/surface modification [42, 43], energy level alignment [44], and development of high-quality charge transport layers [45–47]. As shown in Figure 7.2, the performance rapidly increased in the early stages and slowly to a certified 25.7% recently [19, 48]. Although for fiber-shaped perovskite solar cells, the development is still lag behind that of the planar counterpart, the performance is in a rapidly increasing stage. It is expected to further improve by learning from the planar counterparts. In our opinion, the main limitation in current stage is still the high-quality film deposition process on a curved surface, which will be discussed in detail in the following section.

A fiber-shaped perovskite solar cell is in a one-dimensional structure by configuring the TCE, ETL, perovskite, and HTL layers sequentially on a fiber electrode [49]. As shown in Figure 7.3, there are two typical structures, namely coaxial and twisting structures [8, 13, 50]. For a coaxial structure, the outmost electrode must be transparent to ensure the transmittance of the incident light. For the twisting structure, a highly conductive secondary electrode wire should be applied. However, the contacting area is also reduced and requires a longer diffusion length of the carrier before recombination. More details about the selection of electrodes will be discussed in the following section.

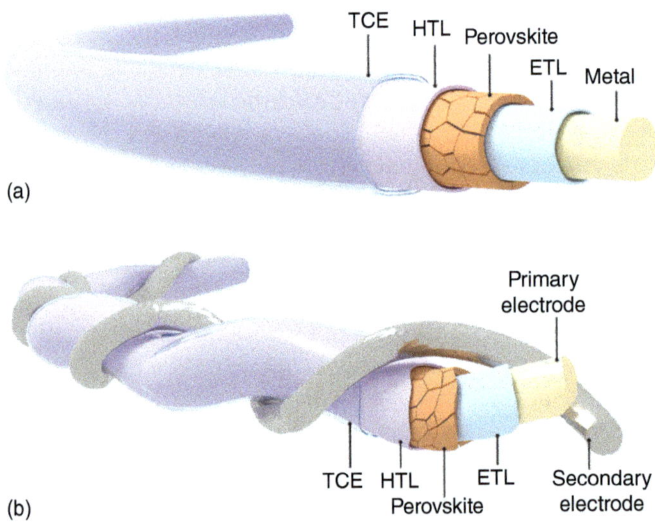

Figure 7.3 The schematic structure of fiber-shaped perovskite solar cell with polycrystalline perovskite photoactive layers. (a) a coaxial structure, and (b) a twisting structure. TCE: transparent conductive electrode; HTL: hole transport layers; ETL: electron transport layers.

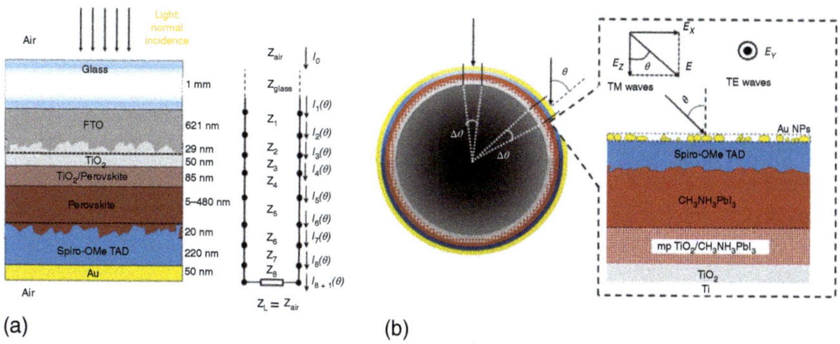

Figure 7.4 The optical modeling of a planar-shaped and a fiber-shaped perovskite solar cell. (a) The structure of planar shaped perovskite solar cells with light incident from the transparent electrode. (b) The structure of fiber shaped perovskite solar cells with disperse Au nanoparticles coating outside. Source: Moshonas et al. [51]/MDPI/CC BY 4.0.

Considering the structure difference between the flat planar surface and curved fiber surface, the optical model for fiber-shaped perovskite solar cells has been investigated (Figure 7.4) [51]. For a fiber with a radius of hundreds of microns, the thickness of the completed cell structure is several order of magnitude lower, which is around 1 μm [8]. The curved surface could be divided into segments, which were taken as a flat structure and had different incident angles (Figure 7.4b). And this curved fiber structure made less use of the direct incident sunlight, while it could benefit more from diffused, scattered, and inclined light [51]. The incorporation of mesoporous layers induced the formation of two active layer stacks, with one perovskite bulk capping layer on top of the mesoporous TiO_2/perovskite layer. Due to the symmetric structure of the fiber-shaped solar cell, the variation of the incident light has neglectable effects on light absorption [8]. Here , sparse Au nanoparticles were considered the top transparent electrode for the fiber solar cells, with a balance of transmittance and conductivity. In this case, a large amount of light was absorbed by the Au layers. At the same time, incorporating semi-transparent dielectric materials filling the gaps between Au nanoparticles could enhance the light absorbance, according to the modeling [51]. On the other hand, the optimization of the top transparent conductive electrode is beneficial for further improvement of fiber solar cells. The above-mentioned model was based on a uniform coating perovskite layer on the curved fiber surface, which is still the main challenge to be solved for obtaining high photovoltaic power conversion performance.

7.3 Deposition of Perovskite Layer on Fiber/Wire Surfaces

7.3.1 Perovskite Formation Strategies

It is obvious that the surface of the fiber electrode is highly curved, which is a grand challenge for the uniform coating of functional layers onto it. The main limitation of high-performance perovskite solar cells comes from the quality of the deposited film, evident from the experience of planar-type perovskite solar cells [36–40]. The coating and formation of the perovskite layer could be categorized into two different processes, namely, one-step and two-step coating processes. For the one-step coating process, the perovskite precursor solution is prepared with a specific composition. One should note that the precursor solution is not a true solution but a mixture of as-formed perovskite colloids and intermediate phases [52, 53]. And the perovskite layer forms directly after the nucleation and crystal growth from the coated precursor solution.

Typically, for a one-step coating process, from the conventional nucleation-growth theory, the concentration will increase with the evaporation of solvent to reach a saturation condition, and the LaMer diagram depicts the nucleation and growth dynamics as shown in Figure 7.5a. However, there is an overpotential barrier to overcome as the nucleation process proceeds [54]. Therefore, a supersaturation

Figure 7.5 Typical nucleation and crystal growth process for perovskite thin film from one-step coating process and two-step coating process on a curved fiber surface. (a) Conventional nucleation-growth theory and LaMer diagram for a film formation from a solution process. (b) The coated perovskite layer outside a fiber electrode with one-step dip-coating process. Source: Reproduced with permission from Li et al. [13], © 2015/John Wiley & Sons, Inc. (c) the dip-coated compact PbI_2 outside a stainless-steel fiber electrode. (d) the formation of $MAPbI_3$ perovskite crystals from a two-step sequential coating process on a stainless-steel fiber electrode. Source: Reproduced with permission from He et al. [10], © 2015/Royal Society of Chemistry.

condition is required ($[M]^*_{min}$). With the nucleation proceeding, the crystal grows simultaneously, which reduces the concentration. When the concentration decreases below a specific point ($[M]^*_{min}$), the nucleation process stops while the solvent evaporation and crystal growth proceed at the same time (in the region between $[M]_{sol}$ and $[M]^*_{min}$). However, depending on the nucleation density and the growth rate, the coverage of the coated film and the size of the crystal grain varies [55]. Specifically for perovskite, the natural evaporation of solvent-induced nucleation density is low, while the crystal growth is much faster [54]. At the early stage, this is the main reason for the poor coverage of the perovskite layer in a planar structure solar cell [55]. With the understanding of this process and the introduction of anti-solvent [36, 37], gas quench process [38], or vacuum-flash process for fast evaporation [56] to induce an instant high concentration for a high nucleation density, high-quality and compact perovskite film with full coverage has been achieved [53]. There are also other strategies to increase the nucleation density. One is to reduce the contact angle through surface treatment, which is beneficial for the reduction of the hetero-nucleation activation energy [57]. And the introduction of additives to induce the hetero-nucleation process could also improve nucleation density and thus promote the film quality [12, 54]. One balance should be considered: with a higher nucleation density, the grain size will be smaller. And it is still a controversial view that a larger grain size is required for higher performance [58].

For the two-step coating process, a sequential coating process was developed according to the chemical reaction process for perovskite [9]. In contrast to the one-step coating of perovskite from the precursor solution, a two-step chemical formation process for perovskite has been developed by first depositing PbI_2 [40]. And the compact PbI_2 is beneficial for compact perovskite formation. However, the crystal structure of the first coated PbI_2 will affect the conversion reaction of perovskite. It has been shown that with an amorphous [59] and porous [60] PbI_2, the complete conversion to high-quality and compact perovskite was obtained. The two-step-coating process is a diffusion-control process; thus, a porous and amorphous structure is beneficial for the interdiffusion of the organic salt coated in the second step [61]. Recently, a certified efficiency of over 25.6% has been reported by adding RbCl into the PbI_2 precursor, which converted PbI_2 into an inactive $(PbI_2)_2RbCl$ compound to stabilize the perovskite phase and enhance the film quality [33].

Figure 7.5 shows the typical perovskite layers that are coated by one-step [13] and two-step [10] processes onto a fiber electrode. Although trying to learn from the planar structure, the anti-solvent and gas-quench processes are difficult to be applied for fiber solar cells. However, the nucleation-crystal growth process and full coverage are the same challenges to be overcome for fiber-shaped perovskite solar cells. It can be observed from Figures 7.5b, d that the coated active layer by different processes is formed by loose packing crystals. For the coating of perovskite in a fiber curve surface, combining experience from a planar counterpart, four strategies have been developed, as shown in Figure 7.6.

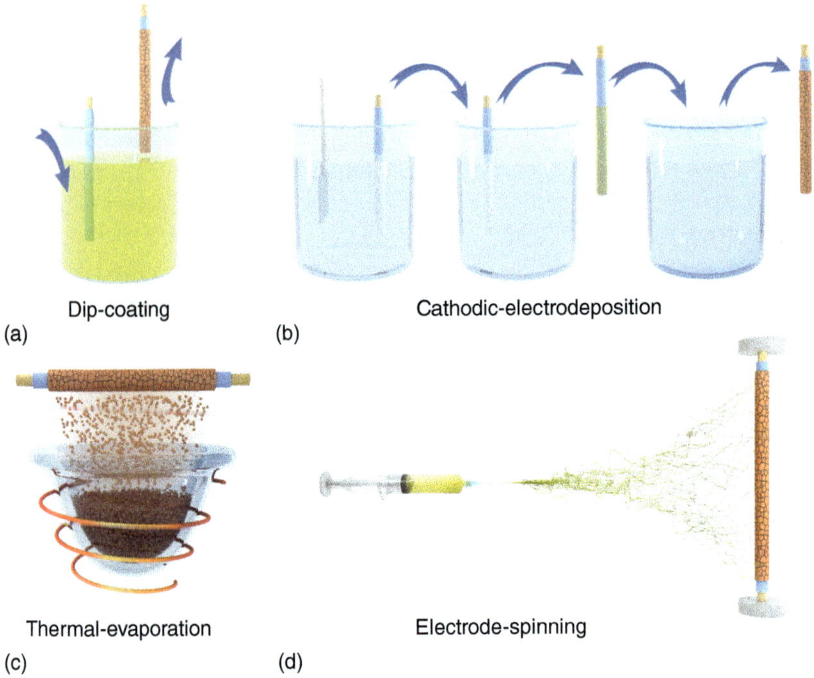

Figure 7.6 The coating methods for perovskite photoactive layers. (a) dip-coating process. (b) cathodic deposition process. (c) thermal evaporation and vapor-assisted deposition process. (d) electrospinning deposition process.

7.3.2 Dip-Coating Process

The first and simplest strategy is the dip-coating process, which is similar to the coating of polymer materials [62]. The first fiber perovskite solar cell was fabricated by this simple method with $MAPbI_3$ precursor in γ-GBL (gamma-Butyrolactone) and Dimethyl sulfoxide (DMSO) solvent (Figure 7.6a) [8]. By inserting the fiber with coated mesoporous TiO_2 ETL structure into the perovskite precursor solution, a wet film could be formed outside the fiber surface. The thermal annealing of the fiber on a hotplate induced the nucleation and crystal growth. However, similar to the nucleation-growth process in a planar surface, the nuclear density is low, and the crystal grew sparsely with a low coverage. Therefore, repeated dip-coating processes were required to improve the coverage [8]. To improve the coverage, mesoporous TiO_2 was coated around the smooth stainless-steel fiber electrode via a dip-coating process (Figure 7.7a). A clear cross-section SEM image of the perovskite fiber solar cell could be observed (inserted in Figure 7.7a). The enlarged surface area by introducing mesoporous TiO_2 enhanced the absorption of the precursor solution and increased the nucleation density. A PCE of 3.3% was achieved for the first work on fiber perovskite solar cells [8]. A twisting structure has been reported by twisting two aligned carbon nanotube fiber electrodes, with one of the electrodes dip-coating of perovskite, and another carbon nanotube fiber functioning as a secondary electrode

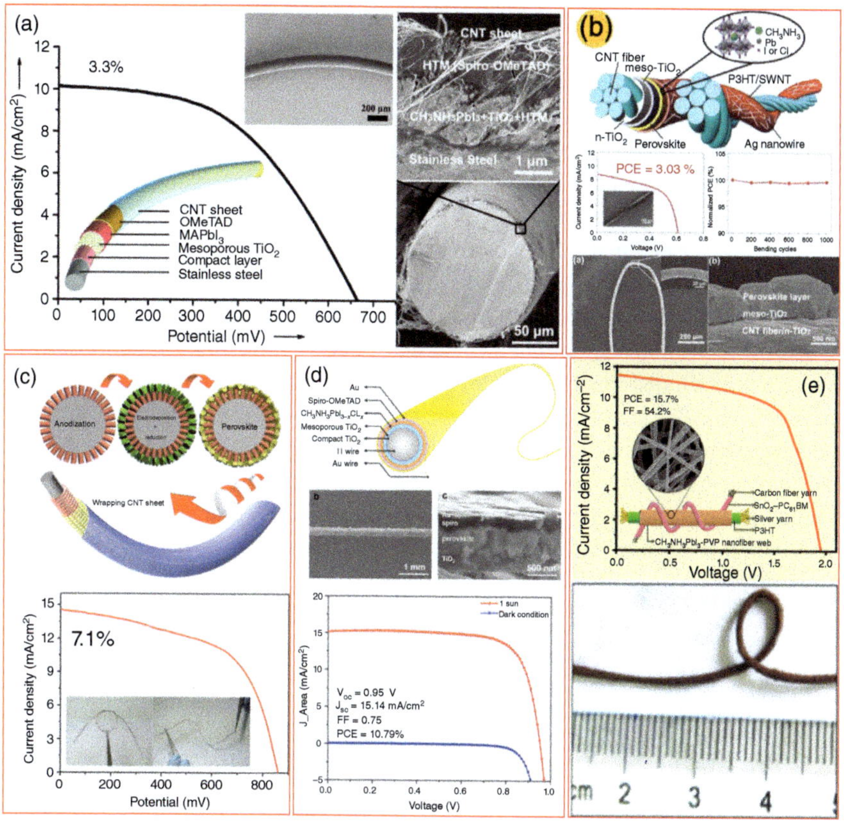

Figure 7.7 Representative fiber-shaped perovskite solar cell structures. (a) a coaxial structure with a dip-coated perovskite layer. Source: Reproduced with permission from Qiu et al. [8]. © 2014/John Wiley & Sons. (b) a twisted structure with carbon nanotube fiber as an electrode and dip-coated perovskite layer. Source: Reproduced with permission from Li et al. [13], © 2015/John Wiley & Sons. (c) a coaxial structure with cathodic electrode-deposited perovskite layer. Source: Reproduced with permission from Qiu et al. [9]. © 2016/John Wiley & Sons. (d) a twisted structure with a secondary Au wire electrode with thermal evaporated perovskite layer. Source: Reproduced with permission from Dong et al. [16]. © 2019/John Wiley & Sons. (e) electrode-spinning perovskite-PVP fiber structure photoactive layer. Source: Reproduced with permission from Li et al. [18]. © 2020/John Wiley & Sons.

(Figure 7.7b) [13, 14]. The coated mesoporous TiO_2 and the high surface area of carbon nanotube fiber both contributed to the deposition of perovskite with a compact interface. A PCE of 3.03% had been achieved (Figure 7.6b) [13].

However, due to the low nucleation density, the coverage of the dip-coating perovskite film is rather limited. To further improve the coverage, the nucleation-growth process needs to be modified by tuning the solution properties. A different precursor component was applied using methylammonium iodide (MAI): $PbCl_2 = 3:1$ dissolving in DMSO. It has been reported that the presence of Cl^- could tune the nucleation process with a higher coverage [55, 63]. This solution has been introduced for dip coating perovskite on fiber electrodes with enhanced coverage. With the further introduction of 1,8-diiodooctane (DIO) [64], the perovskite nucleation and crystal growth could be further modified with a higher nucleation density and slower crystal growth process. A compact perovskite could be achieved on a fiber electrode [12]. However, due to the limitation of the nucleation density, the quality of the perovskite layer is still lower than its counterpart in a planar surface, which was adjusted by anti-solvent, gas-quench, and vacuum-flash processes [53]. With a lead acetate precursor instead of PbI_2, the coverage and quality of the coated perovskite film on the curved surface could be enhanced, and a PCE of up to 7.53% was reported [15].

To overcome the above mentioned challenges, a perovskite quantum dot precursor has been developed for the uniform and compact coating of photoactive layers (Figure 7.8) [65]. The synthesized quantum dots dispersed uniformly in the solvent and deposited around the curved surface. With annealing, the deposited quantum dots crystallized into large grains. With a repeating coating process, the quantum dots filled the gaps between grains and high-quality and compact films. A PCE up to 5.37% had been achieved with the large bandgap perovskite quantum dot precursors [65].

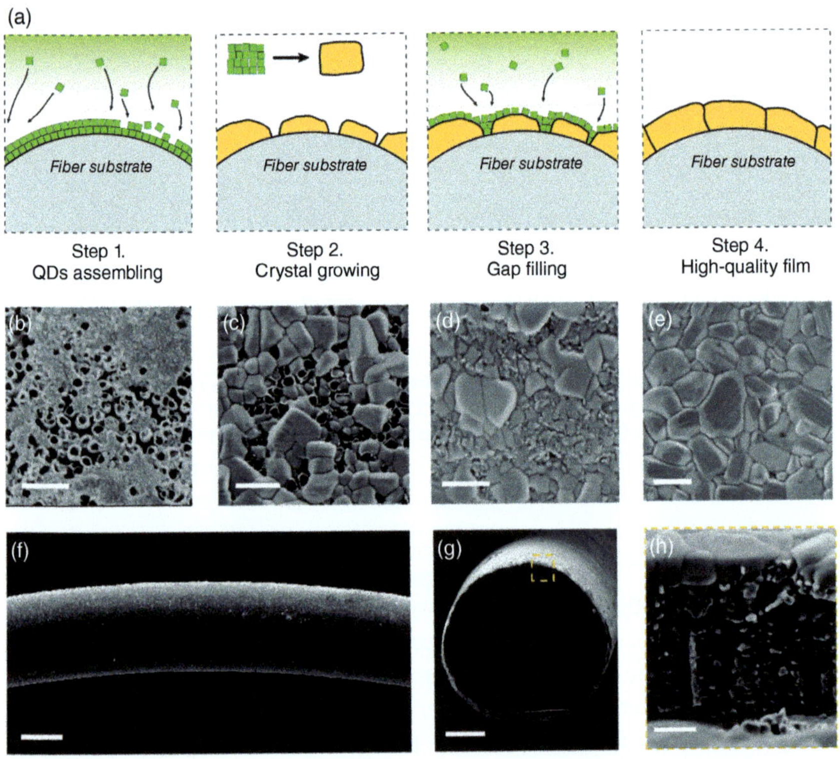

Figure 7.8 The uniform coating of perovskite layer around a curved surface with quantum-dots precursors. Source: Reproduced with permission from Xu et al. [65]. © 2020/The Royal Society of Chemistry.

On the other hand, the two-step coating process could also be applied for the uniform coating of the perovskite layers because the compact and uniform PbI_2 could be much easier coated on the fiber-curved surface (Figure 7.5c) [10]. A PCE of 3.8% had thus been achieved. As mentioned above, the compact layer might prevent the effective infiltration of the second step organic salt and a remaining un-reacted PbI_2 in the interface between perovskite and bottom ETL structures. The PCEs of the fiber-shaped perovskite solar cells based on the dip-coating process are still much lower than the planar structure, and other deposition processes have been developed considering the nature of uniform coating by electro-deposition and thermal evaporation.

7.3.3 Cathodic Electrodeposition

Electrodeposition has been developed for centuries and is potential for uniform coating of nanostructures, metal, and other composite for various applications [66–68]. Although it is hard to directly electrodeposit perovskite materials, the coating process could be divided into several steps, and a porous structured PbI_2 could be designed [9, 69]. The PbI_2 could be converted from electrodeposited PbO, as shown in the following chemical reactions (Figure 7.6b).

$$Pb^{2+} + 2OH^- \rightarrow PbO + H_2O \tag{7.1}$$

$$Pb^{2+} + 2OH^- \rightarrow Pb(OH)_2 \tag{7.2}$$

$$Pb(OH)_2 \rightarrow PbO + H_2O \tag{7.3}$$

$$PbO + 2HI \rightarrow PbI_2 + H_2O \tag{7.4}$$

$$PbI_2 + MAI \rightarrow MAPbI_3 \tag{7.5}$$

This process was developed for the uniform coating of PbI_2 in a mesoporous TiO_2 layer in a planar-type perovskite solar cell [70]. It was reported that with an amorphous or porous PbI_2 layer, high-quality perovskite films could be obtained [59, 60]. In contrast to the dip-coated compact PbI_2, the PbI_2 formed from electrodeposited PbO is porous with thin and flat crystals (Figure 7.7c) [9]. Fully converted PbI_2 and compact $MAPbI_3$ layers were fabricated on a fiber electrode, and a champion PCE of 7.1% was achieved, which was the highest solid-state fiber solar cells at the time reported [9]. This electrodeposition has the potential for large-scale fabrication of long-length solar cells, which are resistant to bending and twisting deformations.

7.3.4 Thermal Evaporation

Besides the solution coating process, vapor deposition is promising for uniform and conformal coating of thin layers [41, 71]. Thermal evaporation of perovskite active layers was developed almost at the same time as the solution coating process, and the first co-evaporation of $PbCl_2$ and MAI for the $MAPbI_xCl_{3-x}$ showed an efficiency of over 15%, which was comparable to the solution-based counterpart [39]. According to the evaporation precursors and reaction steps, thermal evaporation could also be divided into one-step evaporation and sequential vapor deposition process. For the one-step evaporation process it could be further categorized into co-evaporation of reactant [39, 72] and direct thermal evaporation of as-formed perovskite powders [16]. The sequential vapor deposition process is similar to the sequential solution two-step process, with the first step vapor deposition of PbI_2 and the second step vapor deposition of organic salt like MAI and formamidinium iodide (FAI) [73, 74]. This vapor deposition is promising for large-area and uniform deposition and is the typical coating process for commercial thin-film solar modules like CdTe [75]. The area over 100 cm^2 had been reported for perovskite active layers by hybrid chemical vapor deposition process [73]. And at the state of the art, the vapor deposition process for perovskite solar cells has been reported with a PCE of up to 24.42%, approaching its solution-coated counterpart [76].

Among the various vapor deposition processes, the one-step thermal evaporation of the as formed perovskite powders is readily operating [77]. Recently, the direct evaporation of perovskite powder has been introduced into the fabrication of fiber perovskite solar cells (Figure 7.6c). Considering the uniform coating based on vapor deposition, a rotation along the fiber axis is required for uniform coating around the cylinder surface [16]. Using a secondary Au wire electrode and a thin transparent Au electrode, a high PCE of 10.79% had been achieved (Figure 7.7d). However, the performance is still much lower than the planar counterpart. Further advancement in sequential vapor deposition might be important for performance enhancement.

7.3.5 Electrode-Spinning

Other than the above-discussed deposition processes that refer to the planar perovskite solar cells, there are also strategies that are specially developed for fiber solar cells. The electrode-spinning process represents a well-developed technology for nanofiber fabrication. And the deposited film has been widely applied for flexible electronics and energy devices due to its mechanical flexibility and electronic properties [78].

$MAPbI_3$-polyvinylpyrrolidone (PVP) composite nanofibers have been deposited by electrospinning technique (Figure 7.6d). A grain-size tunable perovskite layer with nucleation and growth processes dominated by humidity, spinning voltage, and post-annealing has been deposited on the outer surface of PVP nanofibers. With a polymer nanofiber core and a thin perovskite layer on the nanofiber surface, a flexible $MAPbI_3$-PVP film could be obtained. By replacing the rotating cylinder with a silver yarn coated with HTL P3HT, the $MAPbI_3$-PVP nanofiber composite was deposited. A double-twisted structured fiber solar cell was designed by twisting another carbon fiber yarn coated with ETL SnO_2 and PCBM around the perovskite-deposited silver yarn (Figure 7.7e) [18]. The deposited perovskite exhibited an optical bandgap of 1.55 eV. However, the most absorbed photon energy was 2.1–2.76 eV due to the electrospun nanofiber structure. A champion perovskite fiber solar cell with PCE up to 15.7% had been achieved.

Interestingly, the open-circuit voltage of 1.92 V was much larger than the optical bandgap. Other than double twisting into a fiber perovskite cell, the perovskite-coated silver yarn and ETL-coated carbon yarn could be weaved into a textile to function as a textile solar cell, and a size of 30 mm by 30 mm textile showed an output power density of 1.26 mW/cm^2 [18]. More details about the textile solar cells will be introduced in Chapter 10.

Other than the above-mentioned deposition techniques, which aim at uniform deposition on a highly curved fiber surface, there is a fiber-type perovskite solar cell that is fabricated on the strip substrate. Due to the strip-shaped flat surface, the coating methods for the planar structure counterpart could be applied, such as spin-coating. In this case, a PCE up to 9.49% could be achieved [11].

7.4 Charge Transport Layers for Fiber Perovskite Solar Cells

Another key challenge plagues the high performance of perovskite solar cells lies in the charge extraction and transport, which are related to the interface structure and charge transport layers [79, 80]. The charge transport layers are also important for the stability performance [44]. However, the development of charge transport layers for fiber perovskite solar cells is still lagging behind the planar counterpart.

Currently, TiO_2 nanoparticles, TiO_2 nanotubes, ZnO nanoarray, and SnO_2 have been applied as the ETL layers for perovskite fiber solar cells (Figure 7.9). Spiro-MeOTAD and P3HT are the main materials for HTL layers. TiO_2 and SnO_2

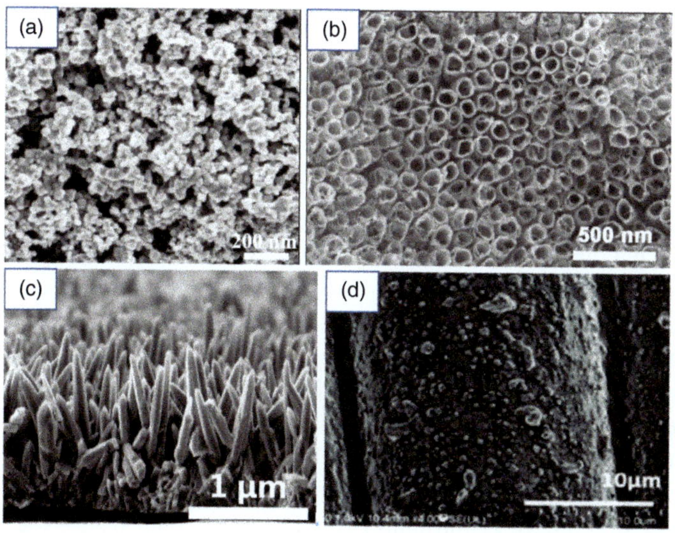

Figure 7.9 The ETL structures that are used in fiber-shaped perovskite solar cells. (a) TiO_2 nanoparticles. Source: Reproduced with permission from Qiu et al. [8], © 2014/John Wiley & Sons, Inc. (b) TiO_2 nanotubes. Source: Reproduced with permission from Qiu et al. [9], © 2016/John Wiley & Sons, Inc. (c) ZnO nanoarray. Source: Reproduced with permission from He et al. [10], © 2015/Royal Society of Chemistry. (d) electrospinning SnO_2-PCBM composite fiber. Source: Reproduced with permission from Balilonda et al. [18], © 2020/John Wiley & Sons, Inc.

have been reported for high-efficiency perovskite solar cells, with efficiency approaching 25% [81, 82]. For TiO_2, a mesoporous layer is required for efficient charge extraction. And a planar compact SnO_2 layer without mesoporous structure is promising for a highly efficient perovskite device due to its better energy alignment and higher carrier mobility [83]. The first report of fiber perovskite solar cells was based on mesoporous TiO_2 nanoparticles, which was beneficial for the charge extraction and high coverage of perovskite film via a one-step dip-coating process (Figure 7.9a) [8]. However, owing to the fragile feature of the mesoporous TiO_2 layer, the thickness needs to be thin enough to prevent breakage. On the other hand, the small thickness will lead to insufficient coverage of the TiO_2 and direct contact between the perovskite and bottom substrate that aggravates the recombination [14]. This resulted in a relatively low PCE of 3.3% for the first fiber perovskite solar cell [8].

To further improve the performance, an aligned TiO_2 nanotube array around the core electrode had been developed (Figure 7.9b). An anode oxidation process was operated to grow aligned TiO_2 nanotube array on Ti wire. And the thickness of the nanotube array could be tuned by the growing time [9]. In addition, the annealing process could help to form a compact TiO_2 layers between the Ti wire and aligned nanotube array, which was beneficial for reducing the interfacial recombination. Combining the aligned TiO_2 nanotube array and the electrodeposition of perovskite with fully coverage and high quality, a high PCE of up to 7.1% was obtained, which significantly outperformed the mesoporous TiO_2 nanoparticles.

ZnO has also been explored for performance enhancement of perovskite solar cells due to its low-temperature annealing process [84]. Uniform coating of ZnO nanoarray was developed by a dip-coating process and applied for fiber perovskite solar cells (Figure 7.9c) [10]. The mild solution-based process enlarged the selection of wire electrodes at the core. And the uniform dip-coating and growing process is also very promising for large-scale coating on textile substrate, which is difficult for the electrodeposition process. Based on the ZnO nanoarray, the fiber perovskite solar cells delivered a PCE of 3.8% [10]. However, due to the stability issue between the perovskite and ZnO, where degradation is prone, ZnO faces challenges for application in perovskite solar cells [85].

Perovskite material is composed of ionic and volatile components, which is the main reason for the defect type of ion vacancy. Due to the ion vacancy and mobility of ions, ion migration is one of the key reasons for performance degradation [86]. The formation and slow movement of the ion vacancy are the origins of the hysteresis phenomenon [87]. The incorporation of PCBM has thus been shown to reduce the hysteresis of perovskite solar cells with high performance [88]. The formation of gradient structure of PCBM onto perovskite layer could reduce the defect density in the interface and help to improve the device performance on a large scale [89]. In addition, SnO_2 with various deposition techniques has been reported for high-efficiency perovskite solar cells, which represent one of the most promising ETL materials [90]. The combination of PCBM and SnO_2 with a nanofiber structure is shown in Figure 7.9d, which is deposited by an electro-spinning process. The nanofiber structure helped to enhance the contact between ETL and perovskite layer on the basis of the double-twisting structure. A remarkable PCE of 15.7% was obtained, the highest at the current stage of fiber perovskite solar cells [18].

There are few HTL materials and structures for fiber perovskite solar cells, and spiro-MeOTAD represents a main player and is also the first solid-state material for high-performance perovskite solar cells [27]. Recently, more attentions have been paid to the HTL structure, and a self-assembled monolayer of MeO-2PACz has shown promising efficiency of over 25% for both inverted single-junction perovskite solar cells and tandem perovskite/Si solar cells [34, 91]. It is expected that there is the potential for incorporation of this newly developed HTL structure to improve the photovoltaic performance of fiber perovskite solar cells.

7.5 Electrode Materials for Fiber Perovskite Solar Cells

As mentioned above, there are two main structures for fiber-shaped perovskite solar cells, i.e. the coaxial and twisting structures. For the coaxial structure, the outer electrode must be transparent for effective absorbance of the incident light. Even for a twisting structure, the charge transport layer should be fully covered on the perovskite layer due to the limited charge carrier diffusion length. For a transparent conductive electrode, the transmittance and conductivity are the most important figures of merit. ITO represents the most popular transparent conductive electrode for thin film solar cells and other optoelectronics, which has a sheet resistance of lower than 10 Ω/□ and a transmittance of over 85% in the visible light range. However, ITO is difficult to be applied for fiber solar cells, which requires other alternatives.

Aligned carbon nanotube sheet is a highly promising candidate owing to its high transparency. However, the maintenance of a high transmittance causes an increase of the electrical resistance, which induces a low fill factor and device performance [8]. The incorporation of a thin Ag layer could be helpful to improve the conductivity while remaining the high transmittance. Improved fill factor and PCE had thus been achieved [9]. The double-twisting structure used highly conductive wire electrodes such as Au wire or Ag/carbon nanotube composite fiber, which overcome the conductivity issue. However, the twisting pitch required further improvement to enhance the light irradiation and charge extraction [13, 16].

7.6 Stretchable and Wearable Fiber Perovskite Solar Cells

For wearable applications that match the body movement while remaining the power generation functions, a stretchable structure is required, which is mainly based on the material design [92]. As mentioned above, perovskite materials exhibit ultrahigh absorbance coefficients. Thus, hundreds of nanometer layer is thick enough for efficient light absorption that simplifies the device fabrication. However, perovskite materials are intrinsically non-stretchable. Therefore, the fabrication of stretchable perovskite solar cells has been a grand challenge for this field. One way

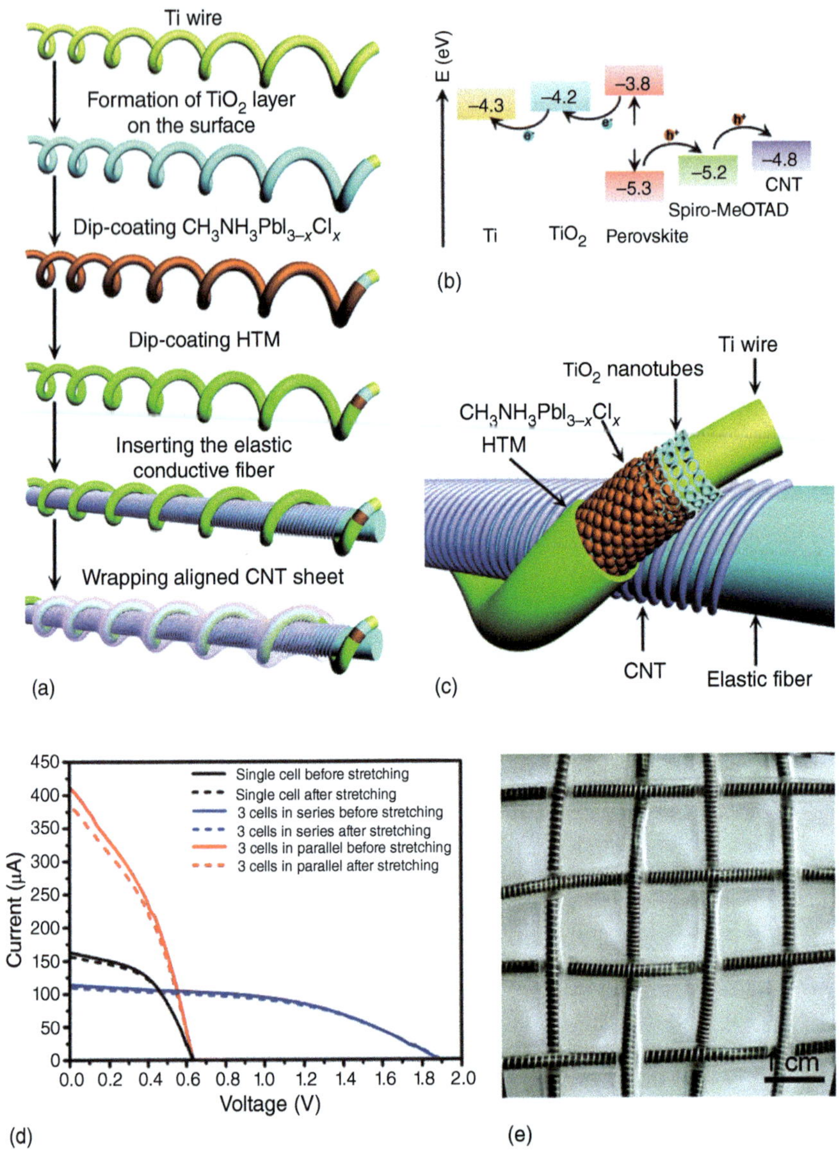

Figure 7.10 Stretchable fiber-shaped perovskite solar cells, structure, and fabrication process. (a) Schematic illustration of the fabrication process for the stretchable perovskite solar cells. (b) Energy level diagram of stretchable perovskite solar cell. (c) The enlarge structure of the stretchable solar cell. (d) The typical current density-voltage curves of the stretchable perovskite solar cells. (e) Optical image showing a textile woven by stretchable perovskite solar cell fiber. Source: Reproduced with permission from Deng et al. [12], © 2015/Royal Society of Chemistry.

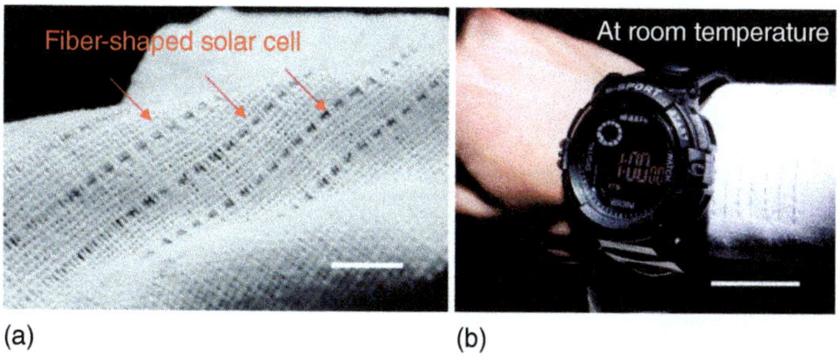

Figure 7.11 Energy textiles based on fiber-shaped perovskite solar cells and wearable applications. (a) Fiber shaped perovskite solar cells weaving into a textile. (b) An electronic watch powering by the energy textile under light illumination. Source: Reproduced with permission from Xu et al. [65], © 2020/Royal Society of Chemistry.

to obtain structurally stretchable perovskite solar cells is to transfer an ultra-thin perovskite solar cell into a pre-strained stretchable substrate. With the release of the pre-strained substrate, a compressed structure could be formed [6]. However, it is difficult to prepare a fiber-shaped device. Impressively, a stretchable fiber-shaped perovskite solar cell had been developed based on a spring structure (Figure 7.10) [12]. The spring structure could tolerate a stretching deformation without any damage to the perovskite layer. The aligned carbon nanotube wrapping outside could protect the inner layer during stretching [12]. After encapsulation, the fiber solar cells could be weaved into breathable textiles for wearable applications, such as powering a watch (Figure 7.11) [65].

7.7 Summary and Perspectives: Challenges for Fiber Perovskite Solar Cells

The past eight years have witnessed the improved PCEs of fiber-shaped perovskite solar cells from 3.3% to over 15%. Compared with dye-sensitized solar cells, fiber-shaped perovskite solar cells show an all-solid-state structure without the usage of liquid electrolytes. And the perovskite material has a much longer carrier lifetime and diffusion length than organic and polymer active materials. All these features make fiber perovskite solar cells a competitive candidate for high-performance and wearable applications. Toxicity, scalability, and stability are currently the main limitations for the commercialization of planar perovskite solar cells, considering that the PCE is approaching that of crystalline Si solar cells [48, 93, 94]. To overcome the toxicity issue, lead-free fiber perovskite solar cells have been realized [17]. However, for fiber perovskite solar cells, the PCE should also be further improved, which remains the fourth limitation. Specifically for fiber perovskite solar cells, the encapsulation issue has not been thoroughly studied yet. A thin encapsulation layer with excellent water and airproof ability is highly required in order to preserve the flexibility and wearability of the fiber perovskite solar cells and prevent the degradation caused by the moisture and oxygen, for which parylene can be a promising candidate [43, 95]. Besides, the

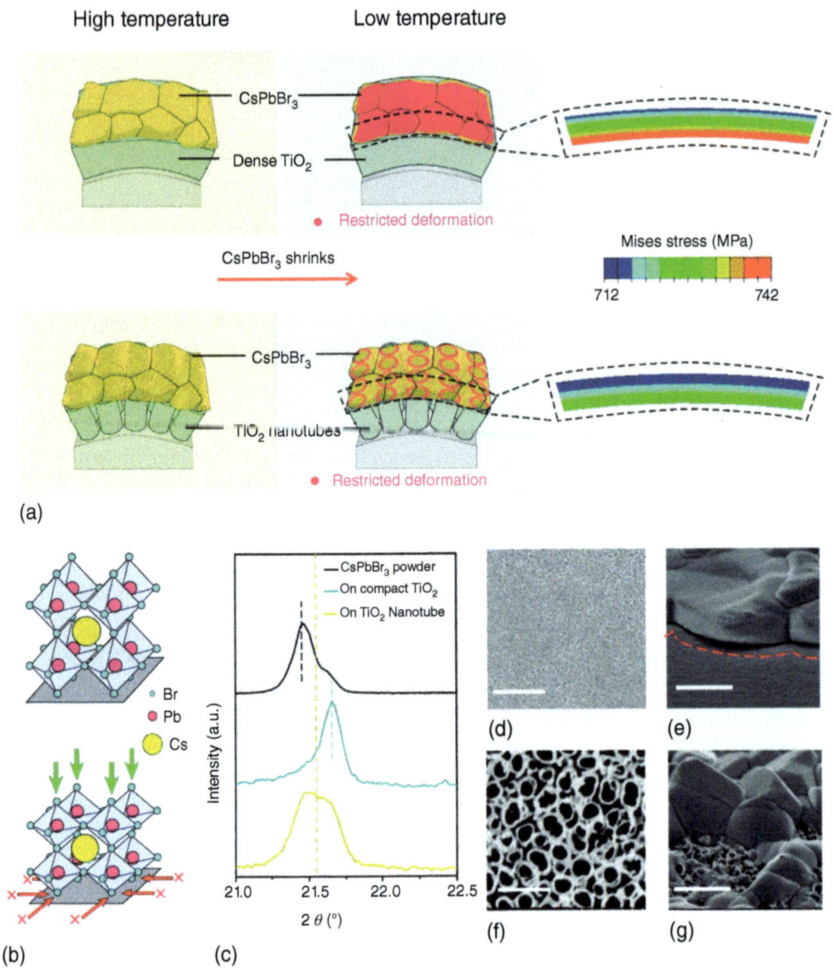

Figure 7.12 The strain properties of perovskite layers on a curved surface and nanoarray structure. (a) Schematic drawing showing the tensile stress of perovskite layer coating on a fiber electrode with TiO_2 nanoparticles or nanotubes. (b) Scheme of the deformation of perovskite layer. (c) XRD spectra showing the lattice plane distance corresponding to deformation condition. (d, e) SEM showing the coating of perovskite on TiO_2 nanoparticle. (f, g) SEM showing the coating of perovskite on TiO_2 nanotube. Source: Reproduced with permission from Xu et al. [65], © 2020/Royal Society of Chemistry.

deposition of crystalline structure on a curved surface might induce compression or stretch strain on the crystals, as shown in Figure 7.12 [65]. It is well known that the strain inside the perovskite crystal will severely reduce the performance [96–98]. Preventing the residual strain during the deposition and formation of a high-quality perovskite layer on the highly curved fiber surface remains a grand challenge. To sum up, fiber-shaped perovskite solar cells can be promising and high-performance candidates in wearable applications, with the rapidly growing device performance and new material/structure innovations.

References

1 Park, N.-G. (2015). Perovskite solar cells: an emerging photovoltaic technology. *Mater. Today* 18 (2): 65–72.

2 Stranks, S.D., Eperon, G.E., Grancini, G. et al. (2013). Electron-hole diffusion lengths exceeding 1 micrometer in an organometal trihalide perovskite absorber. *Science* 342 (6156): 341–344.

3 Miyata, A., Mitioglu, A., Plochocka, P. et al. (2015). Direct measurement of the exciton binding energy and effective masses for charge carriers in organic–inorganic tri-halide perovskites. *Nature Physics* 11 (7): 582–587.

4 Unger, E.L., Kegelmann, L., Suchan, K. et al. (2017). Roadmap and roadblocks for the band gap tunability of metal halide perovskites. *J. Mater. Chem. A* 5 (23): 11401–11409.

5 Lee, M.M., Teuscher, J., Miyasaka, T. et al. (2012). Efficient hybrid solar cells based on meso-superstructured organometal halide perovskites. *Science* 338 (6107): 643–647.

6 Kaltenbrunner, M., Adam, G., Glowacki, E.D. et al. (2015). Flexible high power-per-weight perovskite solar cells with chromium oxide-metal contacts for improved stability in air. *Nat. Mater.* 14 (10): 1032–1039.

7 Lee, G., Kim, M.-C., Choi, Y.W. et al. (2019). Ultra-flexible perovskite solar cells with crumpling durability: toward a wearable power source. *Energy Environ. Sci.* 12 (10): 3182–3191.

8 Qiu, L., Deng, J., Lu, X. et al. (2014). Integrating perovskite solar cells into a flexible fiber. *Angew. Chem. Int. Ed.* 53 (39): 10425–10428.

9 Qiu, L., He, S., Yang, J. et al. (2016). Fiber-shaped perovskite solar cells with high power conversion efficiency. *Small* 12 (18): 2419–2424.

10 He, S., Qiu, L., Fang, X. et al. (2015). Radically grown obelisk-like ZnO arrays for perovskite solar cell fibers and fabrics through a mild solution process. *J. Mater. Chem. A* 3 (18): 9406–9410.

11 Qiu, L., He, S., Yang, J. et al. (2016). An all-solid-state fiber-type solar cell achieving 9.49% efficiency. *J. Mater. Chem. A* 4 (26): 10105–10109.

12 Deng, J., Qiu, L., Lu, X. et al. (2015). Elastic perovskite solar cells. *J. Mater. Chem. A* 3 (42): 21070–21076.

13 Li, R., Xiang, X., Tong, X. et al. (2015). Wearable double-twisted fibrous perovskite solar cell. *Adv. Mater.* 27 (25): 3831–3835.

14 Hu, H., Yan, K., Peng, M. et al. (2016). Fiber-shaped perovskite solar cells with 5.3% efficiency. *J. Mater. Chem. A* 4 (10): 3901–3906.

15 Hu, H., Dong, B., Chen, B. et al. (2018). High performance fiber-shaped perovskite solar cells based on lead acetate precursor. *Sustain. Energy Fuels* 2 (1): 79–84.

16 Dong, B., Hu, J., Xiao, X. et al. (2019). High-efficiency fiber-shaped perovskite solar cell by vapor-assisted deposition with a record efficiency of 10.79%. *Adv. Mater. Technol.* 4 (7): 1900131.

17 Balilonda, A., Li, Q., Bian, X. et al. (2021). Lead-free and electron transport layer-free perovskite yarns: designed for knitted solar fabrics. *Chem. Eng. J.* 410: 128384.

18 Li, Q., Balilonda, A., Ali, A. et al. (2020). Flexible solar yarns with 15.7% power conversion efficiency, based on electrospun perovskite composite nanofibers. *Solar RRL* 4 (9): 2000269.

19 Green, M.A., Dunlop, E.D., Hohl-Ebinger, J. et al. (2022). Solar cell efficiency tables (Version 60). *Progress in Photovoltaics* 30 (7): 687–701.

20 Balilonda, A., Li, Z., Fu, Y. et al. (2022). Perovskite fiber-shaped optoelectronic devices for wearable applications. *J. Mater. Chem. C* 10 (18): 6957–6991.

21 Shan, Q., Wei, C., Jiang, Y. et al. (2020). Perovskite light-emitting/detecting bifunctional fibres for wearable LiFi communication. *Light: Science & Applications* 9 (1): 163.

22 Balilonda, A., Li, Q., Tebyetekerwa, M. et al. (2019). Perovskite solar fibers: current status, issues and challenges. *Adv. Fiber Mater.* 1 (2): 101–125.

23 Ali, M., Ueki, T., Tsurumi, D. et al. (2011). Influence of plasticizer content on the transition of electromechanical behavior of PVC gel actuator. *Langmuir* 27 (12): 7902–7908.

24 Liu, K., Jiang, Y., Jiang, Y. et al. (2019). Chemical formation and multiple applications of organic–inorganic hybrid perovskite materials. *J. Am. Chem. Soc.* 141 (4): 1406–1414.

25 Mitzi, D.B. (2019). Introduction: perovskites. *Chemical Reviews* 119 (5): 3033–3035.

26 Kojima, A., Teshima, K., Shirai, Y. et al. (2009). Organometal halide perovskites as visible-light sensitizers for photovoltaic cells. *J. Am. Chem. Soc.* 131 (17): 6050–6051.

27 Kim, H.-S., Lee, C.-R., Im, J.-H. et al. (2012). Lead iodide perovskite sensitized all-solid-state submicron thin film mesoscopic solar cell with efficiency exceeding 9%. *Sci. Rep.* 2 (1): 591.

28 Giorgi, G. and Yamashita, K. (2015). Organic–inorganic halide perovskites: an ambipolar class of materials with enhanced photovoltaic performances. *J. Mater. Chem. A* 3 (17): 8981–8991.

29 Etgar, L., Gao, P., Xue, Z. et al. (2012). Mesoscopic $CH_3NH_3PbI_3/TiO_2$ heterojunction solar cells. *J. Am. Chem. Soc.* 134 (42): 17396–17399.

30 Xing, G., Mathews, N., Sun, S. et al. (2013). Long-range balanced electron- and hole-transport lengths in organic-inorganic $CH_3NH_3PbI_3$. *Science* 342 (6156): 344–347.

31 Green, M.A., Ho-Baillie, A., and Snaith, H.J. (2014). The emergence of perovskite solar cells. *Nat. Photonics* 8 (7): 506–514.

32 Kim, J.Y., Lee, J.-W., Jung, H.S. et al. (2020). High-efficiency perovskite solar cells. *Chem. Rev.* 120 (15): 7867–7918.

33 Zhao, Y., Ma, F., Qu, Z. et al. (2022). Inactive $(PbI_2)_2RbCl$ stabilizes perovskite films for efficient solar cells. *Science* 377 (6605): 531–534.

34 Jiang, Q., Tong, J., Xian, Y. et al. (2022). Surface reaction for efficient and stable inverted perovskite solar cells. *Nature* 611 (7935): 278–283.

35 Bu, T., Li, J., Li, H. et al. (2021). Lead halide-templated crystallization of methylamine-free perovskite for efficient photovoltaic modules. *Science* 372 (6548): 1327–1332.

36 Jeon, N.J., Noh, J.H., Kim, Y.C. et al. (2014). Solvent engineering for high-performance inorganic–organic hybrid perovskite solar cells. *Nat. Mater.* 13 (9): 897–903.

37 Xiao, M., Huang, F., Huang, W. et al. (2014). A fast deposition-crystallization procedure for highly efficient lead iodide perovskite thin-film solar cells. *Angew. Chem. Int. Ed.* 53 (37): 9898–9903.

38 Huang, F., Dkhissi, Y., Huang, W. et al. (2014). Gas-assisted preparation of lead iodide perovskite films consisting of a monolayer of single crystalline grains for high efficiency planar solar cells. *Nano Energy* 10: 10–18.

39 Liu, M., Johnston, M.B., and Snaith, H.J. (2013). Efficient planar heterojunction perovskite solar cells by vapour deposition. *Nature* 501 (7467): 395–398.

40 Burschka, J., Pellet, N., Moon, S.-J. et al. (2013). Sequential deposition as a route to high-performance perovskite-sensitized solar cells. *Nature* 499 (7458): 316–319.

41 Qiu, L., He, S., Jiang, Y. et al. (2021). Metal halide perovskite solar cells by modified chemical vapor deposition. *J. Mater. Chem. A* 9 (40): 22759–22780.

42 Yang, Z., Babu, B.H., Wu, S. et al. (2020). Review on practical interface engineering of perovskite solar cells: from efficiency to stability. *Solar RRL* 4 (2): 1900257.

43 Liu, Z., Qiu, L., Ono, L.K. et al. (2020). A holistic approach to interface stabilization for efficient perovskite solar modules with over 2,000-hour operational stability. *Nat. Energy* 5 (8): 596–604.

44 Qiu, L., He, S., Ono, L.K. et al. (2020). Progress of surface science studies on ABX_3-based metal halide perovskite solar cells. *Adv. Energy Mater.* 10 (13): 1902726.

45 Foo, S., Thambidurai, M., Senthil Kumar, P. et al. (2022). Recent review on electron transport layers in perovskite solar cells. *International Journal of Energy Research* 46 (15): 21441–21451.

46 Courtier, N.E., Cave, J.M., Foster, J.M. et al. (2019). How transport layer properties affect perovskite solar cell performance: insights from a coupled charge transport/ion migration model. *Energy Environ. Sci.* 12 (1): 396–409.

47 Yin, X., Song, Z., Li, Z. et al. (2020). Toward ideal hole transport materials: a review on recent progress in dopant-free hole transport materials for fabricating efficient and stable perovskite solar cells. *Energy Environ. Sci.* 13 (11): 4057–4086.

48 Qiu, L., He, S., Ono, L.K. et al. (2019). Scalable fabrication of metal halide perovskite solar cells and modules. *ACS Energy Lett.* 4 (9): 2147–2167.

49 Zhai, W., Zhu, Z., Sun, X. et al. (2022). Fiber solar cells from high performances towards real applications. *Adv. Fiber Mater.* 4: 1293–1303.

50 Chen, T., Qiu, L., Yang, Z. et al. (2013). Novel solar cells in a wire format. *Chem. Soc. Rev.* 42 (12): 5031–5041.

51 Moshonas, N., Stathopoulos, N.A., and Pagiatakis, G. (2022). Optical modelling of planar and fibre perovskite solar cells. *Electronics* 11 (13): 2041.

52 Yang, Z., Zhang, W., Wu, S. et al. (2021). Slot-die coating large-area formamidinium-cesium perovskite film for efficient and stable parallel solar module. *Sci. Adv.* 7 (18): eabg3749.

53 Bu, T., Ono, L.K., Li, J. et al. (2022). Modulating crystal growth of formamidinium–caesium perovskites for over 200 cm^2 photovoltaic sub-modules. *Nat. Energy.* 7: 528–536.

54 Liu, C., Cheng, Y.B., and Ge, Z. (2020). Understanding of perovskite crystal growth and film formation in scalable deposition processes. *Chem. Soc. Rev.* 49 (6): 1653–1687.

55 Eperon, G.E., Burlakov, V.M., Docampo, P. et al. (2014). Morphological control for high performance, solution-processed planar heterojunction perovskite solar cells. *Adv. Funct. Mater.* 24 (1): 151–157.

56 Li, X., Bi, D., Yi, C. et al. (2016). A vacuum flash-assisted solution process for high-efficiency large-area perovskite solar cells. *Science* 353 (6294): 58–62.

57 Bi, C., Wang, Q., Shao, Y. et al. (2015). Non-wetting surface-driven high-aspect-ratio crystalline grain growth for efficient hybrid perovskite solar cells. *Nat. Commun.* 6 (1): 7747.

58 Sherkar, T.S., Momblona, C., Gil-Escrig, L. et al. (2017). Recombination in perovskite solar cells: significance of grain boundaries, interface traps, and defect ions. *ACS Energy Lett.* 2 (5): 1214–1222.

59 Wu, Y., Islam, A., Yang, X. et al. (2014). Retarding the crystallization of PbI_2 for highly reproducible planar-structured perovskite solar cells via sequential deposition. *Energy Environ. Sci.* 7 (9): 2934–2938.

60 Yin, J., Qu, H., Cao, J. et al. (2016). Vapor-assisted crystallization control toward high performance perovskite photovoltaics with over 18% efficiency in the ambient atmosphere. *J. Mater. Chem. A* 4 (34): 13203–13210.

61 Xiao, Z., Bi, C., Shao, Y. et al. (2014). Efficient, high yield perovskite photovoltaic devices grown by interdiffusion of solution-processed precursor stacking layers. *Energy Environ. Sci.* 7 (8): 2619–2623.

62 Chen, T., Qiu, L., Li, H. et al. (2012). Polymer photovoltaic wires based on aligned carbon nanotube fibers. *J. Mater. Chem.* 22 (44): 23655–23658.

63 Leyden, M.R., Ono, L.K., Raga, S.R. et al. (2014). High performance perovskite solar cells by hybrid chemical vapor deposition. *J. Mater. Chem. A* 2 (44): 18742–18745.

64 Liang, P.-W., Liao, C.-Y., Chueh, C.-C. et al. (2014). Additive enhanced crystallization of solution-processed perovskite for highly efficient planar-heterojunction solar cells. *Adv. Mater.* 26 (22): 3748–3754.

65 Xu, L., Fu, X., Liu, F. et al. (2020). A perovskite solar cell textile that works at −40 to 160 °C. *J. Mater. Chem. A* 8 (11): 5476–5483.

66 Lelevic, A. and Walsh, F.C. (2019). Electrodeposition of NiP alloy coatings: a review. *Surf. Coat. Technol.* 369: 198–220.

67 Walsh, F.C. and Ponce de Leon, C. (2014). A review of the electrodeposition of metal matrix composite coatings by inclusion of particles in a metal layer: an established and diversifying technology. *Transactions of the IMF* 92 (2): 83–98.

68 Lahiri, A. and Endres, F. (2017). Review—electrodeposition of nanostructured materials from aqueous, organic and ionic liquid electrolytes for Li-ion and na-ion batteries: a comparative review. *Journal of The Electrochemical Society* 164 (9): D597–D612.

69 Al Katrib, M., Perrin, L., and Planes, E. (2022). A way to reach 10% efficiency with carbon-based electrodeposited mixed perovskite solar cells. *Solar RRL* 6 (11): 2200777.

70 Cui, X.P., Jiang, K.J., Huang, J.H. et al. (2015). Electrodeposition of PbO and its in situ conversion to $CH_3NH_3PbI_3$ for mesoscopic perovskite solar cells. *Chem. Commun.* 51 (8): 1457–1460.

71 Jiang, Y., He, S., Qiu, L. et al. (2022). Perovskite solar cells by vapor deposition based and assisted methods. *Applied Physics Reviews* 9 (2): 021305.

72 Ono, L.K., Wang, S., Kato, Y. et al. (2014). Fabrication of semi-transparent perovskite films with centimeter-scale superior uniformity by the hybrid deposition method. *Energy Environ. Sci.* 7 (12): 3989–3993.

73 Qiu, L., He, S., Jiang, Y. et al. (2019). Hybrid chemical vapor deposition enables scalable and stable Cs-FA mixed cation perovskite solar modules with a designated area of 91.8 cm^2 approaching 10% efficiency. *J. Mater. Chem. A* 7 (12): 6920–6929.

74 Qiu, L., He, S., Liu, Z. et al. (2020). Rapid hybrid chemical vapor deposition for efficient and hysteresis-free perovskite solar modules with an operation lifetime exceeding 800 hours. *J. Mater. Chem. A* 8 (44): 23404–23412.

75 Wen, X., Lu, Z., Sun, X. et al. (2020). Epitaxial CdTe thin films on mica by vapor transport deposition for flexible solar cells. *ACS Appl. Energy Mater.* 3 (5): 4589–4599.

76 Li, H., Zhou, J., Tan, L. et al. (2022). Sequential vacuum-evaporated perovskite solar cells with more than 24% efficiency. *Sci. Adv.* 8 (28): eabo7422.

77 Bae, S.R., Heo, D.Y., and Kim, S.Y. (2022). Recent progress of perovskite devices fabricated using thermal evaporation method: perspective and outlook. *Materials Today Advances* 14: 100232.

78 Wang, Y., Yokota, T., and Someya, T. (2021). Electrospun nanofiber-based soft electronics. *NPG Asia Materials* 13 (1): 22.

79 Foo, S., Thambidurai, M., Senthil Kumar, P. et al. (2022). Recent review on electron transport layers in perovskite solar cells. *International Journal of Energy Research* 46: 1–11.

80 Lachore, W.L., Andoshe, D.M., Mekonnen, M.A. et al. (2022). Recent progress in electron transport bilayer for efficient and low-cost perovskite solar cells: a review. *Journal of Solid State Electrochemistry* 26 (2): 295–311.

81 Min, H., Lee, D.Y., Kim, J. et al. (2021). Perovskite solar cells with atomically coherent interlayers on SnO_2 electrodes. *Nature* 598 (7881): 444–450.

82 Hui, W., Chao, L.F., Lu, H. et al. (2021). Stabilizing black-phase formamidinium perovskite formation at room temperature and high humidity. *Science* 371 (6536): 1359.

83 Correa Baena, J.P., Steier, L., Tress, W. et al. (2015). Highly efficient planar perovskite solar cells through band alignment engineering. *Energy Environ. Sci.* 8 (10): 2928–2934.

84 Liu, D. and Kelly, T.L. (2013). Perovskite solar cells with a planar heterojunction structure prepared using room-temperature solution processing techniques. *Nat. Photonics* 8 (2): 133–138.

85 Yang, J., Siempelkamp, B.D., Mosconi, E. et al. (2015). Origin of the thermal instability in $CH_3NH_3PbI_3$ thin films deposited on ZnO. *Chem. Mater.* 27 (12): 4229–4236.

86 Zai, H., Ma, Y., Chen, Q. et al. (2021). Ion migration in halide perovskite solar cells: mechanism, characterization, impact and suppression. *J. Energy Chem.* 63: 528–549.

87 Richardson, G., O'Kane, S.E.J., Niemann, R.G. et al. (2016). Can slow-moving ions explain hysteresis in the current–voltage curves of perovskite solar cells? *Energy Environ. Sci.* 9 (4): 1476–1485.

88 Shao, Y., Xiao, Z., Bi, C. et al. (2014). Origin and elimination of photocurrent hysteresis by fullerene passivation in $CH_3NH_3PbI_3$ planar heterojunction solar cells. *Nat. Commun.* 5: 5784.

89 Wu, Y., Yang, X., Chen, W. et al. (2016). Perovskite solar cells with 18.21% efficiency and area over 1 cm^2 fabricated by heterojunction engineering. *Nat. Energy* 1 (11): 16148.

90 Qiu, L., Liu, Z., Ono, L.K. et al. (2019). Scalable fabrication of stable high efficiency perovskite solar cells and modules utilizing room temperature sputtered SnO_2 electron transport layer. *Adv. Funct. Mater.* 29 (47): 1806779.

91 Mao, L., Yang, T., Zhang, H. et al. (2022). Fully textured, production-line compatible monolithic perovskite/silicon tandem solar cells approaching 29% efficiency. *Adv. Mater.* 34 (40): e2206193.

92 Kim, D.C., Shim, H.J., Lee, W. et al. (2020). Material-based approaches for the fabrication of stretchable electronics. *Adv. Mater.* 32 (15): e1902743.

93 He, S., Qiu, L., Ono, L.K. et al. (2020). How far are we from attaining 10-year lifetime for metal halide perovskite solar cells? *Mater. Sci. Eng. R Rep.* 140: 100545.

94 Qiu, L., Ono, L.K., and Qi, Y.B. (2018). Advances and challenges to the commercialization of organic–inorganic halide perovskite solar cell technology. *Mater Today Energy* 7: 169–189.

95 Jinno, H., Fukuda, K., Xu, X. et al. (2017). Stretchable and waterproof elastomer-coated organic photovoltaics for washable electronic textile applications. *Nat. Energy* 2 (10): 780–785.

96 Zhao, J.J., Deng, Y.H., Wei, H.T. et al. (2017). Strained hybrid perovskite thin films and their impact on the intrinsic stability of perovskite solar cells. *Sci. Adv.* 3 (11): eaao5616.

97 Meng, W., Zhang, K., Osvet, A. et al. (2022). Revealing the strain-associated physical mechanisms impacting the performance and stability of perovskite solar cells. *Joule* 6 (2): 458–475.

98 He, S., Li, S., Zhang, A. et al. (2022). Residual strain reduction leads to efficiency and operational stability improvements in flexible perovskite solar cells. *Mater. Adv.* 3 (15): 6316–6323.

8

Fiber-Shaped Integrated Device

Integration of solar cells with other electronic devices such as batteries, sensors, displays, and actuators can greatly extend the functions and applications. Therefore, fiber-shaped integrated devices based on fiber solar cells have been intensively investigated in various formats [1–14]. In this chapter, we will introduce the development of fiber-shaped integrated devices based on fiber solar cells. First, the classification and mechanism of integrated devices are introduced based on the planar structure. Then fiber-integrated devices with various device structures are discussed. Integrated textiles are further illustrated for practical applications. The summary and perspective for fiber-integrated devices are provided at last.

8.1 Overview of Integrated Device Based on Solar Cells

For solar cells, the generated electric energy is intermittent and is not fully compatible with the requirements of practical applications. Therefore, it is preferred to be stored in energy storage devices, which can be used whenever needed. As early as 1987, Licht et al. developed a photoelectrochemical solar cell with in-situ energy storage capability [15]. As shown in Figure 8.1a, charge separation happens on the surface of a n-type semiconductor $CdSe_{0.65}Te_{0.35}$ under illumination. The generated electron is driven into an external load and can be stored in an electrochemical storage device by reducing SnS to metal Sn. In the dark, the electron generated with the oxidation of Sn and go through the external circuit to complete the discharge process (Figure 8.1b). Under solar illumination of 96.5 mW/cm^2, the photoelectrochemical solar cell generated a current density of 23 mA/cm^2 at a photovoltage of 0.495 V when a load of 1500 Ω was connected, which exhibited a high direct energy conversion efficiency of 11.8% and storage efficiency of 95%. As a result, the overall energy conversion efficiency was calculated to be 11.2% by multiplying the photoelectric conversion efficiency and energy storage efficiency [15].

In 2005, Miyasaka and coworkers developed a three-terminal integrated device by combining a DSSC with a supercapacitor by sharing an internal electrode [2]. As shown in Figure 8.2, for the charging process, terminals A and B were connected. Under illumination, electrons were generated from the dye-sensitized mesoporous TiO_2 layer, transferred along the external circuit and reached and stored at the active

Wearable Solar Cells: Mechanisms, Materials, and Devices, First Edition.
Hao Sun, Zhibin Yang, Longbin Qiu, and Huisheng Peng.

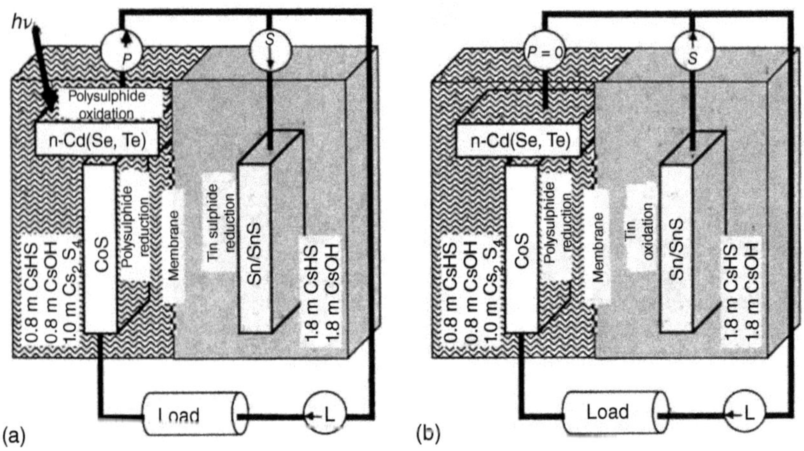

Figure 8.1 Two terminal integrated devices of a n-Cd(Se, Te)/Cs_2S_x/SnS photoelectrochemical solar cell. P, S, and L respond to the direction of electron flow through the photoelectrode, tin electrode, and external load, respectively. (a) Photocharge process under illumination. (b) Discharge process in the dark. Source: Licht et al. [15]/Springer Nature.

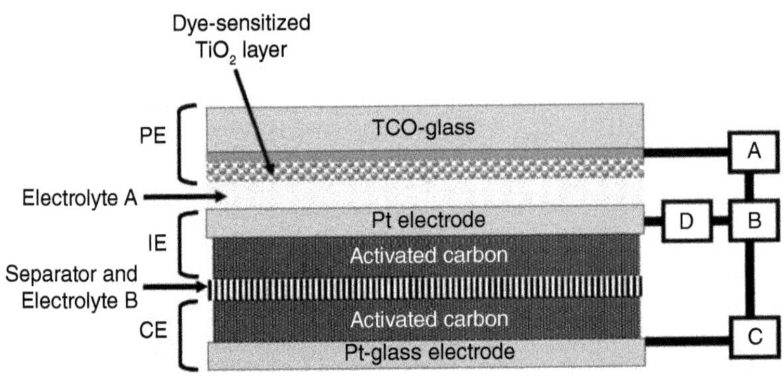

Figure 8.2 A three-terminal integrated device based on a DSSC and a supercapacitor. Source: Murakami et al. [2]/Royal Society of Chemistry.

carbon/platinum internal electrode, which was part of the energy storage device. The dye molecules returned to the ground state when they received electrons from I^- ions in the electrolyte, and the I^- ions were oxidized into I^{3-} ions. For the discharging process, terminals A and B were disconnected, and terminals B and C were connected with a load. The supercapacitor could output electric energy to the load by completing a circuit [2].

Compared to the two-terminal system, the electron transport efficiency in the three-terminal system can be effectively promoted by introducing a third electrode. Taking the discharge process as an example, the charge can be output from the internal electrode B and outer electrode C rather than go through the electrolyte and photoanode in a two-terminal system. In addition, the three-terminal system where the energy conversion and storage part share an internal electrode can realize independent control of photo charging and discharging processes, which is more favorable for practical application.

8.2 Overview of Fiber-Integrated Device Based on Solar Cells

Although the conception of integrated devices has been proposed and studied for more than three decades, fiber-shaped integrated devices had not been realized until 2011. In the beginning, most studies focused on single fiber devices with different structures and compositions. Fiber-shaped DSSCs, PSCs, and nanogenerators were produced for energy harvesting, and fiber-shaped supercapacitors and lithium-ion batteries werc fabricated for energy storage. Since 2015, more efforts have been made to develop integrated textiles for practical applications, focusing on the fabrication process, performance, and stability.

8.2.1 Fiber-Shaped Integrated Device Based on Solar Cells and Supercapacitors

As an important branch of energy storage devices, supercapacitors possess simple device structures and high power densities. A pioneer work by Chen et al. integrated a fiber solar cell with a fiber supercapacitor for the realization of simultaneous photoelectric conversion and storage [5]. For the photoelectric conversion part, a Ti/TiO_2 nanotube/dye photoanode and aligned CNT fiber counter electrodes were twisted to construct a fiber DSSC after introducing an electrolyte (Figure 8.3a–c). For the energy storage part, a Ti/TiO_2 nanotube and aligned CNT fiber electrodes were twisted together to produce a fiber supercapacitor using a gel electrolyte. Once the photoelectric conversion and energy storage components were connected, as shown in Figure 8.3d, the fiber-integrated device would be photocharged under illumination, corresponding to the photovoltaic conversion and electric energy storage in the supercapacitor. When they were disconnected, as illustrated in Figure 8.3e, the supercapacitor can discharge to the circuit. Figure 8.3f shows the voltage evolution of the photo-charge process under AM 1.5 illumination and the discharge process with a discharge current of 0.1 μA. The overall photoelectric conversion and storage efficiency of this integrated device were 1.5% by multiplying photoelectric conversion efficiency ($\eta_{conversion}$) and energy storage efficiency ($\eta_{storage}$).

The integrated device with a twisting structure can be easily damaged under bending or stretching. To overcome this challenge, Yang et al. developed a stretchable fiber-integrated device with a stretchable photoelectric conversion sheath and an energy storage core [10]. Figure 8.4 a,b show the structure of the self-powered energy fiber from a top and cross-sectional view. For the inner stretchable energy storage component, an elastic rubber fiber wrapped with an aligned CNT sheet was used for the inner electrode, and a thin layer of poly(vinyl alcohol) (PVA)-H_3PO_4 gel was used as electrolyte. Then, a second CNT sheet electrode with the same thickness was

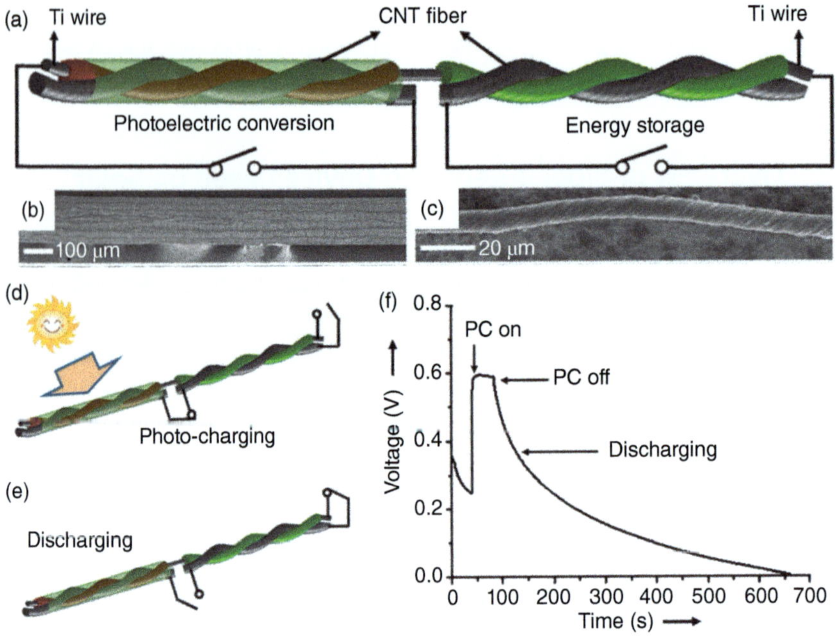

Figure 8.3 (a–c) Schematic illustration and SEM images of the fiber-integrated device. (d) and (e) Schematical circuit connection of the charging and discharging processes, respectively. (f) Photocharge-discharge curve of a typical energy wire. The discharging current is 0.1 μA. Source: Chen et al. [5]/John Wiley & Sons.

wrapped around the fiber to obtain a fiber supercapacitor. The integrated device was then sealed in a plastic tube. Such supercapacitors possessed high stretchability due to the aligned helical structure of the CNT electrode. For the outer stretchable photoelectric conversion component, a third aligned CNT sheet was wrapped onto the supercapacitor as the cathode, and a modified helical Ti fiber served as the photoanode. An integrated device was completed by inserting the cathode into the helical anode. The photo-charging and discharging processes of the integrated energy fiber could be realized by connecting the circuit as Figure 8.4c,d, respectively. It exhibited an overall energy conversion and storage efficiency of 1.83%. Surprisingly, the performance of this integrated fiber can be well maintained under a 20% strain and bending radii of curvature between 5 and 0.5 cm. This work also provided a new approach the fabrication of stretchable integrated devices.

The electrolyte leakage may occur when a traditional DSSC is used for an integrated device. Therefore, Zhang et al. developed an all-solid-state integrated "energy fiber" by using a fiber-shaped PSC [11]. As shown in Figure 8.5a, the fiber-integrated device was constructed on a Ti wire. A PSC was fabricated on one terminal of the Ti wire for photoelectric conversion, and a supercapacitor was prepared on the other side for energy storage [11]. A coaxial structure was realized by winding aligned multi-walled CNT sheets as the outer transparent electrodes. As shown in Figure 8.5b–d, the photoelectric conversion process was achieved, and the charges were stored in a supercapacitor under solar illumination (Figure 8.5b).

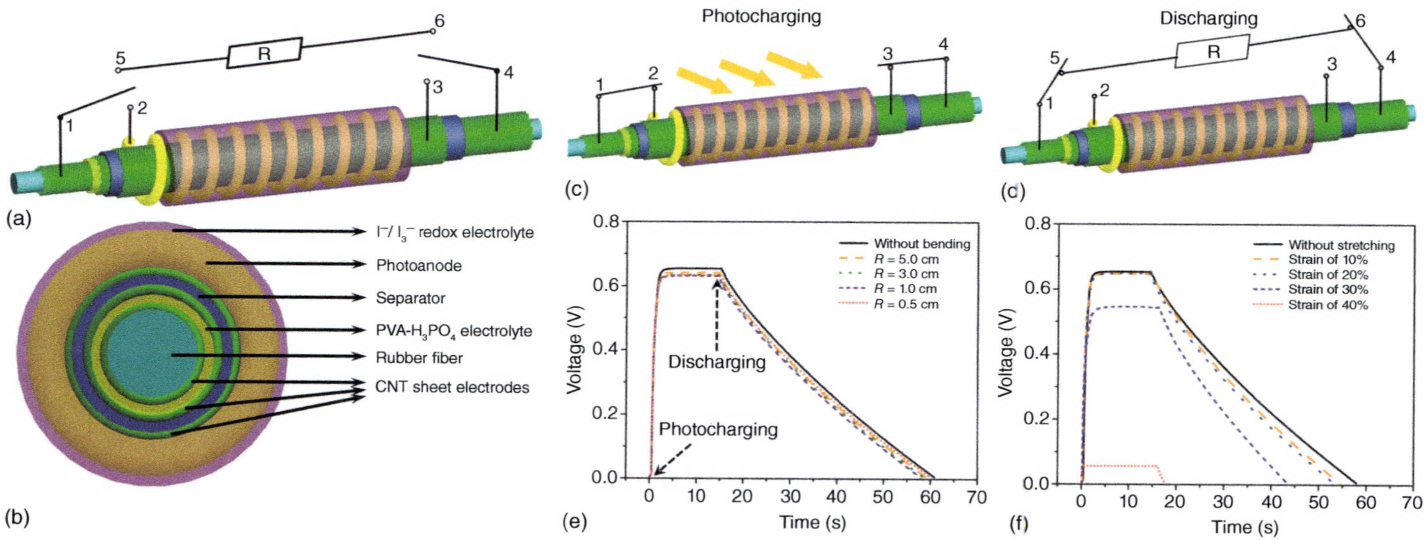

Figure 8.4 (a) and (b), Schematic illustration of the structure of a self-powered energy fiber from a top (a) and cross-sectional view (b). (c) and (d), Schematic circuit connection state of the photo-charging and discharging processes of the self-powering energy fiber. (e) Photo-charging and discharging processes before and after bending with vary radii of curvature. (f) Photo-charging and discharging processes before and after stretching under vary strains. A current density of 0.1 A/g is used for the discharging process. Source: Yang et al. [10]/John Wiley & Sons.

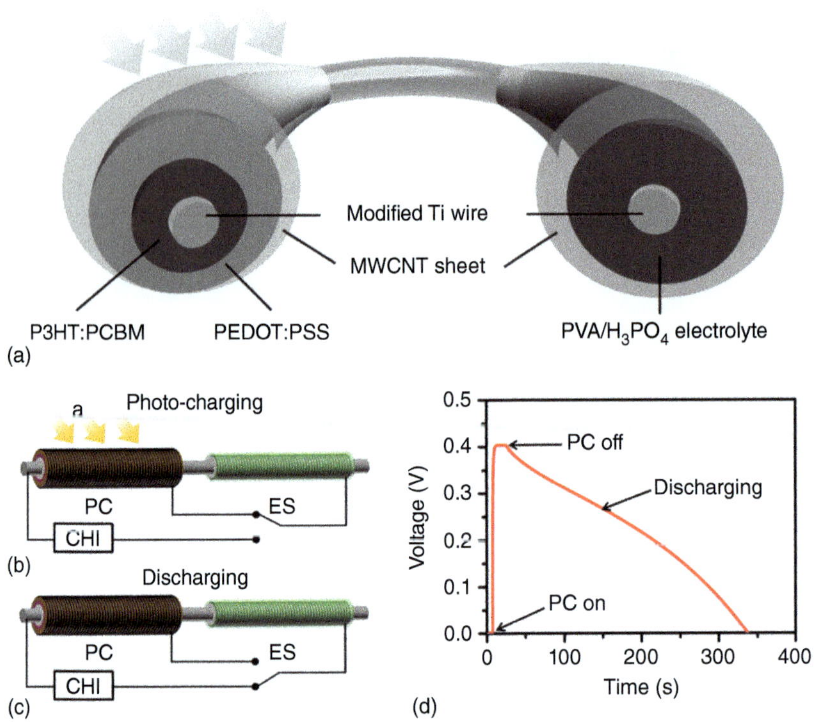

Figure 8.5 (a) Schematic illustration of the all-solid-state fiber integrated device based on a fiber-shaped PSC and a fiber supercapacitor. (b) and (c) the Circuit connection state of the fiber-integrated device in the charging and discharging process. (d) Charging and discharging curve of the fiber-integrated device. An *e* current of 0.1 μA is used in the discharging process. Source: Zhang et al. [11]/John Wiley & Sons.

And the discharge process could be realized by connecting the circuit, as shown in Figure 8.5c. Compared to the fiber-integrated device based on DSSC using a liquid electrolyte, the all-solid-state integrated fiber was expected to be more stable under deformations.

8.2.2 Fiber-Shaped Integrated Device Based on Solar Cell and Lithium-Ion Battery

Many efforts have been made to integrate a fiber supercapacitor with a fiber DSSC or PSC by sharing a common electrode. However, the fiber supercapacitors are always fully charged by the solar cell part within a few seconds due to a mismatch of their energy generation and storage capability. Besides, the output voltages are normally less than 1 V, which are obviously too low to power various electronic devices in practical use. Therefore, Sun et al. developed a fiber-shaped integrated device with a novel core-sheath configuration by integrating a solar cell with a lithium-ion battery [16]. As shown in Figure 8.6a–f, the integrated fiber was composed of a lithium-ion battery and DSSC. This integrated device exhibited excellent performance in photo-charging and discharging processes. It could be photo-charged to

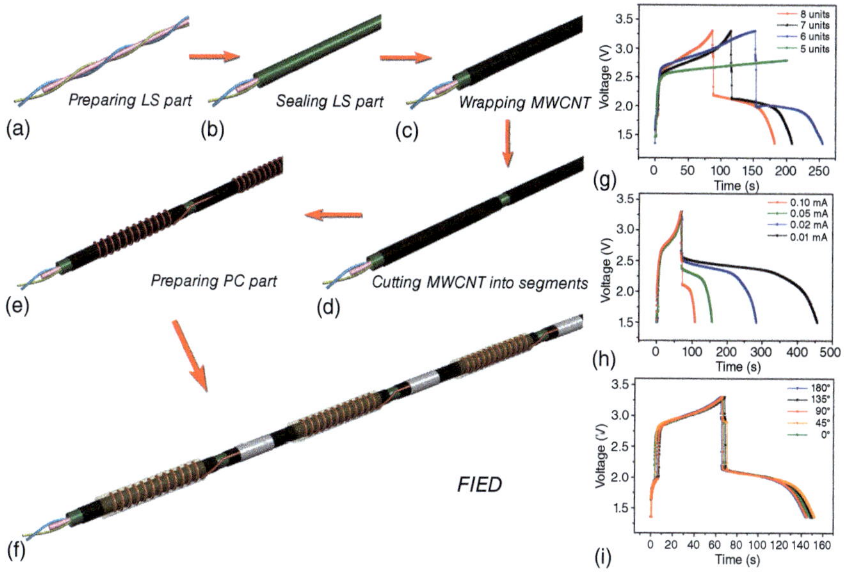

Figure 8.6 (a–f) Schematic illustration of the fabrication process of a fiber-integrated device. (g) Photo-charging and discharging curves with increasing quantity of photoelectric conversion units. (f) Photo-charging and discharging curves with vary discharging currents. (i) Photo-charging and discharging curves with vary incident light angles. All tests were measured under the illumination of simulated AM 1.5 solar light (100 mW/cm^2). Source: Sun et al. [16]/Royal Society of Chemistry.

3.3 V with 5–8 connected solar cells under AM 1.5 illumination and discharged at 0.05 mA (Figure 8.6g). The lithium-ion battery exhibited a high energy density of 22 Wh/kg, which was one to two orders of magnitude larger than the value of previously reported supercapacitors. The fiber-integrated device could discharge at various currents from 0.01 to 0.1 mA after fully charged (Figure 8.6h). In addition, it showed quite stable performance under vary incident light angles from 0 to 180° (Figure 8.6i). The introduction of lithium-ion batteries in fiber-integrated devices could effectively improve the practical potential for wearable applications.

8.2.3 Fiber Integrated Device Based on Nanogenerator and Solar Cell

Apart from solar cells, other energy-harvesting devices can be used for the production of fiber-shaped integrated devices. For example, solar, mechanical, wind, and thermal energy can be harvested and used. Joonho et al. fabricated an energy fiber by integrating solar cells, nanogenerators, and supercapacitors together, which could harvest the environmental solar and mechanical energy (Figure 8.7a) [17]. The assembling process depended on the ZnO nanowire array modified polymethyl methacrylate (PMMA) fiber. Dye, gel electrolyte, and graphene were sequentially deposited on different parts of this fiber to obtain the DSSC, supercapacitor, and nanogenerator. Figure 8.7b shows the *J–V* curves of the DSSC part under illumination of simulated sunlight with a PCE of 0.02%. For the supercapacitor

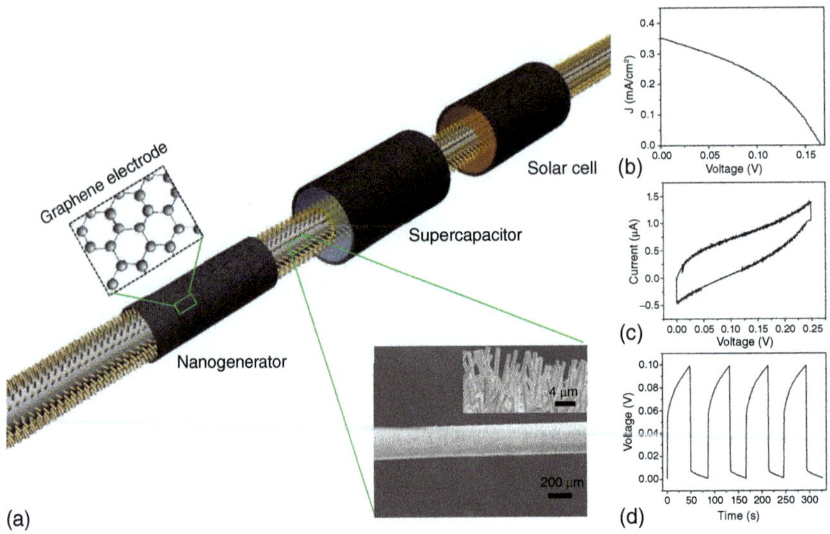

Figure 8.7 (a) Schematic of the multi-energy device containing a nanogenerator (mechanical energy harvesting), solar cell (solar energy harvesting), and supercapacitor (electrochemical energy storage). (b) $J-V$ curve of the DSSC under the illumination of simulated AM 1.5 solar light. (c) Cyclic voltammetry of supercapacitor part using PVA/H_3PO_4 as an electrolyte at 100 mV/s. (d) Galvanostatic charge–discharge curve of supercapacitor part measured with a constant current at 1 μA. Source: Bae et al. [17]/John Wiley & Sons.

part, cyclic voltammetry showed good electrochemical stability and capacitance of this fiber device (Figure 8.7c). The galvanostatic charge–discharge measurement suggested this device has excellent capacitor performance (Figure 8.7d). This work demonstrated the feasibility of multi-function integrated energy-harvesting device and represented an inspirable trial on self-powered integrated devices.

Pan et al. developed an integrated device based on a three-dimensional structure by integrating DSSC and piezoelectric nanogenerator together on a single optical fiber (Figure 8.8a–c) [18]. Owing to the unique property of the optical fiber, light can go through the inner side of the device, which avoids the light shielding by other components, thus improving the PCE of the photovoltaic part (Figure 8.8d). The working mechanism and the energy band diagram of DSSC are illustrated in Figure 8.8e. For the piezoelectric NG part, the pack of ZnO nanowires textured film could directly convert mechanical energy into electricity (Figure 8.8f,g). The integrated device could harvest mechanical energy and photo energy. As shown in Figure 8.8h,i, the single DSSC part showed a V_{OC} of 0.38 V, and the nanogenerator could deliver an output voltage of 2.9 V, which resulted in a total output voltage of 3.3 V for the integrated device. Figure 8.8j shows the output current of the integrated device under a regular cycled compressive strain at the frequency of 1 Hz. The output current reached 7.56 μA when the DSSC and the HG parts were connected in parallel. These results indicated that the integrated device could be available for powering a variety of low-power electronic devices. As a self-powering

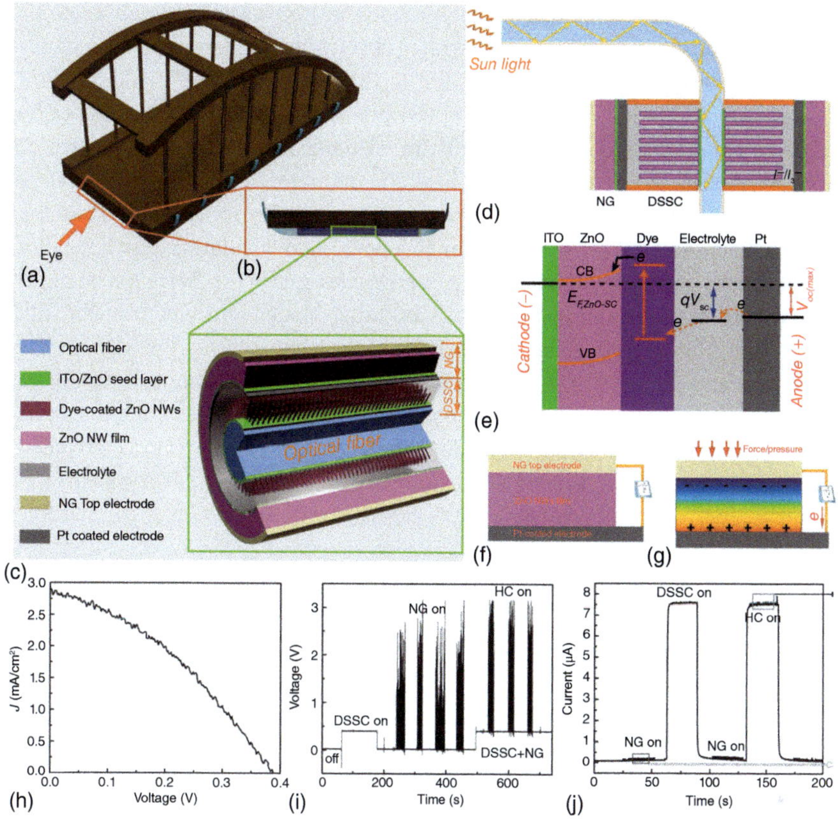

Figure 8.8 Design and principle of a 3D optical fiber-based integrated device consisting of a DSSC and an NG. (a) Taking a smart bridge as a demonstration, integrated devices could be harvesting solar and mechanical energy for driving the self-powering system. (b) Enlarged view of the integrated devices beneath the bridge. (c) 3D schematic of layer-by-layer structure of the integrated device. (d) Detail operating process of integrated device. (e) Electron energy band diagram and working principle of the DSSC part. (f) and (g), Working principle of the NG part. (h) $J-V$ curve of the DSSC. (i) Open-circuit voltage of the integrated device when the NG and DSSC are series connected. (j) Short-circuit current of the integrated device when the NG and DSSC are parallel connected. Source: Pan et al. [18]/John Wiley & Sons.

system, the optical fiber integrated device has great potential in personal electronics, micro-robots, and even smart buildings.

8.2.4 Fiber-Based Integrated Textiles

A flexible lithium-ion battery belt and a textile triboelectric nanogenerator cloth (TENG-cloth) were integrated to form a wearable device by Pu et al. [19]. Based on the electroless plating of the Ni layer, ordinary flexible and insulated polyester fabrics became conductive. They could be used as the current collectors of lithium-ion batteries and TENG. When worn at various positions of the body, the TENG fabric exhibited its potential to transform mechanical energy from various human motions into electricity. The LIB belt also demonstrated decent electrochemical performance even after being severely folded at 180° for 30 times. The LIB belt could power a heartbeat meter strap that remotely communicated with a smartphone powered by the TENG cloth, indicating the practicality for wearable healthcare applications.

It is essential to develop a soft and flexible conductor for wearable TENG as an essential part of the TENG. The common polyester fabric was chosen as the initial substrate, and conductive Ni film (Ni-cloth) and insulating parylene film (parylene-cloth) were sequentially layered on (Figure 8.9a) [19]. Belt-type Ni-cloth and parylene-cloth were employed as the building blocks and were woven into a 5 cm × 5 cm TENG-cloth (Figure 8.9a). One electrode of the TENG was the Ni-cloth belt, and the other was the parylene-cloth belt. The mechanical flexibility, air-breathability, water-washability, and comfortability of the polyester cloth can be well retained in the woven TENG cloth. After the electroless Ni coating, a conformal Ni layer with nanoscale roughness was applied to the polyester fabric's surface. A 0.5 cm × 10 cm Ni-cloth exhibited a low resistance of a few ohms, comparable

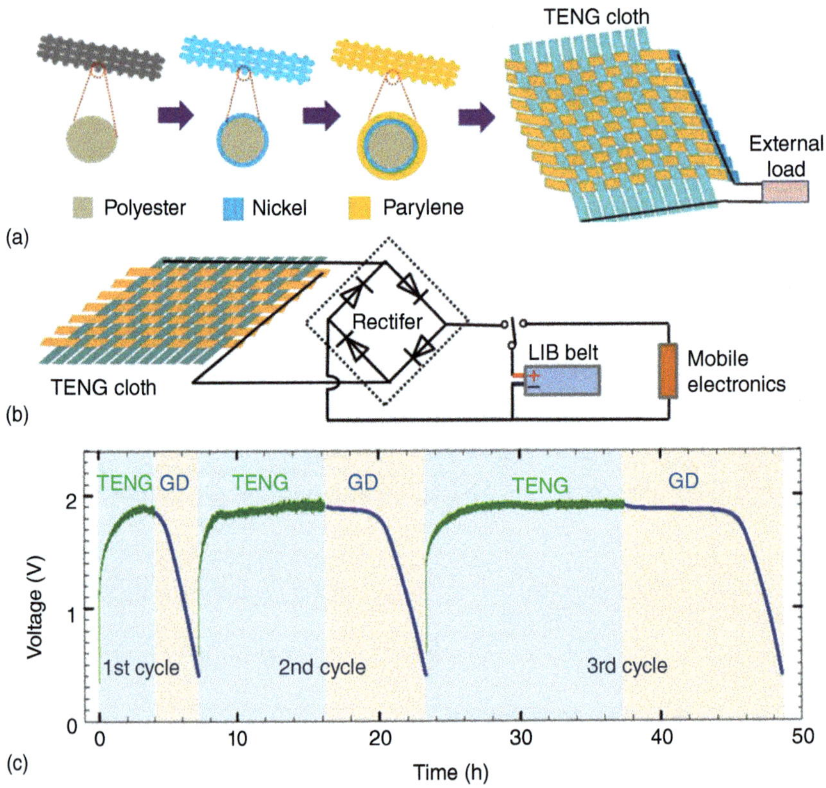

Figure 8.9 The fabrication of TENG cloth. (a) Schematic illustration of the fabrication of TENG-cloth. (b) The equivalent electrical circuit of the self-charging power system. (c) The voltage profiles of the LIB belt were charged by the TENG cloth and galvanostatically discharged at 1 μA for three cycles. Source: Pu et al. [19]/John Wiley & Sons.

to common metal substrates, and outperformed previously reported CNT and graphene papers [20]. By integrating the TENG cloth as an energy harvesting unit and the lithium-ion battery belt as an energy storage unit, the self-powering textile could be obtained (Figure 8.9b). A TENG cloth worn under the arm could be used to power the heartbeat meter strap with the lithium-ion battery belt. The heartbeat meter included a Bluetooth module, which enabled a distant smartphone to receive and analyze the physiological data during wearable use. Figure 8.9b depicts the corresponding electric circuit of the entire electronic system.

A lithium-ion battery belt was charged by the wearable TENG cloth and then discharged for three cycles at 1 μA (Figure 8.9c) [19]. During the charging process, a linear motor was used to simulate human motion at a low frequency of 0.7 Hz. After the lithium-ion battery belt is charged by the TENG cloth, the voltage quickly rises to its working voltage (1.9 V). The duration of the first, second, and third charging cycles were 4, 9, and 14 hours, respectively. When normalized by the area of the TENG cloth, the corresponding discharge capacities are 1.3, 2.8, and 4.4 mAh/m^2, respectively. The self-charging wearable textile makes it possible to reduce or even completely eliminate the battery for miniature healthcare electronics, such as heartbeat meters, pulse meters, and body temperature meters.

In 2016, Chai et al. developed an all-solid-state energy textile by integrating solar cells and supercapacitors [21]. As shown in Figure 8.10a, the evolution of integrated

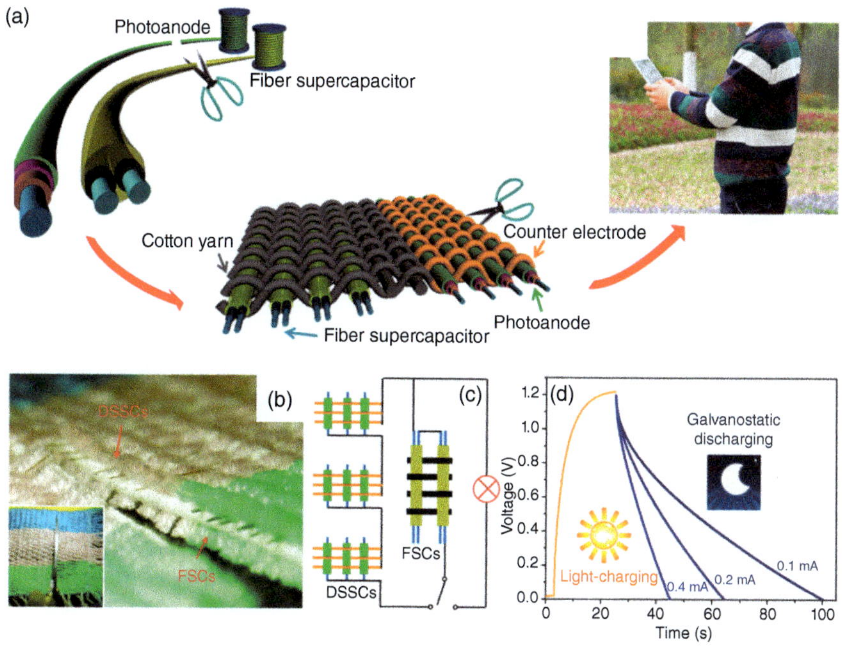

Figure 8.10 (a) Schematic illustration of the structure of an all-solid integrated energy textile. (b) and (c) Photo (b) and equivalent circuit (c) of all-solid integrated energy textile; (d) Photo-charging and discharging processes of the integrated energy textile with vary discharging currents. Source: Chai et al. [21]/American Chemical Society.

devices from thread to cloth has been exhibited. Flexible fiber photoanodes of solar cells and fiber supercapacitors were fabricated, and then, these fiber devices were woven with cotton yarns via an industrial technique (Figure 8.10b). The equivalent circuit of a photo-charging system was shown in Figure 8.10c, i.e. three strings of photoanodes were coupled in parallel for each group, and nine strings of photoanodes were divided into three groups and associated in series. The electricity generated from the DSSC module was stored in the fiber supercapacitor module. When exposed to light, the fiber supercapacitor module could be fully charged by the DSSC module to 1.2 V within 17 seconds (Figure 8.10d). The galvanostatic method showed the discharge capability. To fully discharge the FSC module, it took about 100, 78, and 45 seconds at a discharge current of 0.1, 0.2, and 0.4 mA, respectively. Interestingly, the performance in this DSSC-supercapacitor unit could be readily regulated by increasing the length of photoanodes of the DSSC module. Stylish smart energy cloth could be produced to meet the requirements for energy and aesthetic needs of wearable electronics. This work provided a universal way to fabricate wearable and tailorable textile devices.

Li et al. developed an energy ribbon by integrating a solar cell with a supercapacitor. This unveiled a highly portable self-powering system with potential applications in wearable devices and electric vehicles [13]. Figure 8.11a shows the schematic illustration of the energy textile based on the above energy ribbons. The ribbon

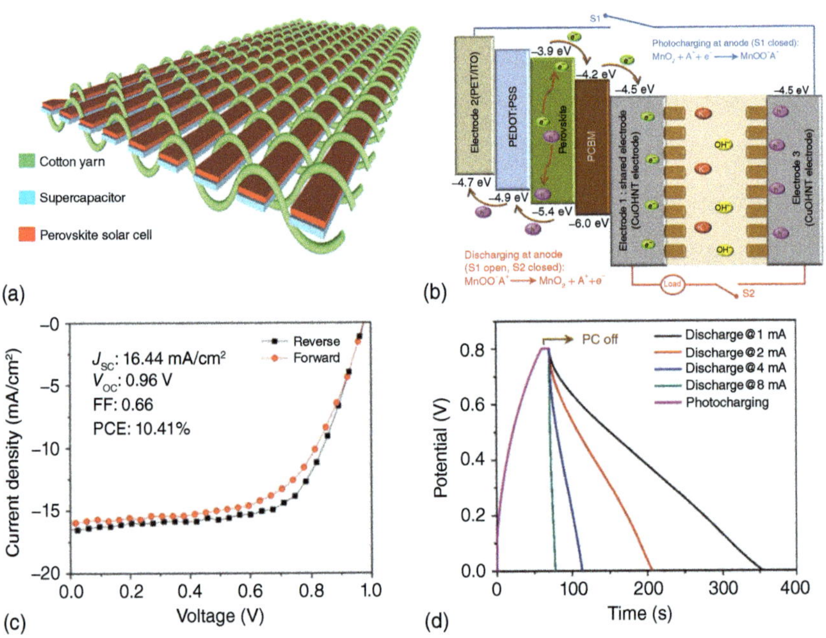

Figure 8.11 (a) Schematic illustration of transportable lightweight cloth with energy ribbon after weaving with the cotton yarn. (b) Charge transfer mechanism of the integrated device. (c) J–V curves of the best device under 1 sun illumination (AM 1.5G). (d) Charge-discharge processes of the energy ribbon. Source: Li et al. [13]/Springer Nature/CC BY 4.0.

included three parts: (i) negative electrode used as a shared electrode between the PSC and supercapacitor; (ii) positive electrode of the PSC; and (iii) positive electrode of the supercapacitor (Figure 8.11b). The solar cell exhibited a high PCE of 10.4% with small hysteresis (Figure 8.11c). To evaluate the photovoltaic conversion and energy storage performance, the energy ribbon was photo-charged using a solar simulator (AM 1.5G) for one minute to achieve an increased potential of 0.8 V, and could discharge at a variety of discharge currents (Figure 8.11d).

8.3 Summary and Perspective

In the past decade, fiber or textile electronics attracted numerous attention for their promising application in wearable electronics. With the fast development of wearable electronics, the demand for multi-functional integrated devices is increasing for better experience in practical applications. In this chapter, we introduced the recent progress of wearable integrated devices in the format of fiber or textile, which is flexible, transferable, integrative, and mechanically stable. The main achievements can be briefly summarized as follows:

(1) A variety of integration strategies have been proposed by combining one or several energy harvesting devices (such as solar, mechanical, and thermal energy) with energy storage devices (such as supercapacitors and Lithium-ion batteries).
(2) The harvested energies from vary sources can be in-situ stored into the supercapacitors or electrochemical batteries, which is a promising approach for portable electronics.
(3) Several studies have demonstrated energy conversion and storage textiles with excellent performance.

These studies provide a theoretical foundation for further innovative design of wearable integrated devices, which can lead to more diverse applications in the wearable electronic fields. However, although great achievement has been made, current integrated fiber devices are mainly focused on energy conversion and storage. More functions can be integrated into fibers in the future, such as sensing, light-emitting, healthcare, sports monitoring, etc. In addition, most of these studies are still in the stage of proof-of-concept. In addition, most reported wearable electronics are prepared by hand in a lab, which locks the development of large-scale fabrication techniques. It is important but challenging to develop a continuous technique to prepare flexible and stable fiber devices and integrate them with stable interface and low electrical resistance automatically. Meanwhile, the performances of fiber-shaped integrated devices decrease with the length and scale due to the increasing electrical resistance of fiber electrodes.

Besides, the safety issue of fiber-integrated electronics is quite important for practical applications. Some fiber-shaped solar cells or batteries contain toxic and flammable organic electrolytes, which could induce risks of fires and explosions once they are damaged during deformation. Therefore, stable and environmentally friendly electrolytes are preferred for practical application. Overall, Future work should pay more attention to new materials, structural designs, safety, and continues fabrication techniques for the commercialization of fiber or textile-integrated devices.

References

1 Miyasaka, T. and Murakami, T.N. (2004). The photocapacitor: an efficient self-charging capacitor for direct storage of solar energy. *Appl. Phys. Lett.* 85 (17): 3932–3934.

2 Murakami, T.N., Kawashima, N., and Miyasaka, T. (2005). A high-voltage dye-sensitized photocapacitor of a three-electrode system. *Chem. Commun.* 26: 3346–3348.

3 Krebs, F.C., Nielsen, T.D., Fyenbo, J. et al. (2010). Manufacture, integration and demonstration of polymer solar cells in a lamp for the "Lighting Africa" initiative. *Energy Environ. Sci.* 3 (5): 512.

4 Wee, G., Salim, T., Lam, Y.M. et al. (2011). Printable photo-supercapacitor using single-walled carbon nanotubes. *Energy Environ. Sci.* 4 (2): 413–416.

5 Chen, T., Qiu, L., Yang, Z. et al. (2012). An integrated "energy wire" for both photoelectric conversion and energy storage. *Angew. Chem. Int. Ed.* 51 (48): 11977–11980.

6 Fu, Y., Wu, H., Ye, S. et al. (2013). Integrated power fiber for energy conversion and storage. *Energy Environ. Sci.* 6 (3): 805.

7 Lee, Y.-H., Kim, J.-S., Noh, J. et al. (2013). Wearable textile battery rechargeable by solar energy. *Nano Lett.* 13 (11): 5753–5761.

8 Zhang, X., Huang, X., Li, C. et al. (2013). Dye-sensitized solar cell with energy storage function through PVDF/ZnO nanocomposite counter electrode. *Adv. Mater.* 25 (30): 4093–4096.

9 Sun, H., You, X., Deng, J. et al. (2014). A twisted wire-shaped dual-function energy device for photoelectric conversion and electrochemical storage. *Angew. Chem. Int. Ed.* 53 (26): 6664–6668.

10 Yang, Z., Deng, J., Sun, H. et al. (2014). Self-powered energy fiber: energy conversion in the sheath and storage in the core. *Adv. Mater.* 26 (41): 7038–7042.

11 Zhang, Z., Chen, X., Chen, P. et al. (2014). Integrated polymer solar cell and electrochemical supercapacitor in a flexible and stable fiber format. *Adv. Mater.* 26 (3): 466–470.

12 Wang, J., Li, X., Zi, Y. et al. (2015). A flexible fiber-based supercapacitor-triboelectric-nanogenerator power system for wearable electronics. *Adv. Mater.* 27 (33): 4830–4836.

13 Li, C., Islam, M.M., Moore, J. et al. (2016). Wearable energy-smart ribbons for synchronous energy harvest and storage. *Nat. Commun.* 7 (1): 13319.

14 Wen, Z., Yeh, M.-H., Guo, H. et al. (2016). Self-powered textile for wearable electronics by hybridizing fiber-shaped nanogenerators, solar cells, and supercapacitors. *Sci. Adv.* 2 (10): e1600097.

15 Licht, S., Hodes, G., Tenne, R. et al. (1987). A light-variation insensitive high efficiency solar cell. *Nature* 326 (6116): 863–864.

16 Sun, H., Jiang, Y., Xie, S. et al. (2016). Integrating photovoltaic conversion and lithium ion storage into a flexible fiber. *J. Mater. Chem. A* 4 (20): 7601–7605.

17 Bae, J., Park, Y.J., Lee, M. et al. (2011). Single-fiber-based hybridization of energy converters and storage units using graphene as electrodes. *Adv. Mater.* 23 (30): 3446–3449.

18 Pan, C., Guo, W., Dong, L. et al. (2012). Optical fiber-based core-shell coaxially structured hybrid cells for self-powered nanosystems. *Adv. Mater.* 24 (25): 3356–3361.

19 Pu, X., Li, L.X., Song, H.Q. et al. (2015). A self-charging power unit by integration of a textile triboelectric nanogenerator and a flexible lithium-ion battery for wearable electronics. *Adv. Mater.* 27 (15): 2472–2478.

20 Hu, L., Pasta, M., La Mantia, F. et al. (2010). Stretchable, porous, and conductive energy textiles. *Nano Lett.* 10 (2): 708–714.

21 Chai, Z., Zhang, N., Sun, P. et al. (2016). Tailorable and wearable textile devices for solar energy harvesting and simultaneous storage. *ACS Nano* 10 (10): 9201–9207.

9

Novel Interfaces in Wearable Solar Cells

9.1 Introduction

This chapter mainly introduces the key interfaces in wearable solar cells. Compared with the traditional planar configuration, solar cells in fiber and fabric formats create a variety of new interfaces that show completely different properties. The resulted electron and ion transfer at these interfaces show novel mechanisms that are important for device performances. In the beginning, the electron transfer between the internal interfaces, as a common problem in various fibrous electronic devices, is introduced. Then, we focus on the separation and transfer of photoelectrons at a variety of interfaces, as well as the developed strategies for interface optimization. The dependence of the photovoltaic performances and structural stability on interfacial properties on are mainly discussed. Finally, recent findings and understandings on multifunctional interfaces and encapsulation structure are analyzed, focusing on the challenges and future directions of interfaces toward device performance improvement.

As mentioned in previous chapters, wearable solar cells woven from fiber-shaped solar cells exhibit higher flexibility and wearability compared with conventional planar counterparts. In addition, they exhibit excellent breathability that can hardly be achieved for solar cells in any other format. Despite these overwhelming advantages, however, the power conversion efficiencies (PCEs) of wearable solar cells cannot compete with their planar counterparts. The main challenge lies in the different interfaces derived from the one-dimensional (1D) device format that requires in-depth investigation and understanding.

The interfaces are critically important for solar cells, which greatly influence the exciton formation, dissociation, and recombination processes of photovoltaic conversion. In particular, the different working mechanisms of dye-sensitized, polymer, and perovskite solar cells, as illustrated in Chapter 2, lead to a variety of different interfaces. Taking polymer solar cells as an example, several key interfaces are formed, including the interfaces between (i) electrode and electron transport material, (ii) electron transport material and photoactive material, (iii) photoactive material and hole transport material, and (iv) hole transport material and electrode (Figure 9.1a). The interfaces in fiber-shaped solar cells undoubtedly become more challenging due to the radial stacking structure on highly curved fiber electrodes

Wearable Solar Cells: Mechanisms, Materials, and Devices, First Edition.
Hao Sun, Zhibin Yang, Longbin Qiu, and Huisheng Peng.

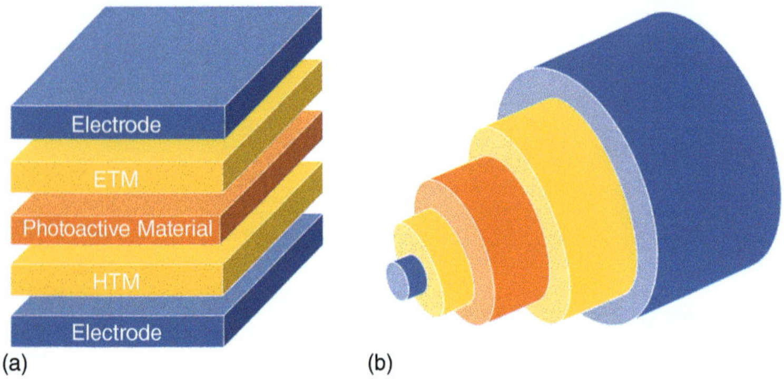

Figure 9.1 (a) Schematic illustration of the general structure of a planar polymer solar cell and a coaxial fiber-shaped polymer solar cell (b).

(Figure 9.1b). Besides, the wearable applications of fiber-shaped solar cells place higher demands on flexibility and shape adaptation, which requires highly stable interfaces under various deformations. Therefore, the design and stabilization of the interfaces in fiber-shaped solar cells are particularly important for their performance and applications.

9.2 Interfacial Charge Transfer

Efficient charge transport is highly important for all types of solar cells, which significantly affects the obtained PCEs. The charge transfer distance in fiber-shaped solar cells can be much longer than in planar solar cells, attributed to the unique 1D configuration, which demands rapid charge transport to reduce the energy loss, ensuring the output voltages and power. Therefore, great efforts have been made to optimize the charge transfer interfaces of fiber-shaped solar cells for high-performance wearable applications.

9.2.1 Charge Transfer Within Fiber Electrode

As mentioned in the previous chapters, fiber electrodes based on carbon nanomaterials, e.g. multi-walled carbon nanotubes (MWCNTs) and graphene, have been widely used in wearable solar cells due to their lightweight, processability, and excellent mechanical properties. The interfaces among numerous MWCNT bundles and graphene sheets undoubtedly play a crucial role in the charge transfer behaviors. For instance, Zhang et al. developed a post-spinning twisting process to prepare MWCNT fibers from spinnable MWCNT arrays [1]. The alignment of MWCNT could ensure the efficient charge transfer along the axial direction of MWCNTs that outperformed the isotropic MWCNT film, thus making a high electrical conductivity of 4.1×10^2 S/cm. They demonstrated the charge transfer mechanism inside the aligned MWCNT fibers via the investigation of the relationship between conductivity and temperature [2]. In general, two mechanisms were proposed to elucidate the conductive behavior among MWCNTs, i.e. variable range hopping (VRH) [3] and tunneling conduction (TC) [4] mechanisms, which are described by the following equations, respectively.

$$\sigma T^{1/2} = \exp(-B/T^{1/4}) \tag{9.1}$$

$$\sigma = \sigma_0 \exp(-A/T^{1/2}) \tag{9.2}$$

where σ is the electrical conductivity, σ_0, A, and B are constants related to the sample, and T is the absolute temperature. The conductivity data at different temperatures are brought into the equations, obtaining the plots of $\ln(\sigma T^{1/2})$ versus $T^{-1/4}$ (Figure 9.2a) and $\ln\sigma$ versus $T^{-1/2}$ (Figure 9.2b). The better linearity in Figure 9.2a indicates that the electron conduction in aligned MWCNT fibers is mainly controlled by the VRH mechanism. The slight deviation at high temperatures is considered to be affected by the electron tunneling during electron transfer.

Efforts have thus been made to acquire an in-depth understanding on the conducting mechanisms of charge transfer in aligned MWCNT fibers. For instance, the relationship between the conductivity and temperature in VRH model can be expressed as $\sigma \propto \exp(-A/T^{[1/(d+1)]})$, where d represents the dimensionality. The linear fit coefficients for the plots of $\ln\sigma$ versus $T^{-1/4}$ ($d = 3$), $T^{-1/3}$ ($d = 2$) and $T^{-1/2}$ ($d = 1$) are calculated as 0.997, 0.995, and 0.992 respectively (Figure 9.2c–e), indicating that the electron transport in MWCNT fibers conforms to a 3Dhopping mechanism [6, 7]. This implies that the electrons are not confined to the 1D channels provided by MWCNTs but hop from one position to another or from one MWCNT to another (Figure 9.2f).

Based on the 3D hopping mechanism, charges can rapidly transfer from the external MWCNTs to internal ones and then efficiently transport along the axial direction of fibers. To this end, the voids and gaps among neighboring MWCNTs will increase the charge transfer resistances. Chen et al. developed a new preparation process by which MWCNTs were first self-assembled into hollow macroscopic cylinders at high temperatures. Then, they condensed in water or ethanol for the formation of microscopic fibers [8] (Figure 9.3a). The obtained MWCNT fibers were further densified through mechanical rolling, which reduced the voids among neighboring MWCNTs and improved the charge hopping efficiencies (Figure 9.3b). This resulted in a remarkable electrical conductivity of 2.24×10^4 S/cm. Apart from interface densification via mechanical and physical strategies, Sun et al. incorporated graphene sheets into MWCNT fibers based on their π–π interaction, which enhanced the charge transfer interface [9] (Figure 9.3c). The graphene sheets served as effective bridges to facilitate the charge transport among neighboring MWCNTs (Figure 9.3d,e), which contributed to an increased electrical conductivity of more than 400 S/cm. The fiber could be further deposited with Pt nanoparticles via electrochemical deposition and used as the counter electrode in fiber-shaped dye-sensitized solar cells. The enhanced charge transfer and electrocatalytic activity resulted in high PCEs of up to 8.5%.

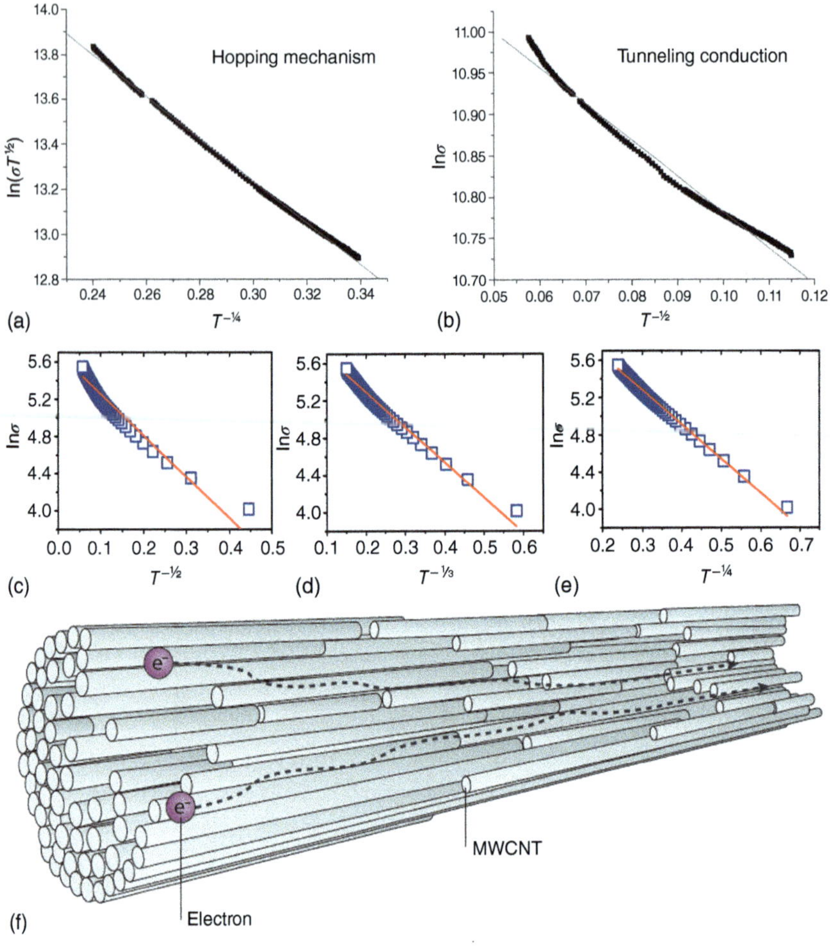

Figure 9.2 Fitting of the conductivity data of our MWCNT fibers with two different conduction mechanisms. Source: Li et al. [2], © 2007/John Wiley & Sons. (a) Plot of $\ln(\sigma T^{1/2})$ versus $T^{-1/4}$, based on the variable range hopping (VRH) mechanism. (b) Plot of $\ln\sigma$ versus $T^{-1/2}$, based on the tunneling conduction (TC) mechanism. (c, d, e) The linear fit curves for the plots of $\ln\sigma$ versus $T^{-1/d}$, where d is the dimensionality, $d = 1, 2, 3$. (f) A model of electron hopping within a conducting fiber made from aligned MWCNTs. The dashed line indicates the transport pathway of electrons. The ability of electrons to hop at the interfaces between neighboring MWCNTs is crucial in determining the electrical resistance of the fiber electrode. Fewer interfaces and shorter distances at the interfaces help to reduce the electron loss during the hopping process. Source: Sun et al. [5], © 2017/Springer Nature.

9.2.2 Charge Transport at the Active Material/Fiber Electrode Interface

The interface between active components and fiber electrodes also plays an important role in the device's performance. For instance, metal wires are difficult to form tight contact with the electron transport layer owing to the poor surface wettability and low surface area, which might cause high interfacial resistances. Moreover, the inferior flexibility of metal wires suffers from severe mechanical property mismatch

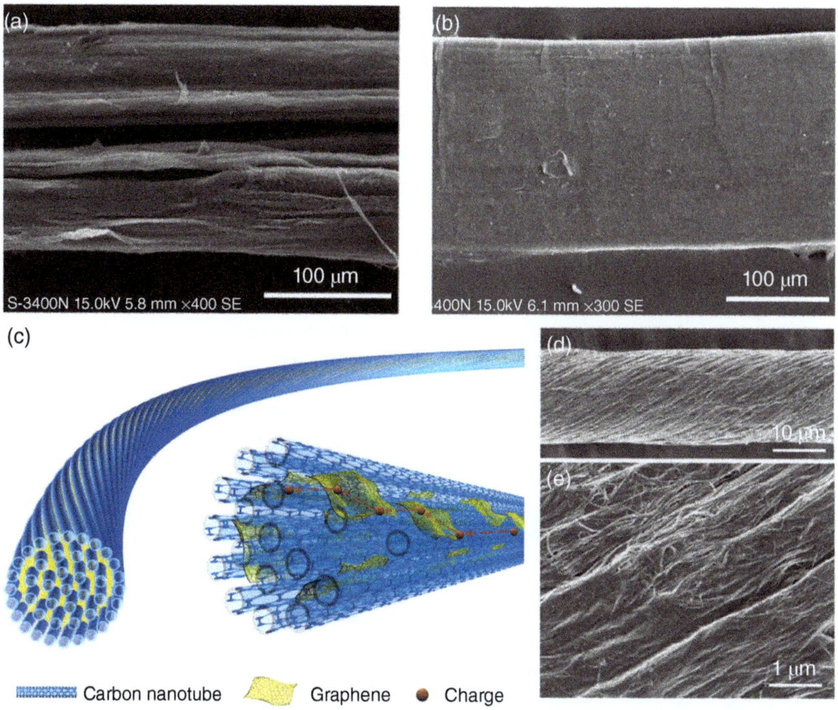

Figure 9.3 (a) Highly conductive carbon-based fibers prepared by modified methods. The SEM image of water-condensed initial fiber. Source: Wang et al. [8], © 2014/Springer Nature. (b) SEM image of the sample after rolling, showing a smooth surface morphology. (c) Schematic illustration to the structure of the graphene/MWCNT composite fiber. Source: Sun et al. [9], © 2014/John Wiley & Sons. Here, the transport routes indicated by the arrows are used to emphasize that the bridged graphene sheets favor the charge transport. (d, e) A graphene/MWCNT composite fiber at low and high magnifications, respectively.

and stress concentration under deformation, which aggravates the instability of the interface layer.

As the electron collection and charge transport material, titanium dioxide (TiO_2) is widely used in solar cells owing to its ideal forbidden bandwidth and high photoelectrochemical stability [13]. Metallic Ti wires loading TiO_2 can serve as the working electrodes of fiber-shaped dye-sensitized, polymer, and perovskite solar cells (Figure 9.4) [10–12]. However, a key challenge lies in the stable interface between the TiO_2 (either in the formats of nanoparticles or nanotubes) and Ti wires. Therefore, multilayered TiO_2 were developed, i.e. a dense and thin TiO_2 film was first coated on a Ti wire, which blocks the reverse migration of holes, and another TiO_2 layer with a higher specific surface area was introduced on the surface, which promoted the surface areas for adsorption of photoactive materials. The laminated TiO_2 structure enabled the efficient electron transfer and suppressed carrier recombination, thus improving the PCEs. Although Ti metal can maintain the interaction with TiO_2 compared to other metal materials, obtaining a uniform coating layer on Ti wire remains challenging, and the dense TiO_2 layer prepared by

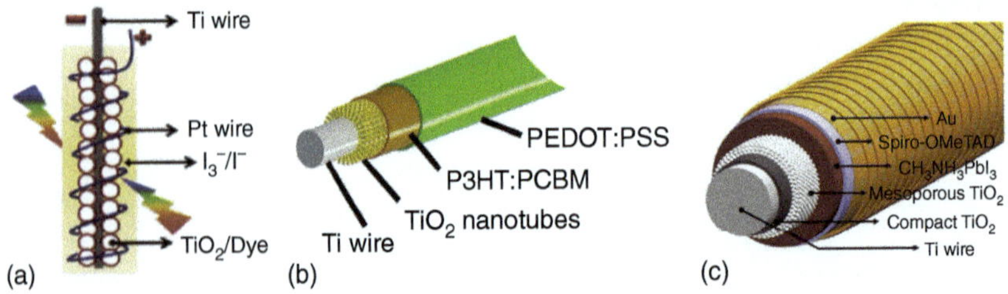

Figure 9.4 (a) Schematic structural diagrams of a fiber-shaped dye-sensitized solar cell. Source: Peng et al. [10], © 2017/Elsevier, (b) a polymer solar cell. Source: Zhang et al. [11], © 2015/John Wiley & Sons, and (c) a perovskite solar cell prepared by using TiO_2-loaded Ti wires. Source: Hu et al. [12], © 2018/Royal Society of Chemistry.

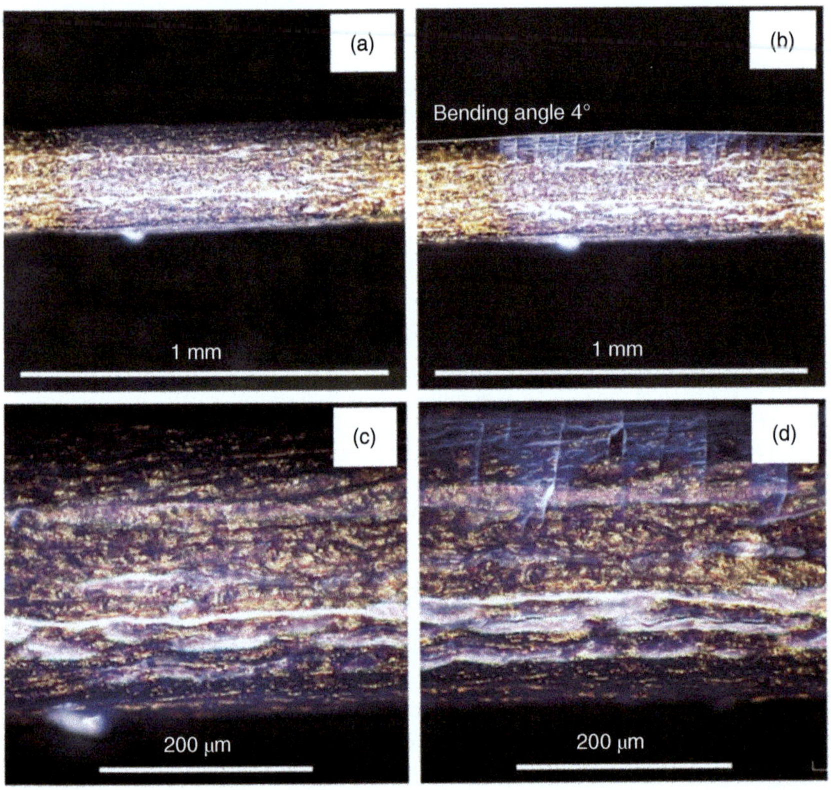

Figure 9.5 Photographs of the TiO_2 coating before (a, c) and after five cycles of bending at 4° (b, d). Cracks can be observed at the edges of the fiber. Source: Sangiorgi et.al. [14], © 2020/Elsevier.

the dip-coating method might peel off under deformation (Figure 9.5) [14]. These issues seriously affected the electron transfer properties at the interface between semiconductor layer and fiber substrate.

To solve the above issue, Song et al. performed atomic layer deposition to prepare a uniform and continuous TiO_2 film with a thickness of 15 nm [15] (Figure 9.6a).

The dense and high-quality TiO_2 film showed no negative effect on the diffusion of photogenerated electrons and effectively inhibited the carrier recombination. The flexibility and stability of the fiber-shaped dye-sensitized solar cell have been significantly improved compared to the devices without the incorporation of such a TiO_2 layer (Figure 9.6b). To overcome the complex production process of atomic layer

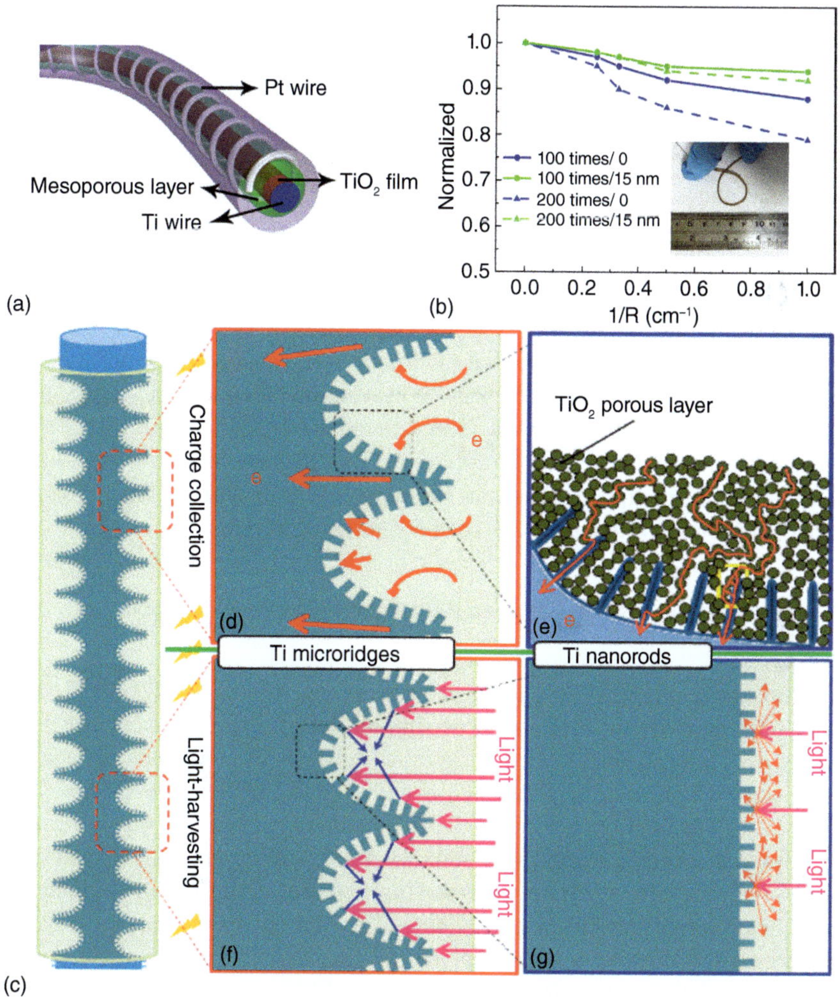

Figure 9.6 (a) Schematic of the fiber-shaped dye-sensitized solar cell structure with the conformal TiO_2 film deposited with atomic layer deposition. Source: Song et al. [15], © 2016/Elsevier. (b) Relationship between the PCEs of the fiber-shaped dye-sensitized solar cells with the atomic layer deposition TiO_2 films of different thicknesses and the $1/R$ (bending radii) after different bending times. The inset image is the optical photo of the bending solar cell. Schematic illustration of the integral Ti wire substrate (c) and the transmission path of electrons on the microridge-modified wire substrate (d, e). Schematic diagram of the light transmission paths on substrate fibers with the microridges (f) and nanorods structures (g). Source: Liu et al. [16], © 2018/Elsevier.

deposition, Lee et al. reported an anodic oxidation approach to in-situ construct micropits on Ti wires [16]. The fiber substrate with microridge/nanorod-modified hierarchical structure was further treated by acid etching. A thin TiO_2 layer could be thus formed on the surface of Ti wire, followed by $TiCl_4$ treatment and sintering (Figure 9.6c–e). It could produce efficient channels to shorten the photoelectron transfer distance from TiO_2 to Ti wire. In addition, the Ti nanorods wrapped by the dense TiO_2 layer suppressed electron recombination and extended the electron lifetime by increasing the contact area between the wire substrate and TiO_2. The novel interface also enhanced the light-harvesting capability, which concentrated light through multiple reflections inside the micropits. It utilized the light-scattering ability of nanorods to improve the utilization of incident light (Figure 9.6f–g). These significantly improved the PCE by 61.6% compared with the simple surface-modified vertical nanorod arrays and microridges.

9.3 Charge Separation

The separation efficiency of excitons at the interface can greatly affect the obtained performances of fiber-shaped solar cells. However, the highly curved surface and limited light-receiving area of fiber devices generate new challenges for charge separation. In this regard, a series of targeted methods and optimizations have been developed, aiming to suppress the recombination of electrons and holes at interfaces, which contributes to promoted PCEs of fiber-shaped solar cells.

9.3.1 Charges Separation Between Photoactive/Electron Transport Materials

The interfacial property between photoactive materials (such as dye N719) and TiO_2 is very important for efficient charge separation during photovoltaic conversion. In the early stage of fiber-shaped solar cells, the TiO_2 layer was simply coated on a metal wire, followed by the incorporation of the photoactive species [17] (Figure 9.7a). The inferior interface between the photoactive and TiO_2 layers could lead to poor device performance. In addition, thick TiO_2 layers with thicknesses above 5 μm were prone to peel off from the fiber substrate, which resulted in low PCEs or even short circuits. Therefore, TiO_2 nanoparticles, nanowires, and nanoporous networks have been explored (Figure 9.7b–d), which significantly improves interfacial contact. However, the obtained PCEs mainly were lower than 1%, originated from the defects that existed in the TiO_2 layer that plague the efficient interfacial charge separation.

Enhancement of the charge separation interface has been verified as an effective strategy. The electron transport in nanotubes might be more efficient than nanoparticles owing to the reduced boundaries among different building blocks. It is thus expected that the photovoltaic performance of TiO_2 nanotubes should exceed that of TiO_2 nanoparticles. However, TiO_2 nanotubes generally exhibited lower PCEs in conventional planar solar cells [21, 22], attributed to the dense

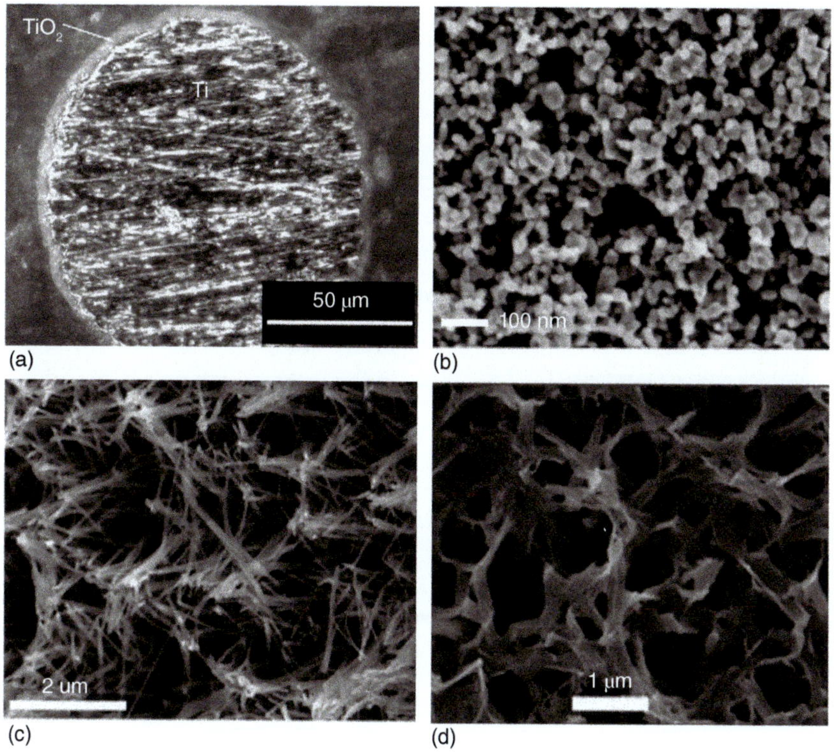

Figure 9.7 (a) SEM images of different TiO_2 morphologies constructed in fiber-shaped solar cells: dense TiO_2 layer. Source: Ramier et al. [17], © 2008/Elsevier, (b) TiO_2 nanoparticles layer. Source: Chen et al. [18], © 2012/American Chemical Society, (c) TiO_2 nanowires layer. Source: Chen et al. [19], © 2013/Royal Society of Chemistry and (d) TiO_2 nanoporous network layer. Source: Tao et al. [20], © 2014/Elsevier.

arrangement of TiO_2 nanotubes that hindered the charge separation and transfer processes (Figure 9.8a). In stark contrast, TiO_2 nanotubes grown on the surface of Ti wire form V-shaped voids (Figure 9.8b), which can facilitate dye adsorption and electrolyte penetration from the wide top void to the narrow bottom [24]. This unique interfacial configuration contributes to high PCEs of fiber-shaped solar cells.

Optimization of the photoactive material/electrolyte interface represents an effective way towards performance improvement. For instance, variations in the density and diameter of TiO_2 nanotubes can enhance electrolyte infiltration and dye adsorption. The *J*–*V* curves show the performances of fiber-shaped dye-sensitized solar cells with different lengths of TiO_2 nanotubes [23] (Figure 9.8e). When the length increased from 10 to 32 μm, the open-circuit voltage (V_{OC}) and filling factor remained almost unchanged, while the short-circuit current density increased from 4.54 to 7.29 mA/cm^2. Longer TiO_2 nanotube significantly increased the surface area of the working electrode, thereby adsorbing more dye molecules with enhanced photoelectron generation and output current densities. However, when the length

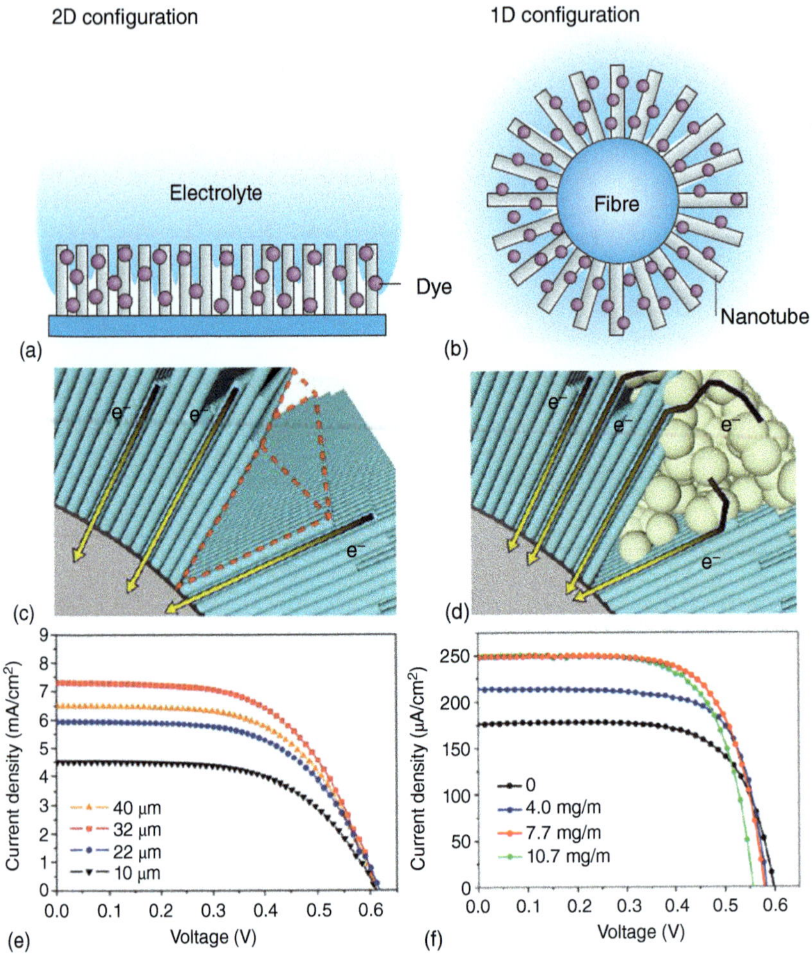

Figure 9.8 Electrolyte infiltration into dye-loaded TiO_2 nanotubes differs in two-dimensional (2D) and 1D configurations. Source: Sun et al. [5], © 2009/Springer Nature. (a) In 2D structures, electrolyte infiltration is hindered by the dense TiO_2 nanotubes. (b) In 1D arrangements, the curved fiber substrate contributes to the formation of V-shaped voids, which favor electrolyte infiltration. Schematic illustration of the electron generation and transport mechanisms in the fiber photoanodes with (c) individual TiO_2 nanotube array and (d) additionally loaded TiO_2 nanoparticles. (e) $J-V$ curves of the fiber-shaped dye-sensitized solar cells with increasing lengths of TiO_2 nanotubes in working electrodes. Source: Sun et al. [23], © 2012/Royal Society of Chemistry; (f) $J-V$ curves of the fiber-shaped dye-sensitized solar cells with different additional nanoparticle loadings.

further reached 40 μm, the charge recombination became severe due to the extended electron diffusion length, which made the short circuit current density reduce to 6.49 mA/cm^2.

Recent progress also demonstrated the optimized TiO_2 nanotube structure toward improved device performance. For instance, enhanced dye molecule distribution can be achieved by the introduction of TiO_2 nanoparticles into TiO_2 nanotube arrays. More photoelectrons can be thus transferred to the fiber electrode after being injected into TiO_2 nanoparticles (Figure 9.8c,d). The PCE and short-circuit current density (J_{SC}) were improved with increased loading of 7.7 mg/m then decreased at higher loadings (Figure 9.8f). TiO_2 nanoparticles can uniformly disperse on the top of nanotube arrays without the aggregation of large-size particles. TiO_2 nanoparticles can also improve the original rough surface morphology in other solar cells, promoting the loading of photoactive materials, accompanied by decreased charge transport resistance and enhanced light scattering.

9.3.2 Charge Separation Based on an Intermediate Layer

In addition to optimizing the interfacial properties of the semiconductor layer with photoactive materials, the construction of an intermediate layer has been used to promote efficient separation and conduction of photogenerated carriers. For dye-sensitized solar cells, the interfacial defects, e.g. dangling bonds and oxygen vacancies, lead to severe photoelectron recombination with oxidized dye molecules or I_3^- in conventional TiO_2 layers. Furthermore, the ester-like bonds are prone to form derived from the interaction between the carboxylic acid group of photoactive dye and the hydroxyl group on TiO_2, which can enhance the dye adsorption on TiO_2 by forming COO–Ti bonds. The generated H_2O as a byproduct further resulted in additional defects and interband electron trap sites on the surface of photoanode, which reduced the photoelectron collection and charge transport injection/extraction kinetics at the TiO_2/dye interface.

To address the above issues, $TiCl_4$ hydrolysis was performed to prepare an ultra-fine TiO_2 layer [25]. However, the rapid and irreversible degradation of $TiCl_4$ under environmental conditions, as well as its acidity and toxicity, has severely hindered large-scale applications [26, 27]. Therefore, it is necessary to construct an interlayer with charge-selective capability, which can simultaneously reduce the interfacial defects and interband electron trap sites in TiO_2 photoanode and improve the charge transport injection/extraction rate. An ideal photoanode interlayer should meet several criteria: (i) it should exhibit strong light absorption, good transparency and high electron injection rate; (ii) it should be able to passivate the TiO_2 photoanode surface area without severe dye coverage; (iii) it should not exhibit competitive adsorption with the dye; (iv) it should prevent dye aggregation and provide sufficient binding sites for promoted dye adsorption on TiO_2 surface; and (v) it should be ideally fabricated via a scale-up and facile method.

Previously, metal oxide interlayers have been widely used for dye-sensitized solar cells, but they usually suffer from poor light absorption and slow electron injection rate. Sönmezoglu et al. introduced deoxyribonucleic acid (DNA) molecule intercalation on the TiO_2 surface for the reduction of the defect sites in TiO_2 photoanodes [28] (Figure 9.9). This interface modification route increased the J_{SC} from 14.86 to 18.89 mA/cm^2, and promoted the V_{OC} from 745 to 816 mV. It represents a promising strategy for fiber-shaped dye-sensitized solar cells through interface optimization.

Polymer-based dipole materials have been introduced into the metal oxide electron transport layer for perovskite solar cells. The high flexibility and processability

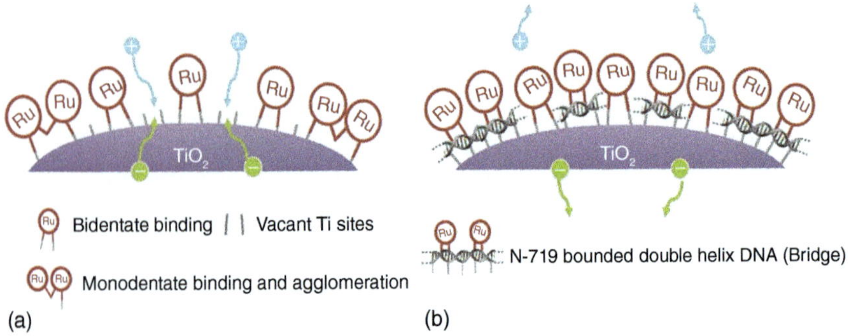

Figure 9.9 The illustration of bare TiO_2 (a) and DNA-modified TiO_2 surfaces (b) sensitized with N-719 dye. DNA molecules are expected to be well-dispersed on the TiO_2 surface, and both the phosphonic end groups of DNA and the remaining vacant sites on TiO_2 were expected to be occupied by N-719 molecules. Hence, both the bare and N-719 bounded double helix DNA molecules adsorbed onto the TiO_2 surface, acting as a bridge for the electrons to move toward the TiO_2. These new fragments inhibited the intermolecular hydrogen bonds of dye molecules, and fast charge recombination via vacant sites reduced due to the surface passivation. Source: Ateş Sönmezoğlu et al. [28], © 2016/John Wiley & Sons.

can adapt to various device structures, including the fiber shape. The permanent dipole moment induced by the polymer-based dipole layer was proved to enhance the kinetics of photoelectron injection/extraction. Furthermore, the free electron generated due to dipole formation can passivate the defect sites on metal oxide. In a recent work, the interfacial properties of TiO_2/dye molecules were significantly improved by incorporating a poly(2-ethyl-2-oxazoline) (PEOx) interlayer [29] (Figure 9.10). The PEOx interlayer improved the photoelectron injection/extraction kinetics by increasing the built-in electric field in the interfacial depletion region, thus increasing the J_{SC} of fiber-shaped dye-sensitized solar cells. In addition, the delocalized free electrons in the PEOx chain exhibited electron-doping-like behavior and passivated the defect sites of TiO_2, which reduced the electron recombination and interfacial ohmic resistance. The polymer layer raised the Fermi energy level of TiO_2 and improved the V_{OC} and fill factor. Moreover, the imine groups in the PEOx chain provided abundant dye anchoring sites that enhanced the J_{SC}.

9.4 Interface and Device Performances

9.4.1 Interfaces in Different Device Structures

As mentioned in previous chapter, fiber-shaped solar cells can be categorized as coaxial, twisting, and parallel structures. The coaxial structure used a conductive fiber electrode as the substrate, and the electron transport layer, photoactive substances, hole transport layer, and outer electrode were incorporated layer by layer onto the fiber substrate. (Figure 9.11a). The twisting structure involves two fiber electrodes with the incorporation of different active components twisted with each other (Figure 9.11b). The parallel structure involves two in-parallel placed

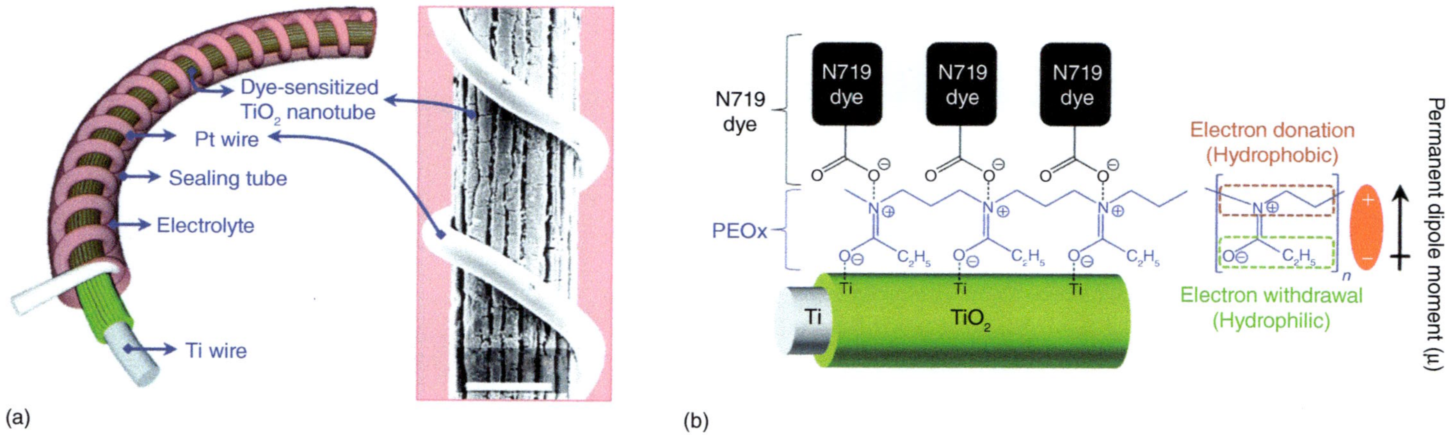

Figure 9.10 Schematic illustration of fiber-shaped dye-sensitized solar cell structure (a) and PEOx-induced permanent dipole moment binding at TiO_2/N719 dye interface (b). Schematic illustration of the expected features resulted from the PEOx interlayer: (1) to enhance the charge-light harvesting capability of fiber-shaped dye-sensitized solar cell by passivating the defect sites of TiO_2 photoanode and increasing the built-in electric field, (2) to enhance the photogenerated electron injection/extraction rates at the TiO_2/N719 dye interface, and (3) to improve dye loading on TiO_2 photoanode surface. Source: Ardhi et al. [29], © 2020/Royal Society of Chemistry.

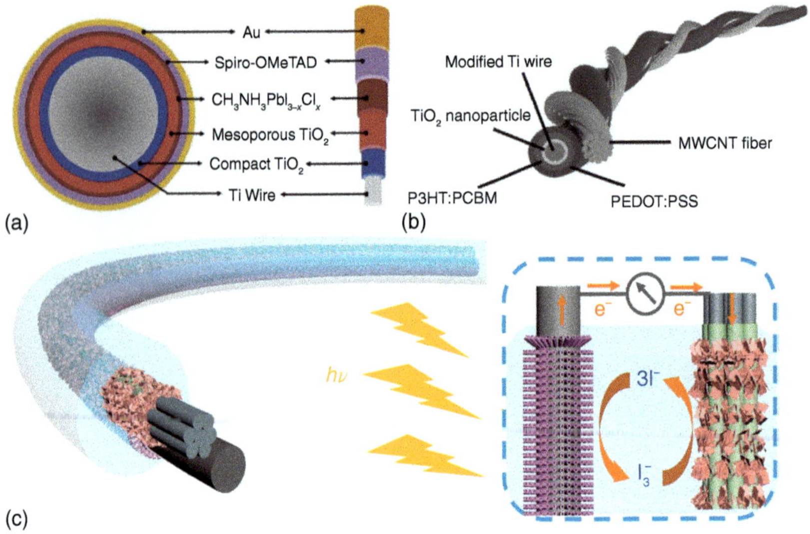

Figure 9.11 Schematic diagrams of typical fiber-shaped solar cells with (a) coaxial structure. Source: Hu et al. [30], © 2016/Royal Society of Chemistry, (b) twisting structure. Source: Zhang et al. [31], © 2014/John Wiley & Sons, and (c) parallel structure. Source: Zhang et al. [32], © 2019/American Chemical Society.

fiber electrodes separated by an electrolyte (Figure 9.11c). Except for the coaxial structure, at least two fiber electrodes are required for a fiber-shaped solar cell. For the construction of the coaxial structure, the photoactive, hole transport layer, and electron transport layers are successively incorporated on one fiber, which poses a higher fabrication requirement to ensure the uniformity and interfacial contact of multilayers. In addition, the mechanical match between different components requires serious consideration in order to avoid structural damage under deformation. As a result, the twisting structure is widely used for the production of fiber-shaped solar cells, owing to the preparation simplicity and interface stabilization.

The deposition of high-quality and uniform active materials on highly curved fiber electrodes is critical for the photovoltaic conversion performance. For example, the quality of the perovskite layer directly determines the PCE and lifetime of the fiber-shaped perovskite solar cells. Due to the smooth surface of the metal wire, the deposited perovskite crystals via the dip-coating method are generally nonuniform [30]. Figure 9.12a,b showed the uneven distribution of the perovskite layer prepared by dip-coating process using the solution comprised of CH_3NH_3I and $PbCl_2$ dissolving in N,N-Dimethylformamide (DMF) as the precursors. Therefore, a series of sedimentation methods have been developed to improve the uniformity of the perovskite layers. For instance, a porous PbO layer was incorporated into the TiO_2 nanotubes by cathodic deposition and subsequently reacted with hydroiodic acid and CH_3NH_3I sequentially to form perovskite $CH_3NH_3PbI_3$ nanocrystals with high coverage and homogeneity, which improved the PCE to 7.1% [33] (Figure 9.12c,d).

Figure 9.12 Surface morphology of perovskite film after (a) dip-coating once (b) dip-coating twice. Source: Hu et al. [30], © 2016/Royal Society of Chemistry. (c, d) SEM images for the perovskite nanocrystal layer by top and side views, respectively. Source: Qiu et al. [33], © 2016/John Wiley & Sons. The nanocrystal layer is loaded on top of the TiO_2 nanotubes.

9.4.2 Interfacial Stability and the Influence on Solar Cell Performance

The structural damage of wearable solar cells caused by external forces should be taken into serious consideration, which generally occurs at the interfaces. A typical example was provided by a fiber polymer solar cell with a twisting structure, in which a sliver wire-based counter electrode failed to create a tight contact at the interface (Figure 9.13a), and the gaps between the electrodes were detrimental for the charge transfer process, which became even worse when the device was bent. This resulted in decreased PCE from 0.48% to 0.14% after 50 bending cycles (Figure 9.13c,d) [34]. In comparison, by using a highly flexible MWCNT fiber as the counter electrode, the obtained fiber polymer solar cell exhibited promoted stability upon bending, which maintained the close contact between the two electrodes. This ensured the high PCE retention of 85% after 1000 bending cycles (Figure 9.13e).

Another sample was provided by coaxial fiber-shaped perovskite solar cells. Stainless steel wire was used as the substrate, which was sequentially coated with various active materials and finally wound with transparent MWCNT sheets as counter electrodes, which exhibited a PCE of 3.3% [35]. Still, the interfacial instability restricted the flexibility, which can only withstand 50 bending cycles (Figure 9.14a). Thus, the uniform coverage of the perovskite layer was prepared on Ti wire via a cathodic deposition solution process, which boosted the PCE

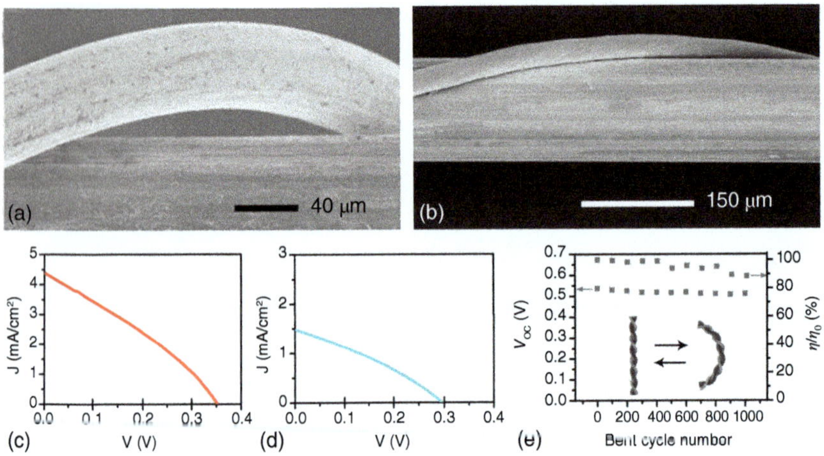

Figure 9.13 Bending properties of fiber-shaped solar cells with twisting structures. Source: Cai et al. [34], © 2012/Royal Society of Chemistry. (a, b) SEM images of fiber-shaped polymer solar cells with silver wires and MWCNT fiber as counter electrodes under different deformation states. (c, d) *J–V* curve of the fiber-shaped polymer solar cells with the silver wire as the fiber electrode before and after bending. (e) Dependence of PCE and voltage on bent cycle number. η_0 and η correspond to the efficiency before and after bending, respectively.

to 7.1% and maintained stable photovoltaic conversion capacity for 400 twisting cycles [33] (Figure 9.14b). Dong et al. further developed an interface layer with fewer defects by vapor-assisted deposition, which demonstrated a remarkable PCE of 10.79% with stable structural stability for over 500 bending cycles [36] (Figure 9.14c). Specifically, the TiO_2 dense layer, mesoporous layer, perovskite layer, spiroOMeTAD layer, and thin gold film were layer-by-layer coated on a Ti wire. The energy level of the outermost gold film was well-matched with the hole transport layer, which benefited the charge transmission at the interface and improved the V_{OC} and fill factor. Cross-sectional scanning electron microscopy (SEM) image verified the close contact between different layers with a total thickness of about 1 μm (Figure 9.14d). These stable interfaces ensure sufficient charge transfer and transport even under large deformation contributing to better device stability.

9.5 Interface of Functional Wearable Solar Cells

Wearable solar devices are also expected to be stretchable during use. Common fiber electrodes like metal wires, polymer fiber, and carbon-based fibers are generally non-stretchable. Therefore, designing functional interfaces with excellent conductivity and stability under stretching is an exciting goal for practical use. Peng et al. prepared stretchable fiber electrodes via a rotation-translation spinning method [37] (Figure 9.15a). Briefly, the two ends of an elastic fiber were fixed on two motors, and a spinnable MWCNT array was placed on a movable platform. The aligned MWCNT sheets were pulled out from the array and continuously wound on the rubber fiber,

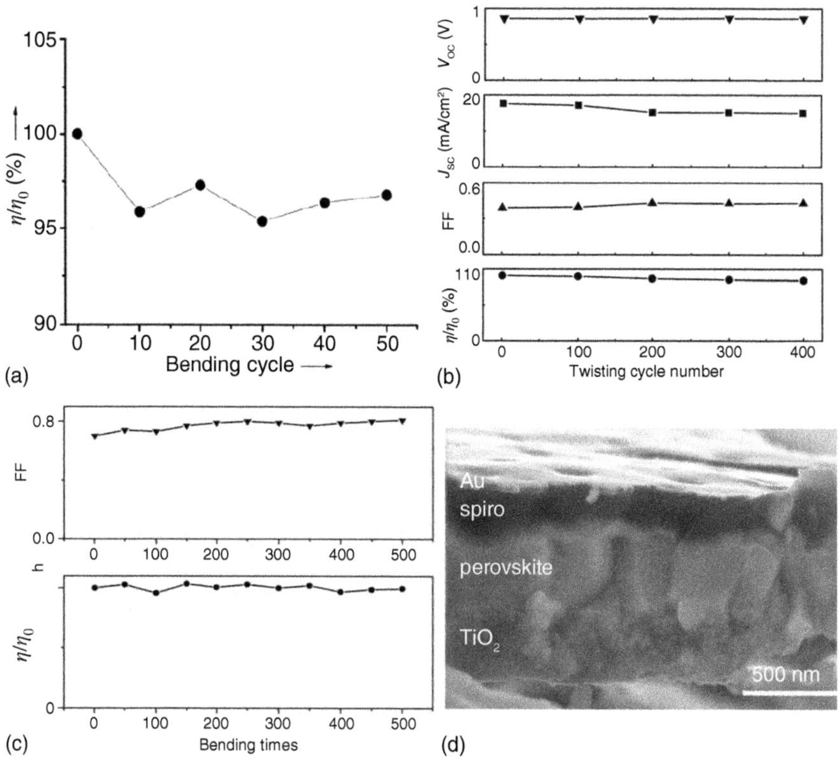

Figure 9.14 (a) Dependence of the PCE on the bending-cycle number. Source: Qiu et al. [35], © 2014/John Wiley & Sons. (b) Dependence of photovoltaic parameters on twisting cycle number. Source: Qiu et al. [33], © 2016/John Wiley & Sons. (c) Changes in fill factor and device efficiency of fiber-shaped perovskite solar cells prepared by vapor-assisted deposition in 500 consecutive bending processes. (d) Cross-sectional SEM image of fiber-shaped perovskite solar cells prepared by vapor-assisted deposition. Source: Dong et al. [36], © 2019/John Wiley & Sons.

thus achieving stretchable fiber electrodes with high conductivity and stretchability. The electrical properties of the stretchable fiber could be tuned by adjusting the helix angle and spinning thickness. The stretchable fiber as the counter electrode could deliver a decent PCE of 7.13%. By coupling with a spring-type photoanode that was also stretchable, the obtained fiber-shaped solar cell could be stretched with a 30% strain for 20 cycles (Figure 9.15b,c). This pioneer work opens a new avenue for the development of stretchable fiber-shaped solar cells that meet the practical requirements.

Undoubtedly, it is critical to integrate energy conversion and storage devices into a single fiber to enhance the integration. A pioneer work by Wang et al. integrated a nanogenerator, a supercapacitor, and a dye-sensitized solar cell on a single fiber [38] (Figure 9.16). By incorporating ZnO nanowires on Kevlar fiber with the deposition of Au, an ideal interface was created with close contact and excellent electrical conductivity. Radially grown ZnO nanowires can simultaneously serve as

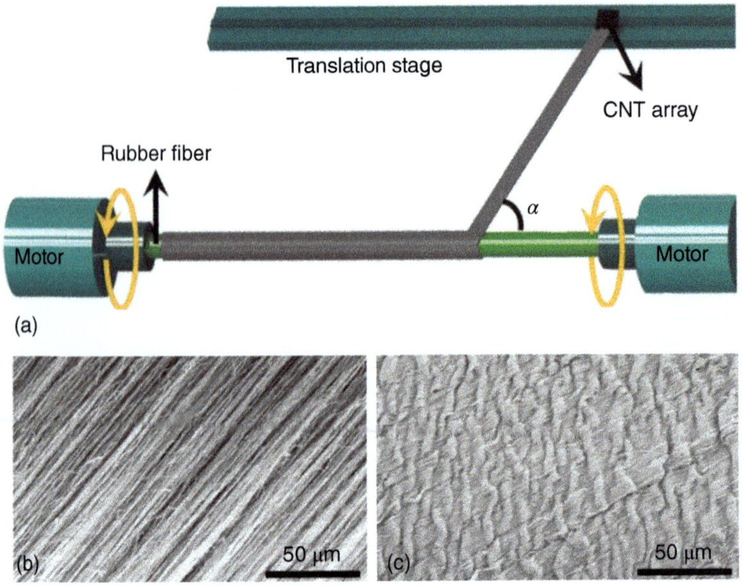

Figure 9.15 (a) Schematic illustration of the fabrication of elastic conducting fibers. High-magnification SEM images of elastic conducting fibers with a helical angle of 30° before (b) and after (c) stretch by 50%. Source: Yang et al. [37], © 2014/John Wiley & Sons.

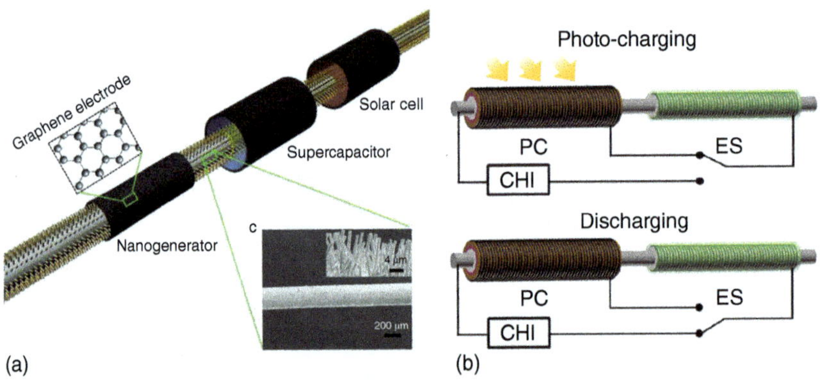

Figure 9.16 (a) Schematic of fiber-based multi-energy device comprising a nanogenerator, solar cell, and supercapacitor. ZnO nanowires are grown on a flexible, thin plastic wire coated with thin Au film. High-quality conductive and transparent graphene on Cu mesh were used as electrodes for each energy device. For solar cells, and supercapacitors, the corresponding electrolytes were filled in between ZnO nanowires and graphene electrodes. Source: Bae et al. [38], © 2011/John Wiley & Sons. (b) Schematic diagram shows the circuit connection state of the integrated polymer solar cell and electrochemical supercapacitor in the process of charging and discharging. Source: Zhang et al. [39], © 2014/John Wiley & Sons.

photoelectrodes for solar cells, piezoelectric layers for piezoelectric nanogenerators, and capacitor electrodes for supercapacitors. This fiber can simultaneously harvest mechanical and solar energy and store it in a supercapacitor. Zhang et al. used a titanium filament grown with TiO_2 nanotubes as a fiber substrate to integrate polymer solar cells and supercapacitors, and coaxially wound MWCNT films were used as counter electrodes, achieving excellent electrochemical performances and flexibility [39]. The total photoelectric conversion efficiency and energy storage efficiency of the integrated device reached 0.82% and decreased by only 10% after 1000 repeated bending cycles.

9.6 Interfaces for Encapsulation

Encapsulation of wearable solar cells in a facile and reliable way is highly important for the stability and safety of fiber-shaped solar cells. For example, effective encapsulation can prevent the leakage of toxic electrolytes, which is critical for the safety of wearable applications. In addition, the water and air-sensitive components, such as perovskite materials, are required for effective encapsulation to enable long-term and high-efficiency operation.

Compared with the planar counterparts, the encapsulation process of fiber-shaped solar cells is more challenging owing to their high curvature interfaces. The wearable applications also put forward stringent requirements on the properties of the encapsulation materials. In the past decades, a variety of high-performance materials have been developed for encapsulation, which will be mainly introduced in this section.

9.6.1 Polymer Tubes

Transparent glassy tubes were used for encapsulation of fiber-shaped solar cells at the early stage [40–42], which unfortunately sacrificed the device's flexibility. As a result, polymer tubes with better flexibility have been employed to accommodate and seal the fiber-shaped solar cells [37, 43–45]. For example, the TiO_2/Ti photoanode and Pt-modified stainless steel counter electrode were injected with liquid electrolyte in a transparent polymer tube with the two ends sealed with wax (Figure 9.17a) [43]. The resulted dye-sensitized solar cell could be bent up to 180° without performance decay (Figure 9.17b,c).

To further accommodate the frequent strain and stress in wearable applications, a stretchable transparent polyethylene (PE) tube served as the encapsulation material to improve the stretchability of the fiber-shaped solar cells [37]. An elastic counter electrode was wound by a spring-like TiO_2/Ti photoanode, and they were inserted into a PE tube followed by infiltration of liquid electrolyte. The obtained solar cells were woven into various textiles, which could be stretched to a certain strain along different directions with stable photovoltaic performances (Figure 9.17d,e). The PCE could be maintained at 90% of the original value after stretching and releasing for 50 cycles at a strain of 20% (Figure 9.17f).

9.6.2 Gel Electrolytes

The use of gel electrolytes can further improve the structure and photovoltaic conversion stability of fiber-shaped solar cells, which have thus been actively pursued in recent years. For instance, poly(vinylidene fluoride-*co*-hexafluoropropene)

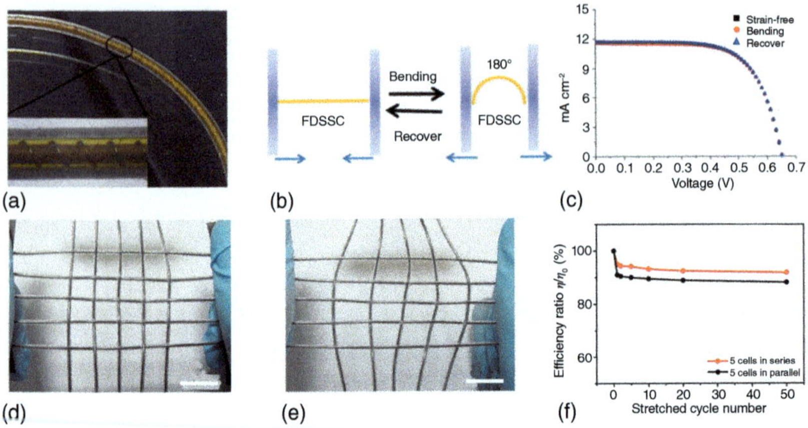

Figure 9.17 (a) Optical image of the flexible fiber-shaped solar cell. (b) Diagram and (c) J–V curves of solar cell in strain-free or bending conditions. Source: Fu et al. [43], © 2013/Elsevier. Photographs of the stretchable solar cell textile before (d) and after stretching (e), respectively. (f) PCE at different stretched cycle numbers at a strain of ~20%. η_0 and η correspond to PCE before and after stretching, respectively. Source: Yang et al. [37], © 2014/Elsevier. Scale bar: (d) and (e), 2 cm.

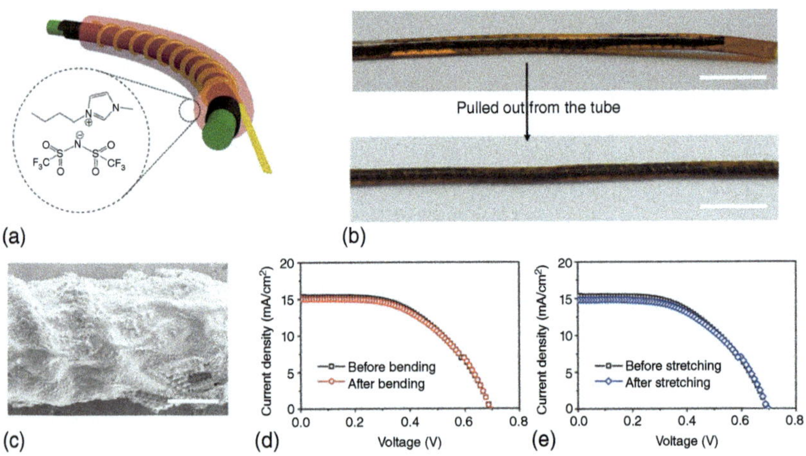

Figure 9.18 (a) Schematic illustration of the structure of the fiber-shaped solar cell with IL gel electrolyte as encapsulation. (b) The fiber-shaped solar cell is pulled from the polymer tube. (c) SEM images of IL gel electrolyte firmly attaching on the electrodes. J–V curves of the DSSC with gel electrolyte after bending (d) or stretching to 25% (e) for 10 cycles. Source: Li et al.[46], © 2013/John Wiley & Sons. Scale bar: (b), 2 cm; (c), 500 μm.

(PVDF-HFP) copolymer and 1-butyl-3-methylimidazolium bis(trifluoromethanesulfonyl) imide (BMImTFSI) ionic liquid has been chosen to prepare gel electrolyte (Figure 9.18a) [46]. A TiO_2/Ti photoanode and stretchable counter electrode were inserted into a polymer tube, and then the heated gel electrolytes were injected at a temperature above the gelation point. After cooling down to room temperature, the obtained fiber-shaped solar cells could be pulled out of the plastic tube

(Figure 9.18b). Long-term stability was achieved for a solar cell with maintaining 90% of the original PCE after 30 days, which benefited from the high stability of gel electrolytes over time, temperature, and moisture. In contrast, the control samples with liquid electrolytes could only maintain 20% of the original PCE. In addition, the good contact between the gel electrolyte and electrodes contributed to improved structural stability under various deformations (Figure 9.18c). The resulted solar cells demonstrated slightly fluctuated PCEs after bending and stretching by 25% for ten cycles, respectively (Figure 9.18d,e).

Moreover, the gel electrolyte configuration of the fiber-shaped solar cell was examined and analyzed by Xiao et al. [47]. Higher PVDF-HFP concentration in the gel electrolyte could produce a more crosslinked network structure, which promoted charge transfer with suppressed recombination. An optimal PCE of 6.02% could be obtained using 9% PVDF-HFP. In addition, the gel electrolyte was found to stably covered the spiral-shaped photoanode (Figure 9.19a). Interestingly, different morphologies were observed for the gel electrolytes at the outer and inner sides of the photoanode. The outer part showed a concrete structure with limited pores (Figure 9.19b), which could block the leakage and evaporation of the liquid

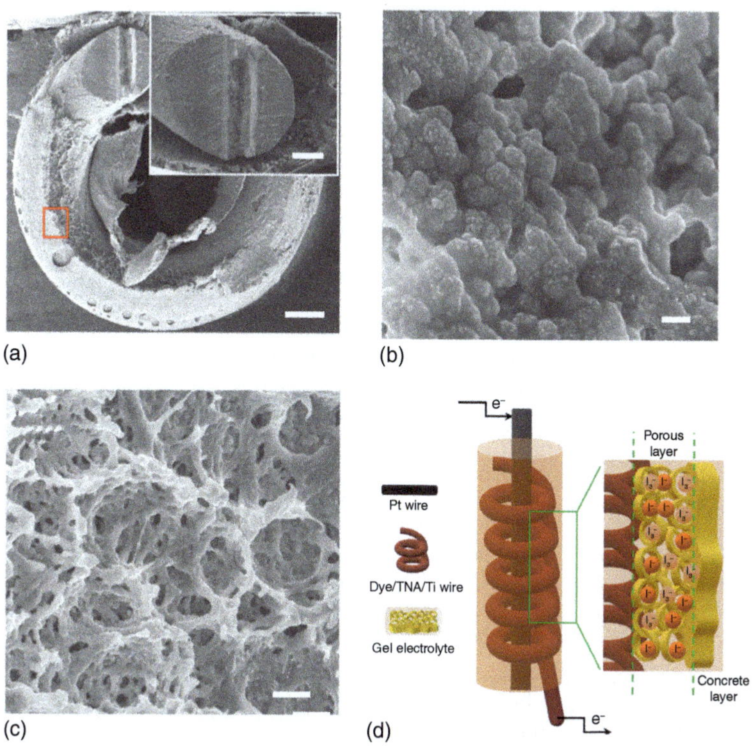

Figure 9.19 (a) The spiral-shaped photoanode with gel electrode covered. The SEM images of the outer (b) and inner (c) parts of the gel electrolyte. (d) The illustration of the fiber-shaped solar cell with the gel electrolyte. Source: Xiao et al. [47], © 2015/Elsevier. Scale bar: (a), 200 μm and inset, 100 μm; (b), 2 μm; (c) 10 μm.

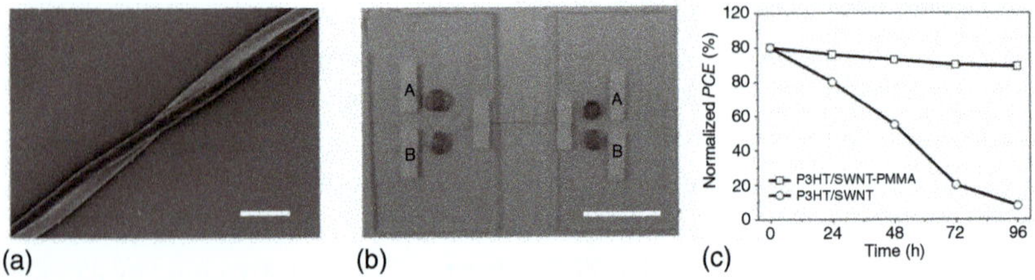

Figure 9.20 (a) SEM images of the twisted $CH_3NH_3PbI_{3-x}Cl_x$/MWCNT photoanode and pristine MWCNT fiber. (b) Optical photograph of the fiber-shaped solar cell with PMMA encapsulating layer. (c) The stability of the fiber-shaped solar cell with and without PMMA encapsulating layers. Source: Li et al. [49], © 2015/John Wiley & Sons. Scale bar: (a), 100 μm; (b), 10 mm.

in gel electrolyte. Numerous pores and strong networks were obtained for the inner part, leading to the accumulation of the liquid electrolyte for rapid charge transfer (Figure 9.19c). This unique configuration significantly improved the photovoltaic conversion performance and long-term stability of the resulted fiber-shaped solar cells (Figure 9.19d).

9.6.3 Polymer Coatings

Direct coating insulating polymer on the electrodes can greatly simplify the sealing process. For example, poly(methyl methacrylate) (PMMA), a hydrophobic material, can be used to inhibit the intrusion of moisture into the perovskite layer[48]. Both $CH_3NH_3PbI_{3-x}Cl_x$/MWCNT photoanode and pristine MWCNT fiber were twisted into a double-twisted architecture (Figure 9.20a), followed by coating PMMA as an encapsulation layer (Figure 9.20b) [49]. With the PMMA modification layer, the PCE could maintained at 89% of the original value after 96 hours of air exposure, while the PCE based on pristine MWCNT dramatically decreased to 8% of the original value without the PMMA protection layer (Figure 9.20c). The resulted fiber-shaped solar cells could tolerate bending with various angles ranging from 0° to 180° without significant PCE decay during 1000 cycles, which indicated that the PMMA layer promoted the structural stability and flexibility of the resulted solar cells.

9.7 Summary and Outlook

The interfaces of fiber-shaped solar cells have been intensively investigated and optimized in the past decade, with significant improvements on mechanism understanding and device performance. However, the highly curved morphology derived from the 1D fiber structure is prone to result in challenging interfaces. For example, the lattice defects caused by uneven and uncontrolled growth of photoactive materials or due to the relatively loose interlayer, especially between twisted fiber electrodes, might result in poor interfacial contact or even structure displacement under deformations. As a result, considerable efforts have been made to regulate the key microstructures e.g. nanorods and nanowires, combined with new processing techniques, e.g. atomic layer deposition (ALD), magnetron sputtering (MS) to improve the quality, uniformity, and stability. Furthermore, an in-depth understanding of charge transport, separation, and recombination has contributed to impressive improvements in PCEs and lifetime. The introduction of the interface layer, such as transition metal oxides and graphene, can effectively mediate the energy barrier between the electrode and photoactive layer, which effectively facilitates carrier transport for higher charge collection efficiencies. Heteroatom doping, monolayer passivation, and material modification can be further used to passivate the internal defects. These strategies are expected to fully meet the requirements in wearable applications with desirable PCEs and structural stability. In addition, simplification of the device structure, such as the exploration of planar heterojunction structure in perovskite solar cells, represents a promising direction to pursue.

It is worth noting that the interfaces created during the weaving process of fiber-shaped solar cells have been rarely investigated, which remains a grand challenge for scale-up production and application. The weaving process involves stress concentration areas, which might cause structural damage or even device failure. To this end, it is highly desirable to focus on the dynamic interface stability during the weaving and wearing process. Experimental and theoretical analysis should be involved to acquire and optimize the above-mentioned interfaces. In addition, the weaving process will inevitably affect the light capture and utilization originating from the light shielding among the neighboring fibers. Therefore, developing transparent materials for fiber-shaped solar cells can be an exciting idea to improve light-harvesting capacity towards higher PCEs.

Last but not least, developing high-performance encapsulation materials with high water/air resistance and mechanical stability is highly important, which can ensure the safety and durability of the obtained devices. Especially for wearable solar cells, safety is the priority for practical applications. Reliable encapsulation with stable interfaces can prevent the erosion of photoactive materials by oxygen and water. Besides, for polymer and perovskite solar cells, in-situ construction of passivation layers can be an effective strategy for the improvement of long-term stability. For dye-sensitized solar cells, the development of solid-state electrolytes can be a promising way to enhance the device stability. Overall, the endeavors on new materials and technologies are highly demanded in future studies to fully unify photovoltaic performance, flexibility performance, and long-term stability.

References

1 Zhang, X., Li, Q., Tu, Y. et al. (2007). Strong carbon-nanotube fibers spun from long carbon-nanotube arrays. *Small* 3 (2): 244–248.

2 Li, Q.W., Li, Y., Zhang, X.F. et al. (2007). Structure-dependent electrical properties of carbon nanotube fibers. *Adv. Mater.* 19 (20): 3358–3363.

3 Zeller, H.R. (1972). Electrical conductivity of one-dimensional conductors. *Phys. Rev. Lett.* 28 (22): 1452–1455.

4 Mataré, H.F. (1984). Carrier transport at grain boundaries in semiconductors. *J. Appl. Phys.* 56 (10): 2605–2631.

5 Sun, H., Zhang, Y., Zhang, J. et al. (2017). Energy harvesting and storage in 1D devices. *Nat. Rev. Mater.* 2 (6): 17023.

6 Peng, H., Sun, X., Cai, F. et al. (2009). Electrochromatic carbon nanotube/polydiacetylene nanocomposite fibres. *Nat. Nanotechnol* 4 (11): 738–741.

7 Peng, H., Jain, M., Peterson, D.E. et al. (2008). Composite carbon nanotube/silica fibers with improved mechanical strengths and electrical conductivities. *Small* 4 (11): 1964–1967.

8 Wang, J.N., Luo, X.G., Wu, T. et al. (2014). High-strength carbon nanotube fibre-like ribbon with high ductility and high electrical conductivity. *Nat. Commun.* 5 (1): 3848.

9 Sun, H., You, X., Deng, J. et al. (2014). Novel graphene/carbon nanotube composite fibers for efficient wire-shaped miniature energy devices. *Adv. Mater.* 26 (18): 2868–2873.

10 Peng, M., Dong, B., Cai, X. et al. (2017). Organic dye-sensitized photovoltaic fibers. *SoEn* 150: 161–165.

11 Zhang, Z., Yang, Z., Deng, J. et al. (2015). Stretchable polymer solar cell fibers. *Small* 11 (6): 675–680.

12 Hu, H., Dong, B., Chen, B. et al. (2018). High performance fiber-shaped perovskite solar cells based on lead acetate precursor. *Sustainable Energy Fuels* 2 (1): 79–84.

13 Que Yaping, W.J., Linhua, H., and Songyuan, D. (2016). Applications of titanium dioxide in perovskite solar cells. *Prog. Chem.* 28 (1): 40–50.

14 Sangiorgi, N., Sangiorgi, A., Dessì, A. et al. (2020). Improving the efficiency of thin-film fiber-shaped dye-sensitized solar cells by using organic sensitizers. *Sol. Energy Mater. Sol. Cells* 204: 110209.

15 Song, W., Wang, H., Liu, G. et al. (2016). Improving the photovoltaic performance and flexibility of fiber-shaped dye-sensitized solar cells with atomic layer deposition. *Nano Energy* 19: 1–7.

16 Liu, G., Wang, M., Wang, H. et al. (2018). Hierarchically structured photoanode with enhanced charge collection and light harvesting abilities for fiber-shaped dye-sensitized solar cells. *Nano Energy* 49: 95–102.

17 Ramier, J., Da Costa, N., Plummer, C.J.G. et al. (2008). Cohesion and adhesion of nanoporous TiO_2 coatings on titanium wires for photovoltaic applications. *Thin Solid Films* 516 (8): 1913–1919.

18 Chen, T., Qiu, L., Cai, Z. et al. (2012). Intertwined aligned carbon nanotube fiber based dye-sensitized solar cells. *Nano Lett.* 12 (5): 2568–2572.

19 Chen, L., Zhou, Y., Dai, H. et al. (2013). Fiber dye-sensitized solar cells consisting of TiO_2 nanowires arrays on Ti thread as photoanodes through a low-cost, scalable route. *J. Mater. Chem. A* 1 (38): 11790–11794.

20 Tao, H., Fang, G.-j., Ke, W.-j. et al. (2014). In-situ synthesis of TiO_2 network nanoporous structure on Ti wire substrate and its application in fiber dye sensitized solar cells. *J. Power Sources* 245: 59–65.

21 Yella, A., Lee, H.-W., Tsao, H.N. et al. (2011). Porphyrin-sensitized solar cells with cobalt (II/III)-based redox electrolyte exceed 12 percent efficiency. *Science* 334 (6056): 629–634.

22 Zhu, K., Neale, N.R., Miedaner, A. et al. (2007). Enhanced charge-collection efficiencies and light scattering in dye-sensitized solar cells using oriented TiO_2 nanotubes arrays. *Nano Lett.* 7 (1): 69–74.

23 Sun, H., Li, H., You, X. et al. (2014). Quasi-solid-state, coaxial, fiber-shaped dye-sensitized solar cells. *J. Mater. Chem. A* 2 (2): 345–349.

24 Chen, T., Qiu, L., Kia, H.G. et al. (2012). Designing aligned inorganic nanotubes at the electrode interface: towards highly efficient photovoltaic wires. *Adv. Mater.* 24 (34): 4623–4628.

25 Yang, S.-C., Yang, D.-J., Kim, J. et al. (2008). Hollow TiO_2 hemispheres obtained by colloidal templating for application in dye-sensitized solar cells. *Adv. Mater.* 20 (5): 1059–1064.

26 Chen, T., Hu, W., Song, J. et al. (2012). Interface functionalization of photoelectrodes with graphene for high performance dye-sensitized solar cells. *Adv. Funct. Mater.* 22 (24): 5245–5250.

27 Knorr, F.J., Zhang, D., and McHale, J.L. (2007). Influence of $TiCl_4$ treatment on surface defect photoluminescence in pure and mixed-phase nanocrystalline TiO_2. *Langmuir* 23 (17): 8686–8690.

28 Ateş Sönmezoğlu, Ö., Akın, S., Terzi, B. et al. (2016). An effective approach for high-efficiency photoelectrochemical solar cells by using bifunctional dna molecules modified photoanode. *Adv. Funct. Mater.* 26 (47): 8776–8783.

29 Ardhi, R.E.A., Tran, M.X., Wang, M. et al. (2020). Chemically tuned, bi-functional polar interlayer for TiO_2 photoanodes in fibre-shaped dye-sensitized solar cells. *J. Mater. Chem. A* 8 (5): 2549–2562.

30 Hu, H., Yan, K., Peng, M. et al. (2016). Fiber-shaped perovskite solar cells with 5.3% efficiency. *J. Mater. Chem. A* 4 (10): 3901–3906.

31 Zhang, Z., Yang, Z., Wu, Z. et al. (2014). Weaving efficient polymer solar cell wires into flexible power textiles. *Adv. Energy Mater.* 4 (11): 1301750.

32 Zhang, J., Wang, Z., Li, X. et al. (2019). Flexible platinum-free fiber-shaped dye sensitized solar cell with 10.28% efficiency. *ACS Appl. Energy Mater.* 2 (4): 2870–2877.

33 Qiu, L., He, S., Yang, J. et al. (2016). Fiber-shaped perovskite solar cells with high power conversion efficiency. *Small* 12 (18): 2419–2424.

34 Cai, X., Hou, S., Wu, H. et al. (2012). All-carbon electrode-based fiber-shaped dye-sensitized solar cells. *Phys. Chem. Chem. Phys.* 14 (1): 125–130.

35 Qiu, L., Deng, J., Lu, X. et al. (2014). integrating perovskite solar cells into a flexible fiber. *Angew. Chem. Int. Ed.* 53 (39): 10425–10428.

36 Dong, B., Hu, J., Xiao, X. et al. (2019). High-efficiency fiber-shaped perovskite solar cell by vapor-assisted deposition with a record efficiency of 10.79%. *Adv. Mater. Technol.* 4 (7): 1900131.

37 Yang, Z., Deng, J., Sun, X. et al. (2014). Stretchable, wearable dye-sensitized solar cells. *Adv. Mater.* 26 (17): 2643–2647.

38 Bae, J., Park, Y.J., Lee, M. et al. (2011). Single-fiber-based hybridization of energy converters and storage units using graphene as electrodes. *Adv. Mater.* 23 (30): 3446–3449.

39 Zhang, Z., Chen, X., Chen, P. et al. (2014). Integrated polymer solar cell and electrochemical supercapacitor in a flexible and stable fiber format. *Adv. Mater.* 26 (3): 466–470.

40 Liu, Z. and Misra, M. (2010). Dye-sensitized photovoltaic wires using highly ordered TiO_2 nanotube arrays. *ACS Nano* 4 (4): 2196–2200.

41 Hou, S., Cai, X., Fu, Y. et al. (2011). Transparent conductive oxide-less, flexible, and highly efficient dye-sensitized solar cells with commercialized carbon fiber as the counter electrode. *J. Mater. Chem.* 21 (36): 13776–13779.

42 Fu, Y., Lv, Z., Wu, H. et al. (2012). Dye-sensitized solar cell tube. *Sol. Energy Mater. Sol. Cells* 102: 212–219.

43 Fu, Y., Peng, M., Lv, Z. et al. (2013). A novel low-cost, one-step and facile synthesis of TiO_2 for efficient fiber dye-sensitized solarcells. *Nano Energy* 2 (4): 537–544.

44 Gu, X.Y., Chen, E.Z., Wei, K. et al. (2021). Design of highly ordered hierarchical catalytic nanostructures as high-flexibility counter electrodes for fiber-shaped dye-sensitized solar cells. *Appl. Phys. Lett.* 118 (5): 053102.

45 Yang, Z., Sun, H., Chen, T. et al. (2013). Photovoltaic wire derived from a graphene composite fiber achieving an 8.45% PCE. *Angew. Chem.* 125 (29): 7693–7696.

46 Li, H.P., Guo, J.J., Sun, H. et al. (2015). Stable hydrophobic ionic liquid gel electrolyte for stretchable fiber-shaped dye-sensitized solar cell. *ChemNanoMat* 1 (6): 399–402.

47 Xiao, B.-C. and Lin, L.-Y. (2020). Tuning electrolyte configuration and composition for fiber-shaped dye-sensitized solar cell with poly (vinylidene fluoride-co-hexafluoropropylene) gel electrolyte. *J. Colloid Interface Sci.* 571: 126–133.

48 Habisreutinger, S.N., Leijtens, T., Eperon, G.E. et al. (2014). Carbon nanotube/polymer composites as a highly stable hole collection layer in perovskite solar cells. *Nano Lett.* 14 (10): 5561–5568.

49 Li, R., Xiang, X., Tong, X. et al. (2015). Wearable double-twisted fibrous perovskite solar cell. *Adv. Mater.* 27 (25): 3831–3835.

10

Textile Solar Cells

10.1 Introduction

With the development of the Internet of Things, daily health monitoring, and the concept of metaverse, wearable electronics are growing at a rapid pace [1–3]. The operation of these wearable electronics relies on a continuous and stable energy supply, which requires sufficient energy harvesting from the environment [4–6]. In previous chapters, the working principles of photovoltaic conversion and various fiber-shaped solar cells have been comprehensively introduced. It is known that the power requirements of these wearable electronics range from several microwatts to milliwatts, which can be achieved by solar cells with a rational increase of the light illumination areas. Figure 10.1 shows the development requirement of wearable solar cells constructed from fiber-shaped solar cells [7]. Due to the challenge in high-quality and uniform deposition of functional and active layers on the fiber substrate, the power conversion efficiencies (PCEs) require further improvement. However, increasing the light illumination area could also fulfill the requirement of the power output. As discussed in previous chapters, the achieving of long devices without serious performance decline relies on the development of highly conductive fiber electrodes and high-quality deposition of the active layers. To this end, wearable photovoltaic textile with large light illumination areas represents a promising strategy for powering wearable electronics, even under weak light for indoor applications.

One facile strategy lies in the weaving of fiber-shaped solar cells into a flexible textile that becomes one part of smart textiles [8, 9]. In this chapter, the weaving methods that obviously differ from the weaving of traditional chemical fibers have been introduced. The photovoltaic and mechanical properties of the obtained textile solar cells are highlighted. Particularly, an emphasis has been made to compare the wearability of textile solar cells and flexible thin-film solar cells. Efforts that have been made to solve the current challenges of textile solar cells are discussed thoroughly. Some practical applications are particularly summarized, including flexible electronics, smart textiles, and biomedical science.

Wearable Solar Cells: Mechanisms, Materials, and Devices, First Edition.
Hao Sun, Zhibin Yang, Longbin Qiu, and Huisheng Peng.

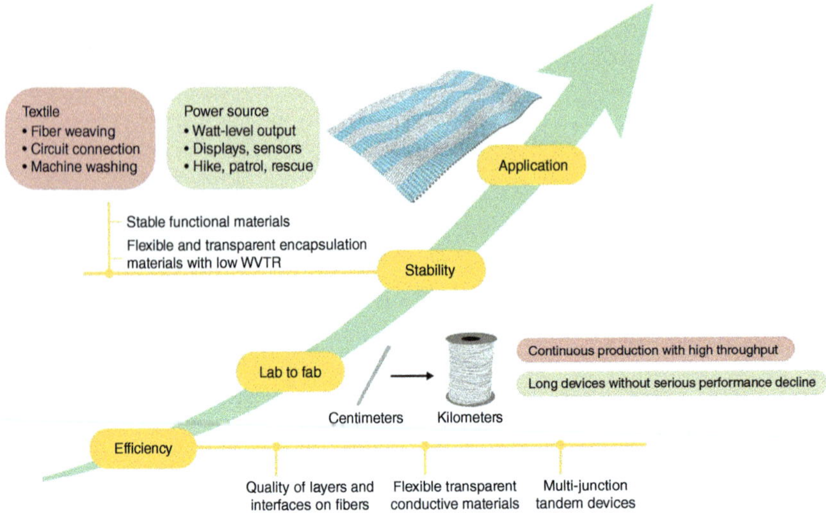

Figure 10.1 The development requirement of wearable textile solar cells are constructed from fiber solar cells. (i) Composed from fiber components with high-quality deposition of functional layers and active layers around the fiber surface. (ii) The scalable fabrication of lab scale centimeters to industrial scale of kilometers. (iii) The wearing and weaving stability, breathability, and washability of flexible textile solar cells. (iv) The integration and application of power-supply textile based on textile solar cells. Source: Reproduced with permission Zhai et al. [7] © 2022, Springer Nature.

10.2 Textile Solar Cell Structures

There are weaving, knitting, and sewing techniques for the construction of the garment and cloth. Several structures have thus been explored for the fabrication of textile solar cells (Figure 10.2). Textiles that are compatible with photovoltaics could be mainly divided into two categories. One is the direct attachment or deposition of thin film solar cells onto the textile, and the other is the weaving of fiber solar cells into a textile. The weaving process could be conducted using industrial shuttle-flying machines [10–14]. However, there is no knitting process reported so far for textile solar cells, possibly due to the mismatched mechanical properties.

Textile solar cells can be mainly categorized into five kinds, as illustrated in Figure 10.2. The first one could be obtained by stitching or sewing a planar-type flexible solar cell to a textile substrate (Figure 10.2a). The second one is the weaving of strip-shaped planar-type solar cells with special design into a textile (Figure 10.2b). The third one involves the deposition of active layers directly on the textile/fabric's electrodes to construct flexible solar cells (Figure 10.2c). The fourth one depends on the incorporation of fiber solar cells into a textile by either weaving or sewing process (Figure 10.2d). And the last one involves an interlaced fiber electrode structure that is specifically developed for textile solar cells (Figure 10.2e). We will introduce and compare the fabrication and performance of these textile solar cells.

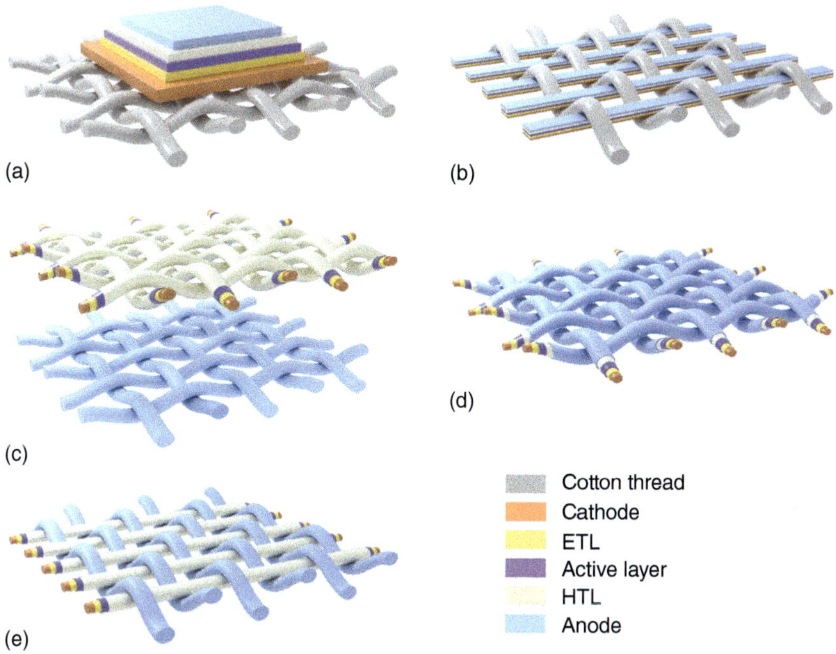

Figure 10.2 Textile solar cells based on various construction substrates and solar cell components. (a) Based on textile substrate and flexible solar cells. (b) Based on planar type strip-shaped solar cells. (c) Based on stacking of textile electrodes. (d) Based on fiber-shaped solar cells in coaxial and twisted structures. (e) Based on interlaced fiber electrodes. The solar cell structure is shown as typical third-generation thin film solar cells as cathode/electron transport layer (ETL)/active layer/hole transport layer (HTL)/anode.

10.2.1 Incorporation of Planar Solar Cells on Textile Substrate

At the early stage of wearable solar cells, silicon-based and thin-film solar cells have been incorporated into clothes, tents, and bags as portable energy supplies for fashion or military usage [6, 15]. However, these solar cells were heavy and rigid. With the development of third-generation solar cells, such as dye-sensitized solar cells, polymer solar cells, and perovskite solar cells, which have ultrahigh absorbance coefficients, flexible and lightweight solar cells could be obtained for wearable applications [16–18]. However, these flexible solar cells based on thin film substrates showed poor breathability, which hinders their wearability. Therefore, the construction of textile solar cells directly on the textile substrates has attracted attention for wearable applications (Figure 10.3) [19, 20].

Polymer solar cells have been actively investigated *via* direct fabrication on textile substrates. Typically, the rough textile substrate is difficult for uniform and continuous deposition of the active material layers. Therefore, it is required to build a bottom planarized electrode on the textile substrate for the subsequent deposition (Figure 10.3a) [19, 22, 23]. The bottom planarized layer also functions as barrier layer for preventing moisture infiltration. The following fabrication is similar to the planar solar cells, through layer-by-layer coating of electrodes, functional, and active layers

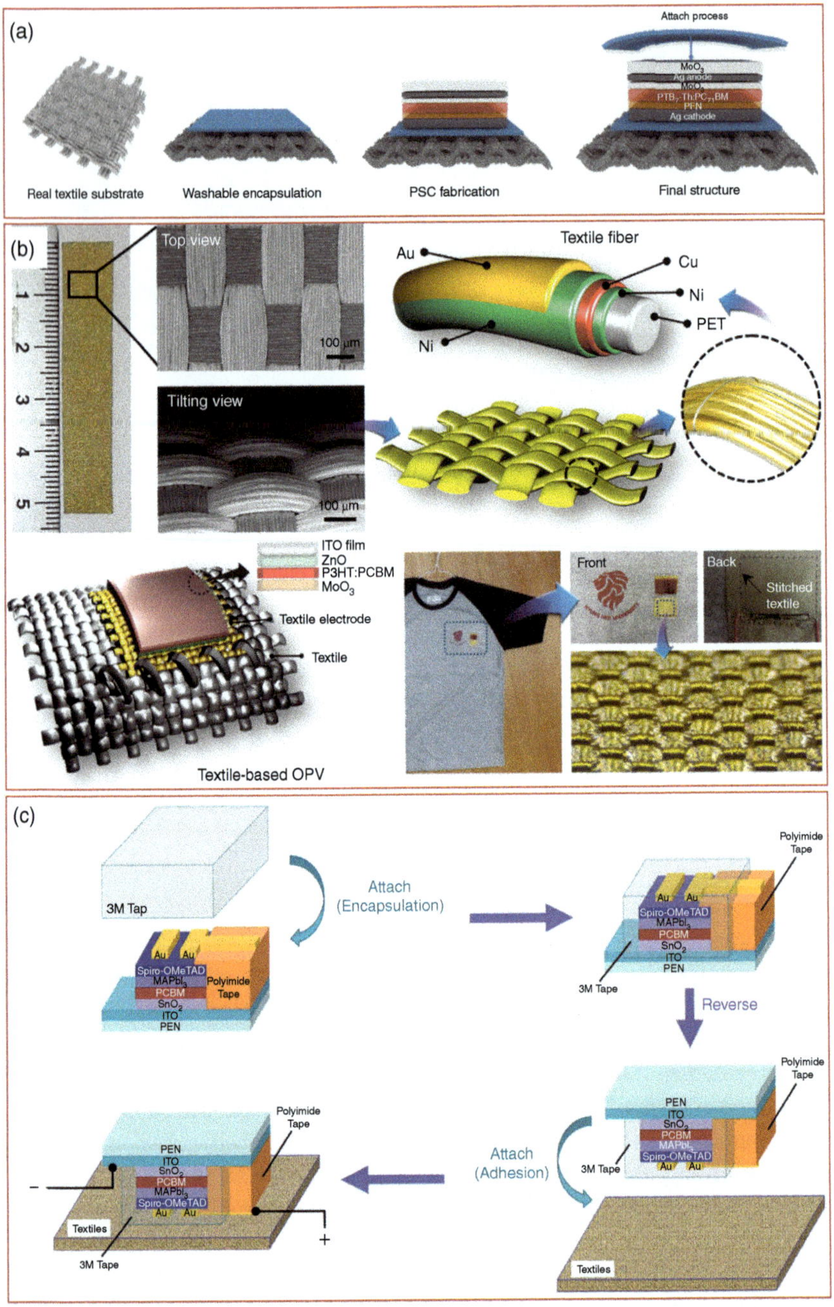

Figure 10.3 Wearable solar cells fabricated based on textile substrate. (a) Flexible polymer fabricated on a planarized textile substrate by depositing washable encapsulation layers. Source: Reproduced with permission from Jeong et al. [19], © 2018/Royal Society of Chemistry. (b) Schematic drawing and optical photos of sewing/stitching a textile solar cell on clothes with textile substrate function as rear electrode. Source: Reproduced with permission from Lee et al. [20], © 2014/Elsevier Ltd. (c) Transferring process of flexible solar cells to a textile substrate. Source: Reproduced with permission from Lam et al. [21], © 2017/Royal Society of Chemistry.

on top of the planarized substrate. In this case, the obtained textile solar cell could exhibit high PCE similar to the planar counterparts [19]. Besides polymer solar cells, perovskite solar cells have been demonstrated on polyurethane (PU) planarized textile substrate, which showed an efficiency of 5.72% [24]. Dye-sensitized solar cells fabricated on polyimide (PI) planarized glass fiber textile or polyester cotton textile substrate were also developed, showing PCEs of 4.04% and 3.24%, respectively [25]. Furthermore, based on the PI planarized heat-resistant glass-fiber textile substrate, a 6 cm^2 textile solar cell was produced with a PCE of 1.8% under 0.2 Sun illumination [26]. Using a solid-state electrolyte, the textile dye-sensitized solar cells using a surface planarized glass fiber textile showed a moderate PCE of 0.4% [27].

To further enhance the performance of textile polymer solar cells, SiO_2 nanoparticles with the plasmonic effect have been introduced to the opaque electrode for light absorption enhancement [28]. After planarization of the rough surface of textile substrate with PU, SiO_2 nanoparticles and ZnO sol-gel precursor were deposited, followed by the deposition of Ag rear electrode. The introduction of SiO_2 resulted in a uniform texture structure on the rear Ag electrode, which was beneficial for the light reflection and reabsorption by the photoactive layers. Based on the plasmonic effect, the PCE enhanced from 7.3% to 8.7%, attributed to the increased short-circuit current density (J_{SC}) enhancement from 13.76 to 15.91 mA/cm^2 [28]. More effective techniques that have been proven in planar solar cells can be further used for textile solar cells based on a planarized substrate.

Other than the direct fabrication of solar cells on a textile substrate, the attachment of free-standing, flexible solar cells to the textile is another approach. In this process, the fabrication of solar cells on a lightweight and thin planar substrate/barrier film layer was performed. Various barrier film layers have been studied that showed comparable performance to traditional PET/ITO substrate [29]. Depending on the functionality of the textile substrate, there are two kinds of textile solar cells that stitch free-standing thin-film solar cells. For the first kind, the textile substrate functions as rear electrodes by intimate contact between the textile and solar cell functional layer (Figure 10.3b) [20]. As a proof-of-concept, layer-by-layer deposition of metal electrodes on PET yarns has been demonstrated. For polymer solar cells or perovskite solar cells, the intimate contact between the rear electrode and functional layer is essential, and the contact area is considered the effective area. However, owing to the rough surface of textile substrate, the contact between the electrode and functional layer is rather limited. Based on the effective area, the textile solar cells, by sewing the half-cell to the textile electrode, showed a J_{SC} of over 13 mA/cm^2, and a PCE of 1.8% [20]. There were also dye-sensitized solar cells based on textile counter electrodes by depositing polypyrrole onto a carbon fabric [30–32]. However, the top electrode was still based on rigid substrates such as glass or FTO, which was not considered as a real textile solar cell.

For the second kind of textile solar cell, the textile only functions as a substrate, and flexible thin-film solar cells were directly transferred onto the textile substrate (Figure 10.3c) [21]. In this case, the resulted photovoltaic performance could outperform other textile solar cells. For instance, such textile solar cells showed a PCE of ~15% and ~8% for perovskite and polymer solar cells, respectively [21, 33].

Considering the structure of dye-sensitized solar cells with a spacer to separate the photoanode and counter electrodes, the textile could also function as the spacer by sewing the photoanode and counter electrode ribbon with the textile in between [34, 35]. A champion efficiency of up to 5.8% has been reported based on this textile structure [34]. Impressively, the high-efficiency flexible GaAs solar cells could also be stitched to a textile substrate [36]. A PCE of 17.49% had been achieved when transferred to a textile substrate, similar to the Kapton substrate of 16.46%.

However, direct sewing of planar solar cells, or incorporating solar cells onto textile substrates has completely vanished the breathability of the resulted solar cells, which is obviously not attractive for wearable applications. Besides, the flexibility of these devices cannot fully meet the requirements for various deformations. Thus, breathable and flexible textile solar cells are in high demand to meet the mechanical and wearable requirements.

10.2.2 Strip-Shaped Solar Cells

To overcome the mechanical and breathable limitations and meanwhile maintain the high performance, strip-shaped solar cells have been fabricated and woven into textile format for wearable use (Figure 10.4) [37–39]. Silicon solar cells and solid-state perovskite solar cells have been investigated. For silicon solar cells, miniature silicon solar cells (3 mm × 1.5 mm × 0.2 mm) were cut from a large wafer (Figure 10.4a). The miniature cells were connected by Cu wires in parallel with a 2 mm gap between each cell, with one Cu wire soldering on the anode side and the other Cu wire soldering on the cathode side. Each miniature cell was encapsulated with a cylindric-shaped transparent resin micro-pod. This cylindric-shaped transparent resin micro-pod is beneficial for light incidence, with an enhanced J_{SC} compared with bare silicon solar cells. For a typical silicon solar yarn, there are ten miniature cells connected to enhance the output current. They can be connected in series or parallel to form minimodules according to the requirements for use. These yarns could be weaved into a textile *via* an industrial shuttle flying process, with the solar yarns as wefts and the other polymer yarns as warps (Figure 10.4a) [37]. The high stability of silicon solar cells is promising for long-term usage of these solar cells [40]. The large area textile with 200 miniature solar cells (44.5 mm × 45.5 mm active area) showed a PCE of 2.15% under one Sun illumination.

Perovskite solar cells show comparably high PCE of 25.7% with the silicon solar cells [41]. More importantly, they can be fabricated via a solution process at low temperatures, which is highly compatible with flexible polymer substrates. To improve the PCEs of fiber-shaped perovskite solar cells, strip-shaped perovskite solar cells have been developed. They were fabricated on a PEN/ITO substrate, and the aligned carbon nanotube sheet served as the anode. The substrate was deposited with the functional layer, and the perovskite active layer was cut into strips with a width of 1 mm. These strip-shaped solar cells showed a champion efficiency of 9.49%, which is much higher than the other solid-state fiber solar cells at that time [42]. They could be woven into a watch band and power an electronic watch under illumination. In addition, they could be woven with polymer yarns to

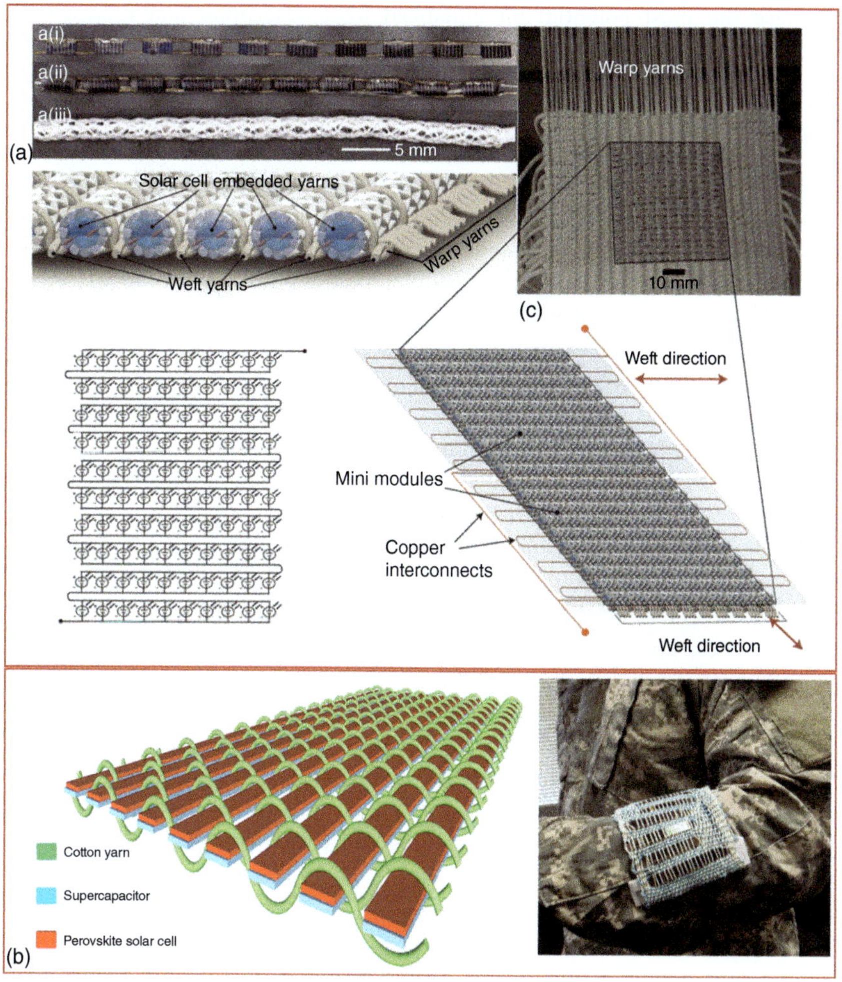

Figure 10.4 Strip-shaped planar-type solar cell weaved with cotton yarn to form a textile solar module. (a) Miniature crystalline Si solar cell integrated into a yarn. Source: Reproduced with permission from Satharasinghe et al. [37], © 2019/John Wiley & Sons, Inc. (b) Textile based on strip-shaped perovskite solar cells. Source: Reproduced with permission from Li et al. [38], © 2016/Springer Nature.

form a flexible textile (Figure 10.4b) [38]. With an inverted cell structure by using Cu ribbon as the cathode and PET/ITO as the anode, the strip-shaped solar cell showed a PCE of 10.41%. Further integration with a strip-shaped supercapacitor realized integrated energy textile with simultaneous energy harvesting and storage capability [38].

Similar with planar solar cells, the photovoltaic conversion performance of the strip-shaped solar cell is also dependent on the angle of incident light. Coupling with the scattering effect of cotton or other polymer yarns, fiber solar cells with a circular-shaped cross-section structure for the construction of textile solar cells might have the most potential for enhanced light absorption and reduce the dependency on the angle of incident light.

10.2.3 Textile Electrode

The construction of textile solar cells based on breathable textile electrodes is another important approach (Figure 10.5) [43, 44]. The low-temperature and solution-based fabrication processes of dye-sensitized [43, 45, 46], polymer [44, 47, 48], and perovskite solar cells [49] enable the preparation of textile solar cells based on textile electrodes.

Obviously, intimate contact between the two textile electrodes is required. For dye-sensitized solar cells, the efficient transport of carriers in between the textile electrodes could be ensured by the liquid electrolyte. As shown in Figure 10.5a, the textile photoanode was composed of a Ti mesh deposited by aligned TiO_2 nanotubes via anodic oxidation for absorption of organic dye N719. The counter electrode was prepared by woven textiles composed of aligned carbon nanotube fibers. After filling liquid electrolytes in between the textile electrodes, textile solar cells based on stacking textile electrodes could be achieved [43]. A champion PCE of 3.1% was obtained based on the novel textile configuration. To prevent the leakage of liquid electrolytes, encapsulation was needed, which made the textile solar cell a compact planar structure with inferior breathability. In addition, the stacking configuration sacrificed the tolerance to shearing deformation during use. Therefore, by replacing the liquid electrode with a solid-state electrolyte, the stability and flexibility could be further enhanced, with a slightly lower PCE of 2.1%. By stacking one more photoanode textile on the other side of the counter electrode textile, a sandwich structure could be formed for facilitated light absorption, which further improved the PCE to 3.62% [43].

Besides textile electrode stacking, weaving two textile electrodes together using interlocking yarns can form a new structure with unexpected performance [46]. In particular, the textile electrodes by weaving ribbons are compatible with the shuttle flying process in the textile industry, thus promising for scale-up fabrication. The photoanode and counter electrode textile were further deposited by printing the corresponding active materials on each side. After electrolyte injection and encapsulation, the textile dye-sensitized solar cells could afford a PCE of 1.7% with decent structure stability [46].

The intrinsically solid-state and flexible polymer solar cells have no leakage concerns. Thus, textile polymer solar cells based on stacking anode sheet to the Ti mesh cathode have been realized (Figure 10.5b) [44]. Similar to the dye-sensitized solar cells, the Ti mesh was incorporated with aligned TiO_2 nanotubes for effective electron extraction and hole blocking. The Ti mesh was transparent with a high electrical conductivity. The following deposition of the photoactive and hole transport layers could be performed by dip-coating method [50, 51]. After stacking with a transparent carbon nanotube sheet, the textile polymer solar cell with bifacial light illumination and absorption potential could be obtained, which exhibited a PCE of 1.08% [44].

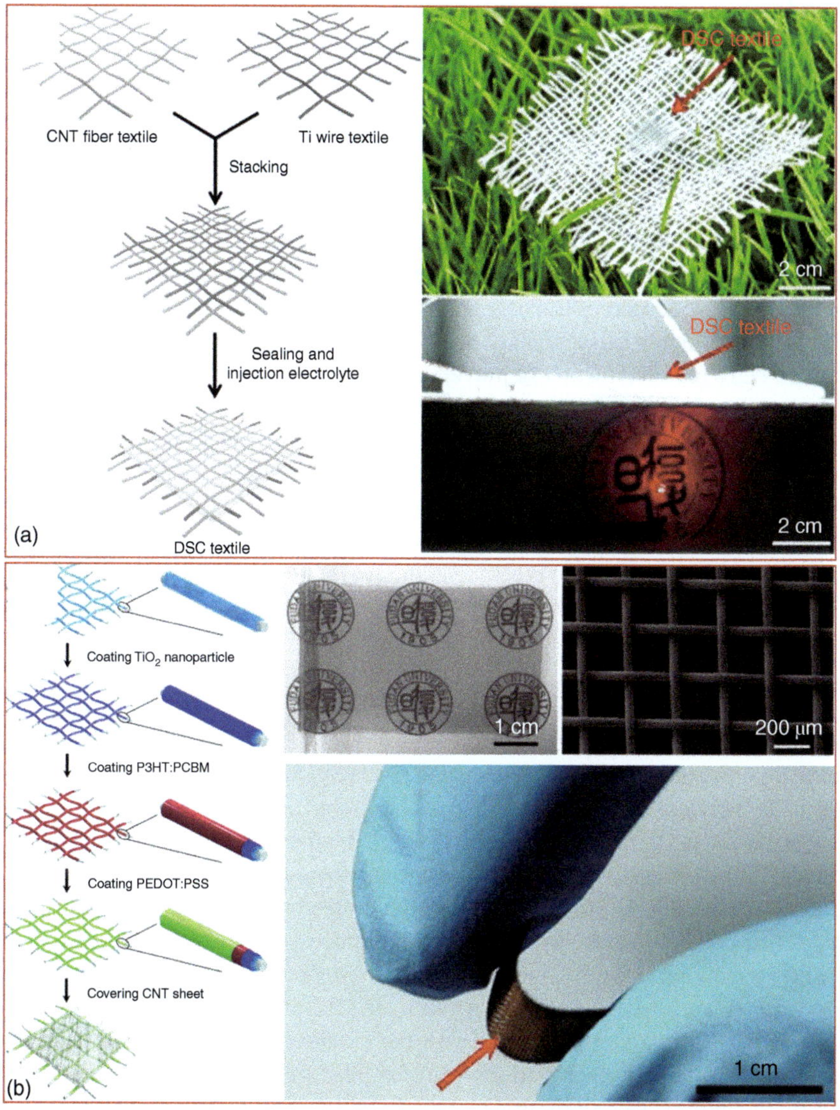

Figure 10.5 Textile solar cells based on stacking textile electrodes. (a) Textile dye-sensitized solar cells by stacking photoanode textiles composed of dye absorbed Ti mesh and counter electrode textiles composed of carbon nanotube fibers. Source: Reproduced with permission from Pan et al. [43], © 2014/John Wiley & Sons, Inc. (b) Textile polymer solar cells by covering modified stainless steel mesh electrodes with transparent carbon nanotube sheet electrodes. Source: Reproduced with permission from Zhang et al. [44], © 2014/John Wiley & Sons, Inc.

Besides the metal mesh, polymer-based textile electrode could be applied based on low-temperature processing. A transparent textile anode was fabricated using polyester/Ag nanowire/graphene core-shell yarns [47]. A blade-coating process was applied to deposit HTL and photoactive layer of bulk heterojunction. After coating the ETL and cathode on top of the active layer, a textile polymer solar cell with PCEs of up to 2.27% was obtained. Due to the lightweight and flexibility of polyester textiles, the textile polymer solar cell exhibited excellent flexibility and wearing compatibility [47]. Besides the dip coating [48] and blade coating processes, spray coating has also been applied for the fabrication of textile polymer solar cells [22]. For instance, perovskite solar cells had been produced by stacking stainless-steel textile electrodes with highly transparent carbon nanotube sheet electrodes, although the obtained PCEs required further improvement [49].

It should be noted that the mechanical deformation and abrasion may lead to relative shifting of the electrodes, which reduces the effective contact between different components and thus decreases the power output. In addition, the above-discussed strategies, namely planar solar cells sewing on the textile substrate and stacking textile electrodes, both suffer from poor breathability. Thus, solar cells in a fiber format that can be woven into highly breathable and wearable textiles are promising and have been actively pursued.

10.2.4 Textile Solar Cells Based on Fiber-Shaped Solar Cells

Textile solar cells woven from fiber-shaped solar cells can afford high mechanical properties, breathability, and wearability [50, 52]. Fiber-shaped dye-sensitized solar cells, polymer solar cells, and perovskite solar cells have been also woven into textiles, as shown in Figure 10.6 [53–55]. Figure 10.6a exhibits the typical structure of coaxial and twisting structures of fiber-shaped solar cells [50, 53]. Up to date, weaving, knitting, and sewing processes have been demonstrated for textile solar cell fabrication, while the knitting process has not been reported yet. For the weaving process, the fiber solar cells could function as both the weft and warp yarns (Figure 10.6b) [54]. Owing to the high flexibility of fiber solar cells that can be twisted into a knot (Figure 10.6a), the obtained textile solar cells showed high structure stability upon various deformations. Furthermore, by making the photoanode into a spring shape, the textile can be intrinsically stretchable [56].

The sewing process represents another way for fabrication of textile solar cells based on fiber solar cells (Figure 10.6c) [55, 57]. In this case, the encapsulated fiber solar cells could be sewed into desirable patterns. A photovoltaic textile with a large area of 17 cm × 22 cm was demonstrated, which presented an energy output of 22.7 mW and could power a smart textile sensor and display system (Figure 10.6c) [55].

Several issues are required to be further addressed. For instance, the effective illuminated areas of textile solar cells have rarely been discussed. On the other hand, the mechanical stability during the weaving/sewing process has not been fully studied yet. Furthermore, due to the challenges for the fabrication of long fiber solar cells, the weaving of fiber-shaped solar cells into a textile with high performance

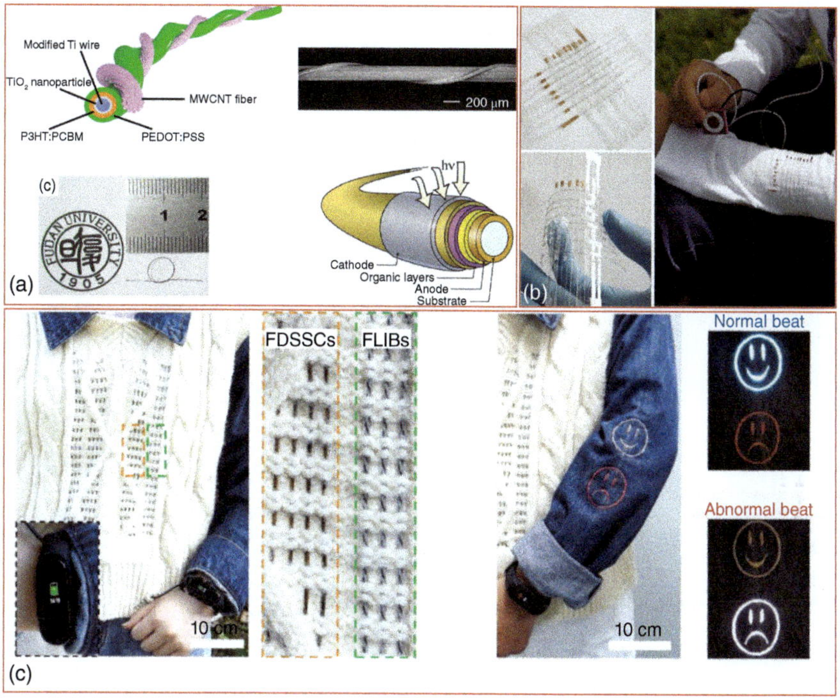

Figure 10.6 Fiber-shaped solar cells weaving into a flexible textile solar cell structure. (a) Typical coaxial and twisted structure fiber-shaped solar cells. Source: Reproduced with permission from Pan et al. [53], © 2015/Elsevier (b) Textile solar cells based on fiber polymer solar cells. Source: Reproduced with permission from Zhang et al. [54], © 2014/John Wiley & Sons, Inc. (c) Textile based on interweaving of fiber dye-sensitized solar cells and colorful yarns. Source: Reproduced with permission from Kang et al. [55], © 2022/John Wiley & Sons, Inc.

remains a grand challenge. Therefore, one possible solution lies in the following contexts on interlaced fiber electrodes weaving in weft and wart directions.

10.2.5 Interlaced Fiber Electrodes

Dye-sensitized and polymer solar cells have been designed into the interlaced anode and cathode fiber electrode structure [11, 58]. However, the use of liquid electrolytes in dye-sensitized solar cells should be avoided. There were several reports regarding the interlaced fiber electrode based on solid-state dye-sensitized solar cells [10, 12, 14, 59] and polymer solar cells [11, 13]. Even though fiber perovskite solar cells demonstrated higher PCEs, up to 15.7% [60], the interlaced fiber electrode structure has not been realized for perovskite solar cells.

To construct the interlaced fiber electrode structure, separated fiber electrodes for anode and cathode are woven by an industrial loom using a shuttle-flying process (Figure 10.2e). Typically, for the reported interlaced fiber electrode structure, a cathode with an electron transport layer, active layer, and hole transport layer is inserted in the weft direction, while an anode with a metal-coated polymer core is fixed in

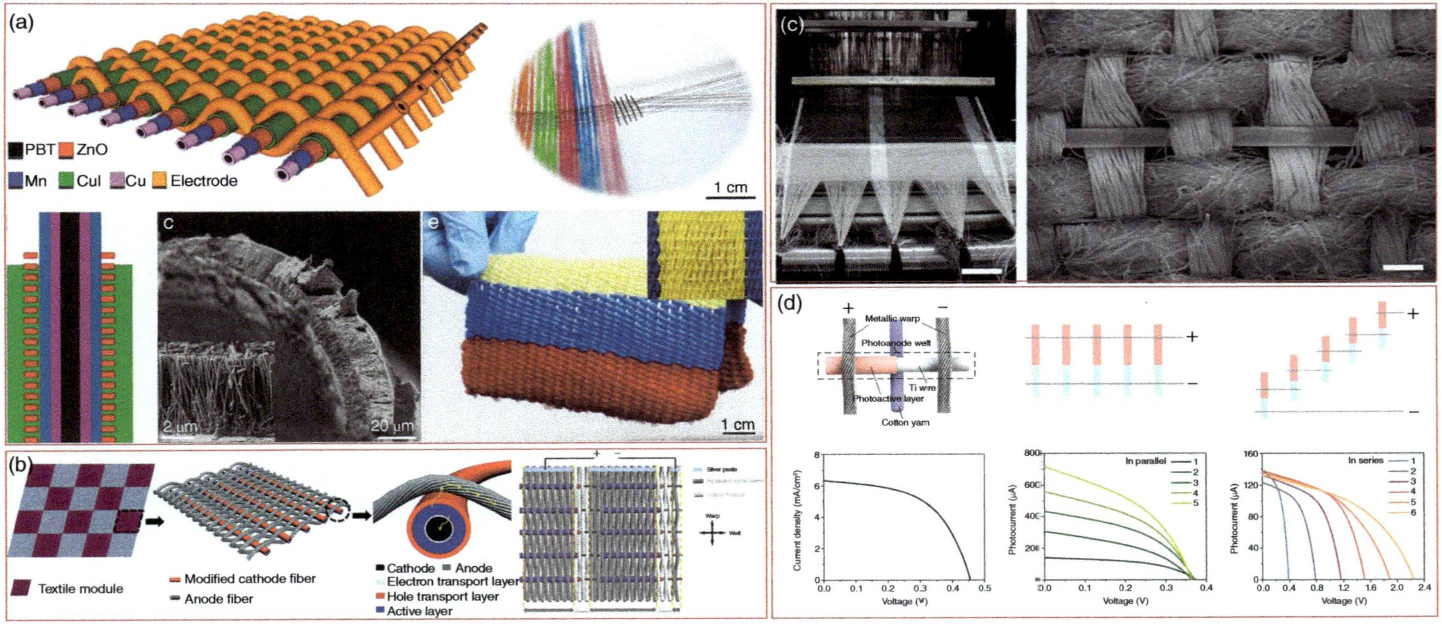

Figure 10.7 Interlaced fiber electrode structured textile solar cells. (a) Dye-sensitized solar cells with weft fiber cathode (photoanode) structure of PBT/Cu/Mn/ZnO/N719/CuI and warp fiber anode (counter electrode) of PBT/Cu. Source: Reproduced with permission from Zhang et al. [10], © 2016/John Wiley & Sons, Inc. (b) Polymer solar cells with weft fiber cathode structure of Ti/ZnO/PTB7:PC71BM/PEDOT: PSS and warp fiber anode of silver-coated nylon. The right part showed the parallel connection of five cathode fiber electrodes and then further connected in series with two identical parts. (c) The industrial loom with shuttle flying process for the weaving process of interlaced fiber electrode structured textile solar cells. Source: Reproduced with permission from Liu et al. [11], © 2018/Royal Society of Chemistry. (d) The performance of the interlaced fiber electrode solar cells with single cathode and schematic drawing showing the series and parallel connection structure. Source: Reproduced with permission from Shi et al. [13], © 2021/Springer Nature.

the warp direction [10, 11]. In this way, the tension-induced degradation of the active layer is minimized. When the cathode is modified with an active layer fixed in the warp direction, it shows enhanced interfacial tension in the active layer and thus reduces the device performance [10].

For the construction of textile dye-sensitized solar cells based on interlaced fiber electrode structure, polymer fibers coated with Cu from a solution bathing process serve as both anode (counter electrode) and cathode (photoanode) (Figure 10.7a) [10, 12, 14, 59]. Prior to the cathode modification, Mn was electrodeposited for the following process. Due to the use of polymer, the TiO_2 coating at high temperatures was not applicable, so a low-temperature dip-coating was performed for ZnO with the absorption of dye N719. With the coating of solid-state hole transport layer, the obtained cathode served as the weft in the weaving process. And the metal-coated anode was fixed as the warp in the weaving process.

The single cathode structure showed a PCE of around 1%, which was much lower than those of fiber-shaped solar cells. To further improve the device's performance, textile polymer solar cells with interlaced fiber electrodes were developed (Figure 10.7b). A Ti wire served as a cathode and was deposited with the ETL ZnO via a sol-gel process [11]. The metal electrode was then deposited with the blends of Poly({4,8-bis[(2-ethylhexyl)oxy]benzo[1,2-b:4,5-b′]dithiophene-2,6-diyl}{3-fluoro-2-[(2-ethylhexyl)carbonyl]thieno[3,4-b]thiophenediyl}) (PTB7) molecule with [6,6]-phenyl C71 butyric acid methyl ester (PC71BM) blend and HTL poly(3,4-ethylenedioxythiophene): polystyrene sulfonate (PEDOT: PSS) from a dip coating process. With the optimization of the ZnO and active layer by tuning the precursor concentration, the PCE could be improved from 0.0067% to 1.67%. The weaving process is shown in Figure 10.7c using an industrial loom based on the shuttle flying process.

Weaving interlaced fiber electrode textiles can realize efficiency in series or parallel connections in order to achieve desired current and voltage (Figure 10.7d) [13]. There are several ways to realize series and parallel connections. For an in-parallel connection, the cathode was placed in parallel. While for in-series connection, the cathode and anode should be shifted, as shown in Figure 10.7d, although this could make the weaving process very difficult. An alternative approach is shown in Figure 10.7b [11], in which each string can serve as an independent unit separated by the cotton thread that realizes the in-series connection [10, 12, 14, 59]. The energy output could be readily tuned to match the expected applications, such as powering an electronic watch and mobile phone.

10.3 Weaving Patterns and Incidence Angle for Textile Solar Cells

To enhance the output power of the interlaced fiber structure solar cells, the weaving process was optimized with promoted projected photoactive area through optimization of the weaving pattern (Figure 10.8). Here, the area used for PCE calculation is defined as the projection area (Figure 10.8a). The voids between the interlaced

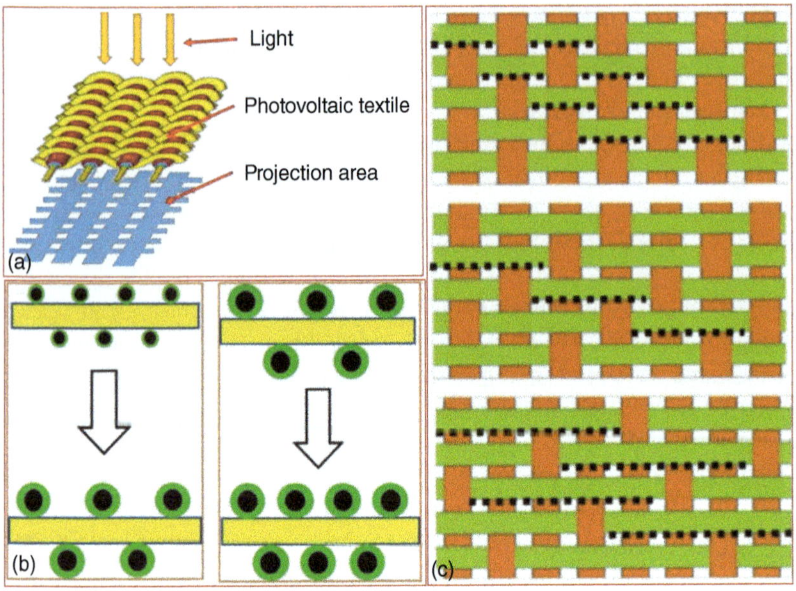

Figure 10.8 Weaving pattern and effect of shading on the illumination area. (a) The definition of the projection area of textile solar cells. The voids that formed from the weft and warp fiber during the weaving process are not included in the calculation of the projection area. (b) The shading effect of waft anode fiber from the interval distance and diameter on the ratio of photoactive area on the projection area. (c) The plain, twill, and strain (from top to bottom) weaving patterns for interlaced fiber electrodes. Source: Reproduced with permission from Zhang et al. [10]. © 2016/Wiley-VCH Verlag GmbH & Co. KGaA, Weinheim.

cathode and anode are not included in the projection area. The diameter and the intervals of the anode fiber electrodes can affect the active area ratio. In addition, the effective contact area between the cross-fiber electrodes is also limited for fiber electrodes with small diameters, which causes increased electrode resistances and reduced carrier extraction and transportation. And for a longer interval distance, the carrier transport distance along the axial direction increases, which also reduces the carrier extraction and transportation with severe recombination. Therefore, a balance between the electrode diameter, interval distance, photoactive area ratio, and the efficiency of carrier extraction and transportation needs serious consideration.

The intervals between the anode fiber electrodes can be modified by using different weaving patterns, as shown in Figure 10.8c, namely the plain, twill, and strain weaving patterns [10]. The weaving patterns were obtained by tuning the up and down moving frequencies of each warp fiber electrode. And the intervals between the anodes could be several times of the diameter of the warp fiber electrodes. For solid-state dye-sensitized solar cells, as illustrated in Figure 10.7a, the diameter of the warp fiber electrode is smaller than the weft-modified fiber electrode [10]. And the effective area ratio with respect to projection area for the three weaving patterns of plain, twill, and strain ranges from 87.3% to 96.1%, due to the three-dimensional cross-contact structure. The plain weaving patterns exhibited the highest effective illumination area of 96.1% ,the strain weaving patterns had the smallest effective illumination area, and the photocurrent density reduced from 6.35 to 5.8 mA/cm^2 when changed the weaving patterns from plain to strain [59]. Further investigation on the weaving pattern with the consideration of effective illumination area and carrier collection is expected to improve the performance of the textile solar cells.

Another factor that might influence light illumination and absorption is the incidence angle with respect to the textile solar cell plane. For wearable applications of textile solar cells, the movement is random, and the incidence angle will change all the time, unlike the solar panel that is fixed in the power station and can even track the sunlight. As shown in Figure 10.9a, there are perpendicular illumination and inclined illumination, and also direct and indirect illumination from other yarns' scattering [37]. For planar-type solar cells, the incident angle determines the effective projection area, which decreases with the increasing incidence angle. And typically, for planar-type solar cells, only the front side of the cell is transparent for light absorption, while the rear side is basically opaque [61]. It was observed that when the incident angle was higher than 90 degrees, there was almost no electricity generated for the planar solar cells (Figure 10.9b). However, when weaving into a textile

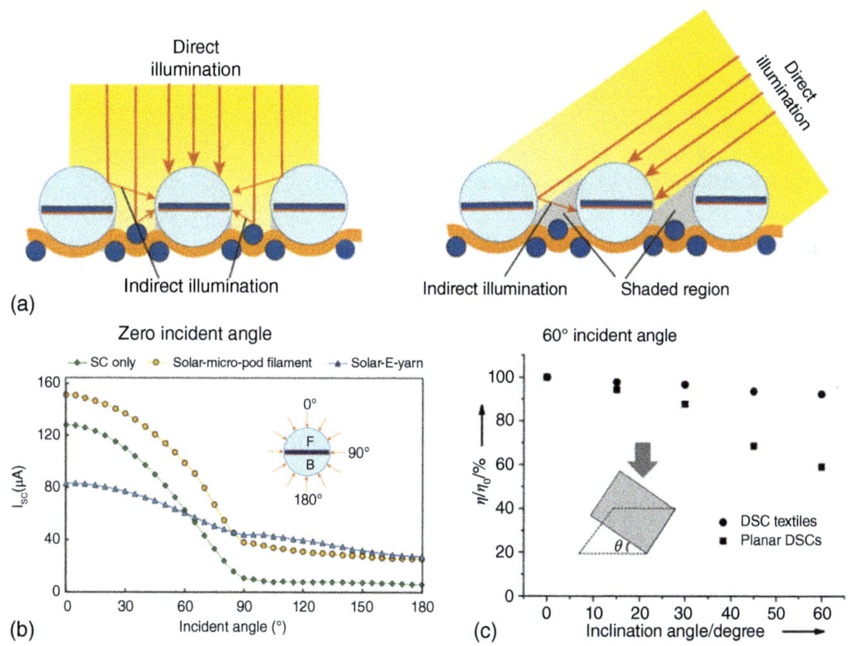

Figure 10.9 The effect of illumination light incidence angle on the device performance. (a) Schematic drawing showing the light incidence angle of 0° and 60°. There is scattering of light from the other weaving materials to enhance light absorption. (b) For a textile solar cell based on planar-type solar cells. Source: Satharasinghe et al. [37], © 2019/John Wiley & Sons/CC BY 4.0. (c) The comparison between a textile solar cell based on fiber-shaped textile electrodes and planar-type solar cells. Source: Reproduce with permission from Pan et al. [43], © 2014/Wiley-VCH Verlag GmbH & Co. KGaA, Weinheim.

or sealed into a transparent cylinder, there is current generated even illuminated from the rear side due to the indirect illumination scattering from the other yarns, which showed great potential in wearable applications (Figure 10.9b) [37].

As discussed in the previous chapters, the PCEs of fiber-shaped solar cells showed negligible sensitivity related to the incident angle. Similarly, in the textile solar cells, the incident light angle did not show much influence on the output power (Figure 10.9c) [43]. With optimization of the fiber interval, the shading effect could be minimized. The scattering light could also be fully utilized for the maximal power output. Furthermore, with the circular encapsulation and the refractive index difference between each layers, the effective illumination area could be, in fact, larger than the projection area, which achieved higher current density than expected [55]. All these results showed that a fiber electrode-based textile solar cell has the potential for enhanced light absorption with less dependence on the incidence angle.

10.4 Mechanics and Breathability of Textile Solar Cells

The textile solar cells for wearable applications need to be flexible, bendable, twistable, and waterproof. For the waterproof and washability, it is attributed to the encapsulation structure and materials, which will be discussed in the following section. Among the various textile solar cell structures discussed above, the planar type solar cells are not tolerant to twisting deformation. For solar cells based on textile electrodes, the encapsulation also limited the twisting deformation properties of the textiles. Furthermore, these two structures reduced or even eliminated the breathability of the formed textiles. When weaving from fiber-shaped solar cells or interlaced fiber electrodes, the twisting stability of the resulted textile could be well maintained.

As shown in Figure 10.10a, bending deformation is common for all types of flexible solar cells, including textile solar cells. It normally did not deteriorate the performance of textile solar cells. Besides bending, there are also shearing, twisting, and stretching deformations (Figure 10.10b). Systematic studies have proved the high stability upon various deformations for textile solar cells based on fiber-shaped solar cells or interlaced fiber electrodes. For wearable applications, the abrasion should be carefully considered. Without the encapsulation and protection layers, the textile solar cells are prone to be worn out during abrasion due to the easy abrasion of tens or hundreds of nanometers thick functional layer and photoactive layer. Up to date, an abrasion study has been conducted using textile solar cells woven from solar cell yarns (Figure 10.10c). The solar cell yarns consisted of miniature silicon solar cells encapsulated by UV-curable resin (Figure 10.4a). Furthermore, the encapsulated solar cell yarns are protected and decorated by knitted-braid shells. In the abrasion test, the protection layer showed slightly worn-out surface. However, the textile solar cells showed almost unchanged photovoltaic parameters after 6000 abrasion cycles.

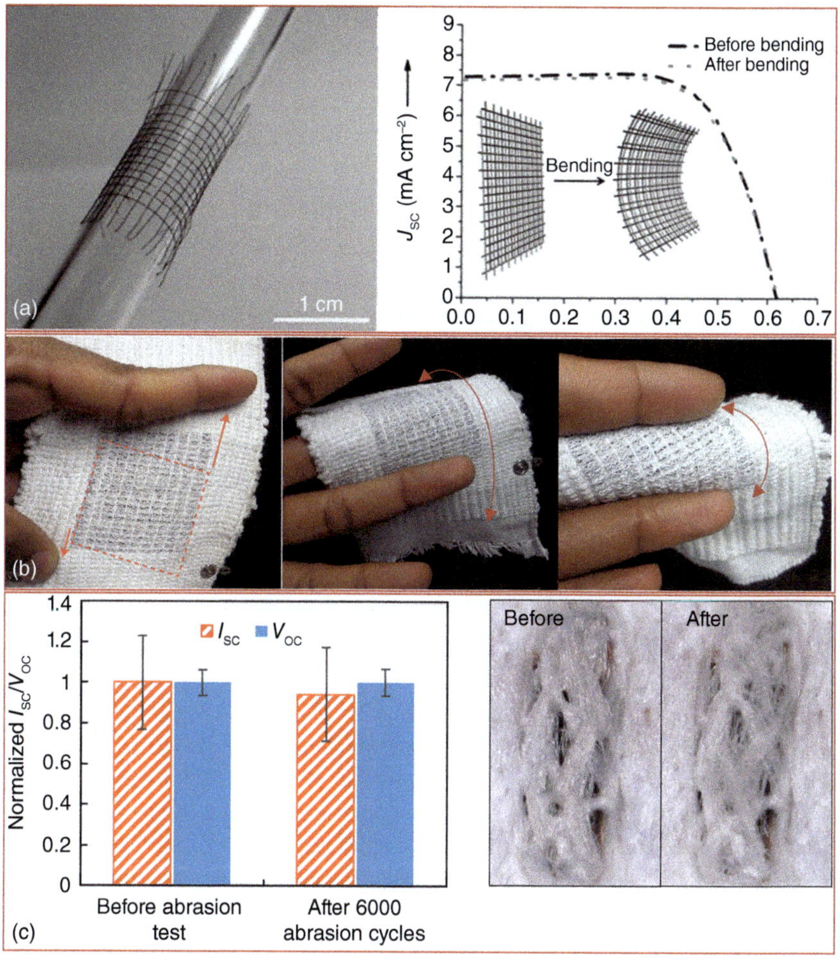

Figure 10.10 Mechanical properties of textile solar cells. (a) Bending deformation of a textile solar cell and the typical J–V curves before and after bending deformation. Source: Reproduced with permission from Pan et al. [43], © 2014/John Wiley & Sons, Inc. (b) The shear deformation, draping deformation, and bending deformation for textile solar cells. (c) abrasion test for a textile solar cell based on miniature silicon solar cells and the V_{OC} and J_{SC} values before and after 6000 abrasion cycles. Source: Reproduced with permission from Satharasinghe et al. [37], © 2019/John Wiley & Sons, Inc.

The moisture filtration and the transfer of heat are related to the breathability of a textile, which is important for wearing compatibility. For breathability measurement, it was based on the Gravimetric Absorbance Test System (GATS) [37]. In specific, the testing textile solar cells and a control textile sample were compared, with the rear side of the textile covered with a thin polyethylene film, aiming to expose the textile solar cell area to the porous plate of the GATS. The absorption test was conducted until each textile was saturated with distilled water. It was not surprising that the textile solar cells had similar moisture management properties with almost the same saturated water capacity [37]. The mechanics and breathability are vital for the wearing comfortability of textile solar cells. Furthermore, by weaving with other colorful yarns, a nice garment with textile solar cells could be obtained, which was very attractive for daily wearing applications.

10.5 Encapsulation Structure and Materials for Textile Solar Cells

Encapsulation is essential for long-term stability under ambient environments and wearing abrasion. Particularly for a wearable textile solar cell, it should be tolerant to a machine-washing process, which is an industrial standard that needs to be passed [37]. Most of the above-discussed textile solar cells are not encapsulated, except for the dye-sensitized solar cells that use liquid electrolytes (Figures 10.5a and 10.6c) and silicon-based solar cells (Figure 10.4a). Without encapsulation, the electrode and active materials might be worn out during use. As a result, the encapsulation of textile solar cells should be taken into serious consideration.

There are currently several encapsulation processes and materials that have been reported for textile solar cells. These encapsulation strategies were developed for the structure of textile solar cells based on textile substrates (Figures 10.11a,b) [19, 33], textile electrodes stacking together (Figure 10.5a) [43], planar solar cells (Figure 10.4a) [37], and fiber-shaped solar cells (Figure 10.6c) [55]. However, there is no encapsulation for the interlaced fiber electrode structure yet. After encapsulation, the device stability obviously improved especially for dye-sensitized solar cells. For textile electrode-based dye-sensitized solar cells, surlyn films were applied to encapsulate the photoanode and counter electrode textile electrodes between two poly(ethylene terephthalate) (PET) membranes. However, such a thin film structure has sacrificed breathability. With the use of solid-state electrolyte, the encapsulated textile could stably work for more than 300 hours without obvious performance degradation. The fiber-shaped solar cells encapsulated in a flexible and transparent fluorinated ethylene propylene tube (Figure 10.6c), it showed stable operation duration for more than 1000 bending cycles with negligible performance degradation [55].

One of the most important issues for the encapsulation is the machine-washing or hand-washing tolerance, which has been proved for several textile solar cells (Figure 10.11) [19, 33, 37, 39]. To improve water tolerance during the washing process, several encapsulation materials and structures have been explored. Ultrathin solar cells, like polymer solar cells and perovskite solar cells, have been developed in the past decade [17, 62–64]. They were fabricated on an ultrathin substrate with a thickness of around 2 μm, like PET and parylene. Based on the ultrathin structure and effective capping layer with ideal thickness and modulus, the active and functional layers could be well maintained under deformations [65]. And the thin-film solar cells could be crumped and then stretched back to a thin film. However, it was not tolerant to water penetration, as verified by the high water vapor transmission rate [66]. One way to endure the water penetration was the

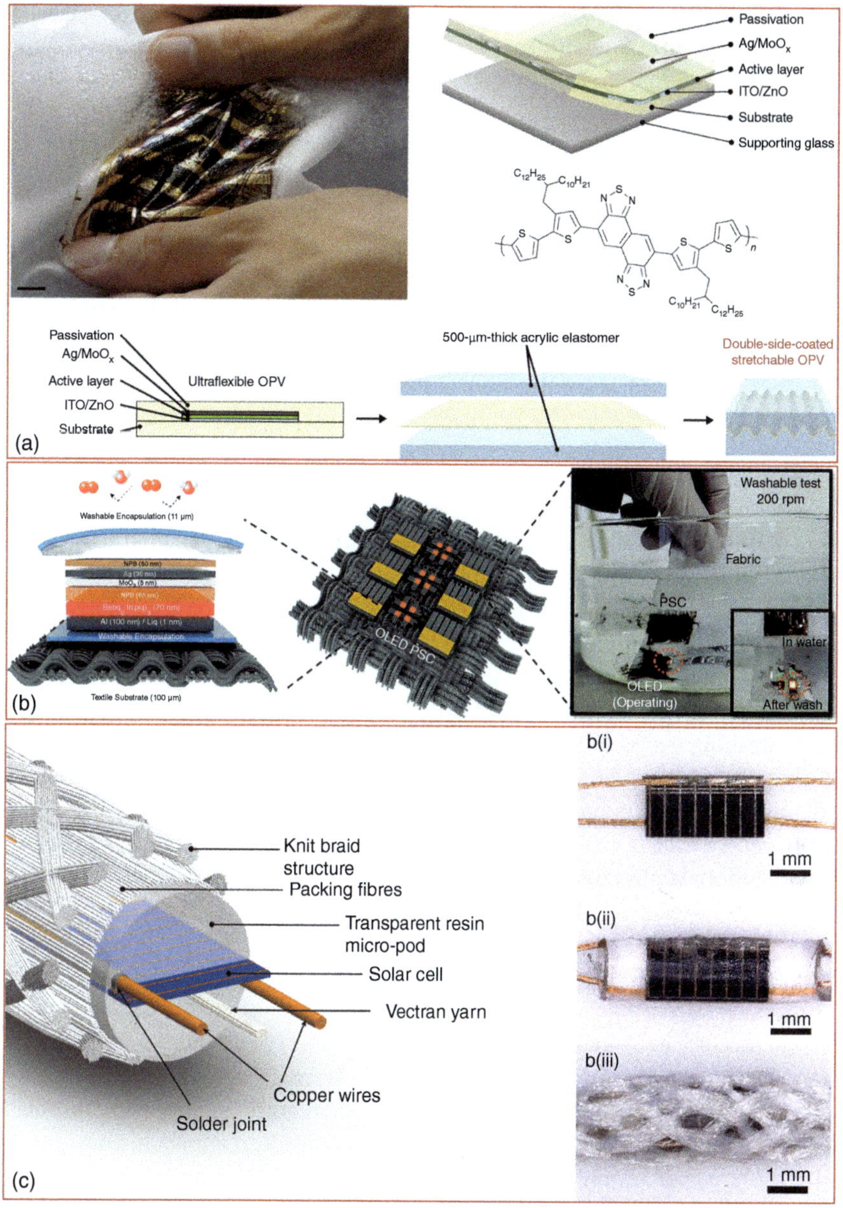

Figure 10.11 Encapsulation structure and washing stability. (a) Encapsulation of ultra-thin polymer solar cells by chemical vapor deposited parylene and sealing by pre-stretched acrylic elastomer. Source: Reproduced with permission from Jinno et al. [33], © 2017/Springer Nature. (b) Encapsulation of polymer solar cells based on textile substrate with atomic layer deposited multilayer structure that tolerant to washing. Source: Reproduced with permission from Jeong et al. [19], © 2019/Royal Society of Chemistry. (c) Encapsulation of miniature silicon solar cells with UV-curable transparent resin micro-pod. Source: Reproduced with permission from Satharasinghe et al. [37], © 2019/John Wiley & Sons, Inc.

incorporation of a waterproof elastomer (Figure 10.11a). Ultrathin organic solar cells on parylene substrate with a PCE of 7.9% had been demonstrated by coating on sides with waterproof elastomer [33]. The double-side waterproof elastomer encapsulated organic solar cells showed high stretchability and stability in water, which could maintain the photovoltaic conversion performance after immersing in water for 120 minutes (Figure 10.11a).

Deposition of nanometers thick metal oxide layer represents another encapsulation strategy. It enabled a low water vapor transmission rate that was comparable to a glass substrate (Figure 10.11b). For instance, a 30 nm layer composed of 10 alternative Al_2O_3 and ZnO layers (each for 3 nm) was developed [19]. After bending and washing for 20 cycles, the performance of the encapsulated textile solar cells and organic light-emitting diode showed negligible difference, which confirmed the effectiveness of the encapsulation layer. A lighting diode could be observed during and after the washing process. However, the compact encapsulation layer again makes the textile un-breathable.

To make fully textile-compatible solar cells with high breathability and washability, a UV-curable resin was applied for miniature solar cells (Figure 10.11c) [37]. Due to the high stability of silicon solar cells, there are more options for the UV-curable resin that have no reaction with the silicon. Based on the transparent resin micro-pod encapsulation and the protection of polymer knit braid structure, the textile solar cells based on solar yarns are tolerant to the machine-washing process, with 20 g of detergent and line dried inside of a wash bag for 25 cycles [37]. After 25 washing cycles by machine, the J_{SC} and open-circuit voltages (V_{OC}) all maintained over 95% of their initial values. The successful usage of transparent UV curable resin showed its promising applications to other fiber-shaped solar cells, like polymer solar cells, solid-state dye-sensitized solar cells, and perovskite solar cells, with rational optimization of the UV curable resin.

10.6 Application of Textile Solar Cells

Textile solar cells can be ideal power sources for a variety of wearable applications. Various wearable and portable electronics have been developed for intelligent communication and healthcare monitoring. They require reliable and flexible power supplies with output power ranging from microwatt to milliwatt. Due to the intermittency power supply features of solar cells, the integration with wearable energy storage devices such as batteries and supercapacitors is more applicable (Figure 10.12a). The fiber-shaped energy storage devices could be woven with textile solar cells to make a self-powered energy textile (Figure 10.12b). This could greatly overcome the intermittency of sunlight and enhance the practicality. For instance, the applications of the energy textile for health monitoring and intelligent communication were exhibited in Figure 10.12c by integrating with the recently developed textile display and other smart sensors. The textile solar cells could charge a variety of wearable devices, including a mobile phone, an electronic watch, and an integrated heartbeat, humidity, and temperature sensor (Figure 10.12d).

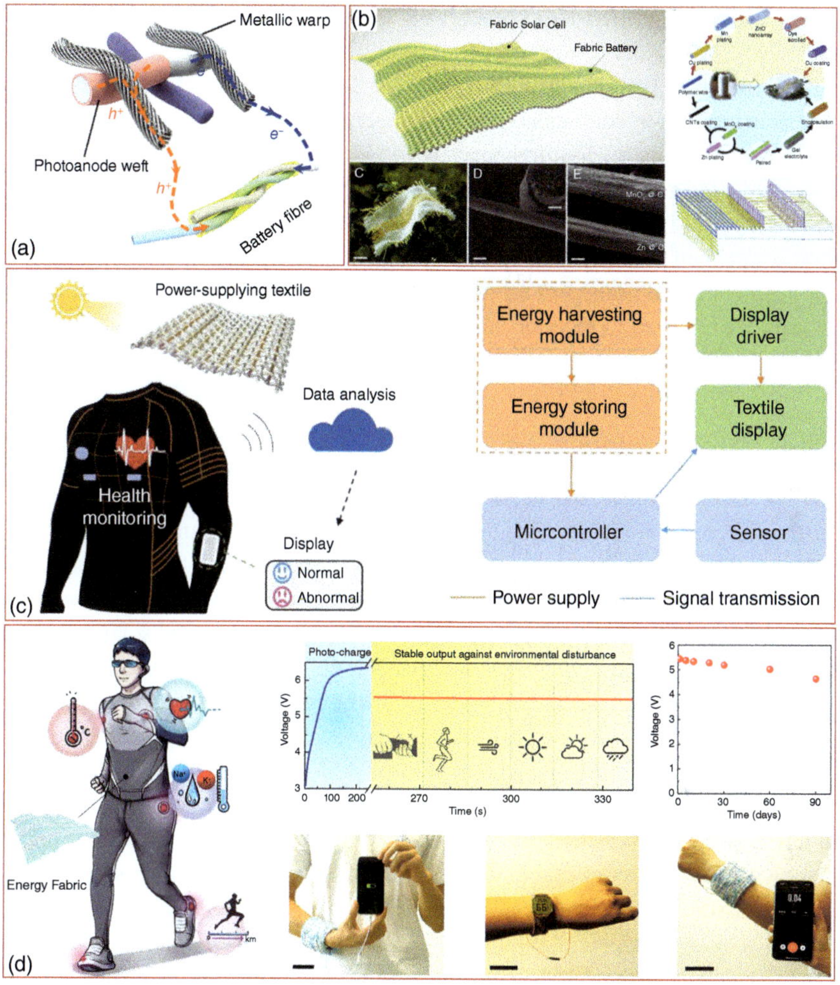

Figure 10.12 Wearable application of textile solar cells by integrating with other functional devices. (a) Schematic drawing showing how the textile solar cells with fiber electrodes connected with fiber-shaped batteries for the photo-charging process. Source: Reproduced with permission from Shi et al. [13], © 2021/Springer Nature. (b) The energy textile is composed of textile solar cells and barriers that are weaved by a shuttle flying process. Source: Reproduced with permission from Zhang et al. [12], © 2020/Elsevier. (c) The working process of a functional textile that incorporated power supplying textile, microcontroller, sensor, and textile display. Source: Reproduced with permission from Kang et al. [55], © 2022/Wiley-VCH Verlag GmbH & Co. KGaA, Weinheim. (d) The demonstration of data collection from the functional textile, which powers a normal phone, electronic watch and recording heartbeat. Source: Reproduced with permission from Zhang et al. [12], © 2020/Elsevier.

10.7 Summary and Perspectives

The development of textile solar cells is still in its early stages, which mainly focuses on the construction of solar cells in a textile format. Several structures have been developed and discussed comprehensively in this chapter. However, each design has its own pros and cons. New structures should be further developed to meet the requirements for flexibility, breathability, washability, and wearable comfortability. And effective encapsulation approaches are highly demanded to maintain the above-mentioned properties for practical use. On the other hand, measurement standards are required to be established for the sustainable development of textile solar cells. Photovoltaic textile is promising for wearable applications. Considering the device's performance with large light illumination areas, high mechanical properties, and real textile structures that are breathable and comfortable for wearing, considerable investigation and improvement are urgently needed in future studies. When constructed from textile substrate or textile electrode, the performances could be further improved by referring to the effective technology of planar solar cells.

Strip and fiber solar cells are the potential to fulfill the requirement of breathability and comfortability. However, strip-shaped ones are very sensitive to the incident light angle. Currently, the performance and stability of fiber-shaped solar cells are still the main challenges that plague textile solar cells. Due to the special curved surface of the fiber electrodes that are used for the textile solar cells, the uniform coating of high-quality active and functional layers remains challenging, and the subsequent surface/interface engineering for the performance enhancement should be pushed forward, which can greatly improve their practicability.

In terms of the type of textile solar cells, dye-sensitized solar cells have intimate contact with the electrodes due to the use of liquid electrolytes, which suffer from the leakage issue. Polymer solar cells are light-weight and flexible with an ultra-thin photoactive layer. However, the photovoltaic conversion performance is relatively limited, hindered by the harsh requirements for the fabrication parameters and processes. Perovskite solar cells with all-solid-state and high-efficiency features are promising candidates for the production of textile solar cells. However, the use of toxic Pb elements should be avoided for better-wearing safety. All these issues need to be further considered and optimized in order to truly apply textile solar cells to wearable electronics.

References

1 Wang, C., Wang, C., Huang, Z. et al. (2018). Materials and structures toward soft electronics. *Adv. Mater.* 30 (50): e1801368.
2 Sanderson, K. (2021). Electronic skin: from flexibility to a sense of touch. *Nature* 591 (7851): 685–687.
3 He, S., Hong, Y., Liao, M. et al. (2021). Flexible sensors based on assembled carbon nanotubes. *Aggregate* 2 (6): 143.

4 Libanori, A., Chen, G., Zhao, X. et al. (2022). Smart textiles for personalized healthcare. *Nat. Electron.* 5 (3): 142–156.

5 Wei, X., Zhang, W., and Fan, X. (2022). Nonprinted IC textiles for wearable electronics. *Acc. Mater. Res.* 3 (12): 1201–1205.

6 Schubert, M.B. and Werner, J.H. (2006). Flexible solar cells for clothing. *Mater. Today* 9 (6): 42–50.

7 Zhai, W., Zhu, Z., Sun, X. et al. (2022). Fiber solar cells from high performances towards real applications. *Adv. Fiber Mater.* 4 (6): 1293–1303. https://doi.org/10.1007/s42765-022-00184-1.

8 Hatamvand, M., Kamrani, E., Lira-Cantú, M. et al. (2020). Recent advances in fiber-shaped and planar-shaped textile solar cells. *Nano Energy* 71: 104609.

9 Satharasinghe, A., Hughes-Riley, T., and Dias, T. (2020). A review of solar energy harvesting electronic textiles. *Sensors (Basel)* 20 (20): 5938.

10 Zhang, N., Chen, J., Huang, Y. et al. (2016). A wearable all-solid photovoltaic textile. *Adv. Mater.* 28 (2): 263–269.

11 Liu, P., Gao, Z., Xu, L. et al. (2018). Polymer solar cell textiles with interlaced cathode and anode fibers. *J. Mater. Chem. A* 6 (41): 19947–19953.

12 Zhang, N., Huang, F., Zhao, S. et al. (2020). Photo-rechargeable fabrics as sustainable and robust power sources for wearable bioelectronics. *Matter* 2 (5): 1260–1269.

13 Shi, X., Zuo, Y., Zhai, P. et al. (2021). Large-area display textiles integrated with functional systems. *Nature* 591 (7849): 240–245.

14 Chai, Z., Zhang, N., Sun, P. et al. (2016). Tailorable and wearable textile devices for solar energy harvesting and simultaneous storage. *ACS Nano* 10 (10): 9201–9207.

15 Li, Q. and Zanelli, A. (2021). A review on fabrication and applications of textile envelope integrated flexible photovoltaic systems. *Renew. Sust. Energy Rev.* 139: 110678.

16 Yu, K., Rich, S., Lee, S. et al. (2019). Organic photovoltaics: toward self-powered wearable electronics. *Proc. IEEE* 107 (10): 2137–2154.

17 Kaltenbrunner, M., Adam, G., Glowacki, E.D. et al. (2015). Flexible high power-per-weight perovskite solar cells with chromium oxide-metal contacts for improved stability in air. *Nat. Mater.* 14 (10): 1032–1039.

18 Ehrmann, A. and Blachowicz, T. (2019). Recent coating materials for textile-based solar cells. *AIMS Materials Science* 6 (2): 234–251.

19 Jeong, E.G., Jeon, Y., Cho, S.H. et al. (2019). Textile-based washable polymer solar cells for optoelectronic modules: toward self-powered smart clothing. *Energy Environ. Sci.* 12 (6): 1878–1889.

20 Lee, S., Lee, Y., Park, J. et al. (2014). Stitchable organic photovoltaic cells with textile electrodes. *Nano Energy* 9: 88–93.

21 Lam, J.-Y., Chen, J.-Y., Tsai, P.-C. et al. (2017). A stable, efficient textile-based flexible perovskite solar cell with improved washable and deployable capabilities for wearable device applications. *RSC Adv.* 7 (86): 54361–54368.

22 Arumugam, S., Li, Y., Liu, J. et al. (2018). Optimized process of fully spray-coated organic solar cells on woven polyester cotton fabrics. *Mater. Today Proc.* 5 (5): 13745–13752.

23 Kylberg, W., de Castro, F.A., Chabrecek, P. et al. (2011). Woven electrodes for flexible organic photovoltaic cells. *Adv. Mater.* 23 (8): 1015–1019.

24 Jung, J.W., Bae, J.H., Ko, J.H. et al. (2018). Fully solution-processed indium tin oxide-free textile-based flexible solar cells made of an organic–inorganic perovskite absorber: toward a wearable power source. *J. Power Sources* 402: 327–332.

25 Liu, J., Li, Y., Li, M. et al. (2019). Processing of printed dye sensitized solar cells on woven textiles. *IEEE J. Photovolt.* 9 (4): 1020–1024.

26 Opwis, K., Gutmann, J.S., Lagunas Alonso, A.R. et al. (2016). Preparation of a textile based dye sensitized solar cell. *Int. J. Photoenergy* 2016: 1 11.

27 Liu, J., Li, Y., Yong, S. et al. (2019). Flexible printed monolithic-structured solid-state dye sensitized solar cells on woven glass fibre textile for wearable energy harvesting applications. *Sci. Rep.* 9 (1): 1362.

28 Cho, S.H., Lee, J., Lee, M.J. et al. (2019). Plasmonically engineered textile polymer solar cells for high-performance, wearable photovoltaics. *ACS Appl. Mater. Interfaces* 11 (23): 20864–20872.

29 Güler, E.N., Distler, A., Basu, R. et al. (2022). Fully solution-processed, light-weight, and ultraflexible organic solar cells. *Flexible and Printed Electron.* 7 (2): 025003.

30 Saberi Motlagh, M., Mottaghitalab, V., Rismanchi, A. et al. (2022). Performance modelling of textile solar cell developed by carbon fabric/polypyrrole flexible counter electrode. *Int. J. Sustainable Energy* 41 (8): 1106–1126.

31 Müller, S., Wieschollek, D., Juhász Junger, I. et al. (2019). Back electrodes of dye-sensitized solar cells on textile fabrics. *Optik* 198: 163243.

32 Kang, Z., Tan, X., Li, X. et al. (2016). Self-deposition of Pt nanoparticles on graphene woven fabrics for enhanced hybrid Schottky junctions and photoelectrochemical solar cells. *Phys. Chem. Chem. Phys.* 18 (3): 1992–1997.

33 Jinno, H., Fukuda, K., Xu, X. et al. (2017). Stretchable and waterproof elastomer-coated organic photovoltaics for washable electronic textile applications. *Nat. Energy* 2 (10): 780–785.

34 Yun, M.J., Cha, S.I., Seo, S.H. et al. (2014). Highly flexible dye-sensitized solar cells produced by sewing textile electrodes on cloth. *Sci. Rep.* 4: 5322.

35 Yun, M.J., Cha, S.I., Seo, S.H. et al. (2015). Insertion of dye-sensitized solar cells in textiles using a conventional weaving process. *Sci. Rep.* 5: 11022.

36 Yun, Y., Moon, S., Kim, S. et al. (2022). Flexible fabric-based GaAs thin-film solar cell for wearable energy harvesting applications. *Sol. Energy Mater. Sol. Cells* 246: 111930.

37 Satharasinghe, A., Hughes-Riley, T., and Dias, T. (2019). An investigation of a wash-durable solar energy harvesting textile. *Prog. Photovolt. Res. Appl.* 28 (6): 578–592.

38 Li, C., Islam, M.M., Moore, J. et al. (2016). Wearable energy-smart ribbons for synchronous energy harvest and storage. *Nat. Commun.* 7 (1): 13319.

39 Ilén, E., Elsehrawy, F., Palovuori, E. et al. (2022). Washable textile embedded solar cells for self-powered wearables. *Res. J. Textile Apparel.*

40 He, S., Qiu, L., Ono, L.K. et al. (2020). How far are we from attaining 10-year lifetime for metal halide perovskite solar cells? *Mater. Sci. Eng. R Rep.* 140: 100545.

41 Green, M.A., Dunlop, E.D., Hohl-Ebinger, J. et al. (2022). Solar cell efficiency tables (Version 60). *Prog. Photovoltaics* 30 (7): 687–701.

42 Qiu, L., He, S., Yang, J. et al. (2016). An all-solid-state fiber-type solar cell achieving 9.49% efficiency. *J. Mater. Chem. A* 4 (26): 10105–10109.

43 Pan, S., Yang, Z., Chen, P. et al. (2014). Wearable solar cells by stacking textile electrodes. *Angew. Chem. Int. Ed.* 53 (24): 6110–6114.

44 Zhang, Z., Li, X., Guan, G. et al. (2014). A lightweight polymer solar cell textile that functions when illuminated from either side. *Angew. Chem. Int. Ed.* 53 (43): 11571–11574.

45 Bandara, T.M.W.J., Hansadi, J.M.C., and Bella, F. (2022). A review of textile dye-sensitized solar cells for wearable electronics. *Ionics* 28 (6): 2563–2583.

46 Yun, M.J., Sim, Y.H., Cha, S.I. et al. (2019). Three-dimensional textile platform for electrochemical devices and its application to dye-sensitized solar cells. *Sci. Rep.* 9 (1): 2322.

47 Wu, C., Kim, T.W., Guo, T. et al. (2017). Wearable ultra-lightweight solar textiles based on transparent electronic fabrics. *Nano Energy* 32: 367–373.

48 Borazan, İ., Bedeloglu, A.C., and Demir, A. (2020). A photovoltaic textile design with a stainless steel mesh fabric. *J. Ind. Textiles* 51 (10): 1527–1538.

49 He, S., Qiu, L., Fang, X. et al. (2015). Radically grown obelisk-like ZnO arrays for perovskite solar cell fibers and fabrics through a mild solution process. *J. Mater. Chem. A* 3 (18): 9406–9410.

50 Chen, T., Qiu, L., Yang, Z. et al. (2013). Novel solar cells in a wire format. *Chem. Soc. Rev.* 42 (12): 5031–5041.

51 Chen, T., Qiu, L., Li, H. et al. (2012). Polymer photovoltaic wires based on aligned carbon nanotube fibers. *J. Mater. Chem.* 22 (44): 23655–23658.

52 Wen, Z., Yeh, M.H., Guo, H. et al. (2016). Self-powered textile for wearable electronics by hybridizing fiber-shaped nanogenerators, solar cells, and supercapacitors. *Sci. Adv.* 2 (10): e1600097.

53 Pan, S., Zhang, Z., Weng, W. et al. (2014). Miniature wire-shaped solar cells, electrochemical capacitors and lithium-ion batteries. *Mater. Today* 17 (6): 276–284.

54 Zhang, Z., Yang, Z., Wu, Z. et al. (2014). Weaving efficient polymer solar cell wires into flexible power textiles. *Adv. Energy Mater.* 4 (11): 201301750.

55 Kang, X., Zhu, Z., Zhao, T. et al. (2022). Hierarchically assembled counter electrode for fiber solar cell showing record power conversion efficiency. *Adv. Funct. Mater.* 32 (51): 2207763.

56 Deng, J., Qiu, L., Lu, X. et al. (2015). Elastic perovskite solar cells. *J. Mater. Chem. A* 3 (42): 21070–21076.

57 Qiu, L., He, S., Yang, J. et al. (2016). Fiber-shaped perovskite solar cells with high power conversion efficiency. *Small* 12 (18): 2419–2424.

58 Zhang, L., Shi, E., Ji, C. et al. (2012). Fiber and fabric solar cells by directly weaving carbon nanotube yarns with CdSe nanowire-based electrodes. *Nanoscale* 4 (16): 4954–4959.

59 Chen, J., Huang, Y., Zhang, N. et al. (2016). Micro-cable structured textile for simultaneously harvesting solar and mechanical energy. *Nat. Energy* 1 (10): 16138.

60 Li, Q., Balilonda, A., Ali, A. et al. (2020). Flexible solar yarns with 15.7% power conversion efficiency, based on electrospun perovskite composite nanofibers. *Solar RRL* 4 (9): 2000269.

61 Liu, Z., Qiu, L., Ono, L.K. et al. (2020). A holistic approach to interface stabilization for efficient perovskite solar modules with over 2,000-hour operational stability. *Nat. Energy* 5 (8): 596–604.

62 Park, S., Heo, S.W., Lee, W. et al. (2018). Self-powered ultra-flexible electronics via nano-grating-patterned organic photovoltaics. *Nature* 561 (7724): 516–521.

63 Fukuda, K., Yu, K., and Someya, T. (2020). The future of flexible organic solar cells. *Adv. Energy Mater.* 10 (25): 2000765.

64 Kaltenbrunner, M., White, M.S., Głowacki, E.D. et al. (2012). Ultrathin and lightweight organic solar cells with high flexibility. *Nat. Commun.* 3 (1): 770.

65 Lee, G., Kim, M.-c., Choi, Y.W. et al. (2019). Ultra-flexible perovskite solar cells with crumpling durability: toward a wearable power source. *Energy Environ. Sci.* 12 (10): 3182–3191.

66 Cheacharoen, R., Boyd, C.C., Burkhard, G.F. et al. (2018). Encapsulating perovskite solar cells to withstand damp heat and thermal cycling. *Sustain. Energy Fuels* 2 (11): 2398–2406.

11

Summary and Outlook

11.1 Introduction

Wearable solar cells have attracted broad attentions, attributed to the great demand for wearable electronics with a variety of applications. Among various structures designed for wearable applications [1], fiber and textile solar cells developed in the past decades have emerged, with significantly improved power conversion efficiencies from lower than 1% to over 15% now [2] and are expected to be further enhanced to over 20% in the near future [3]. In previous Chapter 2, the working mechanisms of solar cells, specifically for dye-sensitized solar cells [4], polymer solar cells [5], and perovskite solar cells [6] have been introduced. The design of fiber solar cells and textile solar cells on these three kinds of solar cells has been reviewed and discussed. As a proof of concept, these fiber and textile solar cells have been verified to power a variety of wearable devices with high flexibility and wearing comfortability. In this chapter, we will briefly summarize the development of wearable solar cells, particularly in the formats of fiber and woven textiles, focusing on the pros and cons. In addition, the current challenges and future directions are mainly discussed, from the pursuit of high power conversion efficiency to high stability and scalability from a practical application point of view. We also discuss the key challenges for fiber solar cells for real-world applications in the near future, including continuous production technology, stability, safety, and cost-effectiveness. The interdisciplinary cooperation involving microelectronics and biomedical engineering is also briefly discussed in the end.

11.2 Development of Planar Solar Cells for Wearable Applications

Planar solar cells demonstrate wearability to some extent (Figure 11.1) [7]. At early stage, the concept of wearable solar cells was demonstrated by flexible planar-type solar cells. Even for crystalline silicon solar cells, when the thickness is thin enough, it can exhibit certain flexibility [8]. There are also some commercial products for portable applications that one can easily access on the websites of Alibaba or Amazon. These flexible solar cells are typically sewn into garments, tents, and bags

Wearable Solar Cells: Mechanisms, Materials, and Devices, First Edition.
Hao Sun, Zhibin Yang, Longbin Qiu, and Huisheng Peng.

Figure 11.1 Wearable solar cell applications based on flexible planar-type solar cells. Source: Reproduced with permission from Hashemi et al. [7], © 2020/Royal Society of Chemistry.

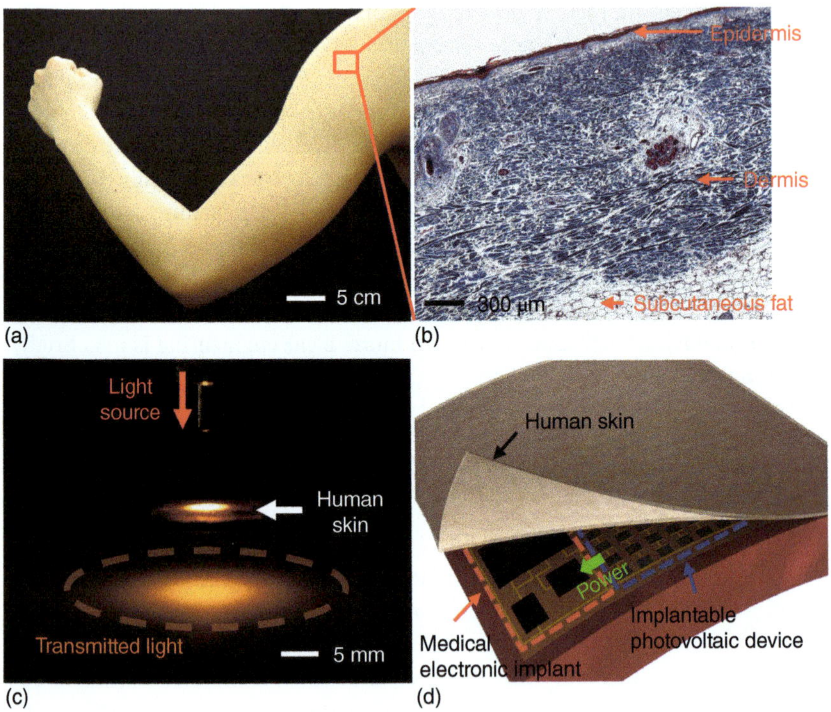

Figure 11.2 Implantable photovoltaic solar cells for wearable applications. Source: Reproduced with permission from Song et al. [9], © 2017/Elsevier.

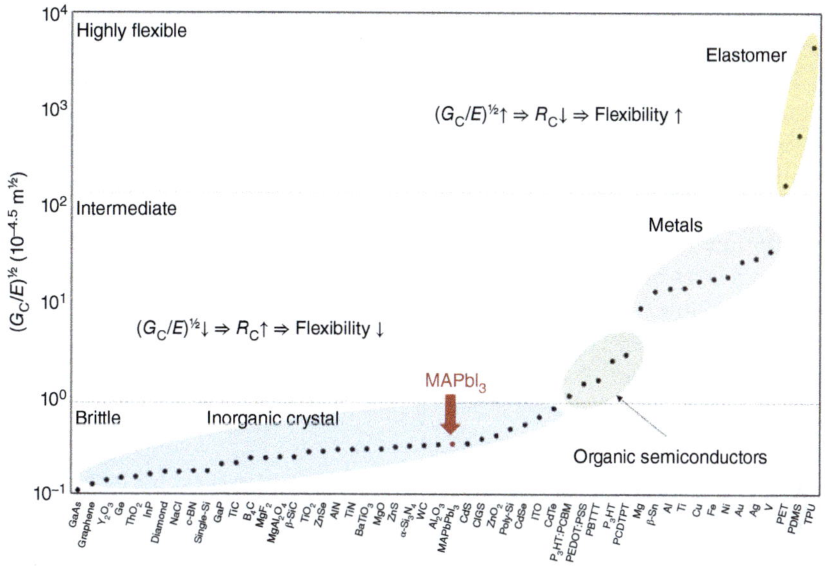

Figure 11.3 The mechanical properties of typical functional layer and photoactive layer materials in flexible and wearable solar cells. Source: Seong et al. [13]/Optica Publishing Group/CC BY 4.0.

for portable powering systems. However, they are generally too heavy to carry and cannot fully meet the requirements for flexibility and wearability in practical use.

There are also reports on epidermal polymer solar cells for powering wearable electronics [9, 10]. They exhibit a planar structure with the employment of ultrathin film substrate for device production [10–12]. The obtained solar cells can be implanted under the human skin for implantable applications (Figure 11.2) [9]. However, the light transmission through the skin is severely reduced, and the cell performance is rather limited. More importantly, the implanting of such solar cells through surgery might cause comfortability and infection issues.

Regarding the practical wearable applications, the key challenges include mechanical stability, power conversion efficiency, long-term stability, and large-scale productions. We compare the mechanical brittleness of typical functional and photoactive materials for solar cells (Figure 11.3) [13]. To this end, polymer semiconductors and perovskite photoactive materials exhibit great potential for flexible and wearable solar cells. Among various configurations, fiber-shaped solar cells are the most promising candidates for wearable applications due to their excellent tolerance to mechanical deformation and high breathability and wearability after being woven into textiles [14].

11.3 Performance Improvement of Fiber and Textile Solar Cells

As the primary building block, the individual solar cell determines the photovoltaic performance of the connected modules. To improve the photovoltaic power conversion performance, high-quality and uniform layers, such as the functional and photoactive layers, are highly important [15]. However, the high radius of curvature makes the uniform coating of these layers rather challenging, particularly for ultrathin film coating with thicknesses below 1 μm. To this end, developing effective coating techniques is highly required. In addition, the incorporation of the passivation reagent/process to suppress the recombination process should be taken into serious consideration. The interface engineering for fiber and textile solar cells remains further improved. Current fabrication strategies mainly involve bottom-up processes via layer-by-layer coating techniques. It might be also possible to develop an inverse top-down process by reshaping a planar solar cell into a fiber form, which has been proven in wearable energy storage devices [16] but has not been realized for solar cells yet. This represents a promising direction that deserves future investigation.

11.4 Continuous Production of Fiber and Textile Solar Cells

For fiber-shaped solar cell shooting for weaving into a textile format, a suitable length is vital for the continuous weaving process. However, due to the conductivity issue for the fiber electrodes, the power conversion efficiencies of fiber-shaped solar cells cannot be well maintained with increasing lengths. With the use of metal wire as highly conductive electrodes, fiber dye-sensitized solar cells could maintain cell performance with a long length [17]. The liquid electrolyte can benefit the fast and efficient transfer of carriers by redox reaction. For polymer solar cells and perovskite solar cells, however, the carrier diffusion lengths are short, which requires intimate contact of transparent electrode to the core layer. The area without intimate contact is considered to be unavailable, or worse, as a recombination center. A twisting structure coupling with the use of high-conductive wire electrodes might solve the above issue. With a small twisting pitch, the efficient illumination area is limited. While with a large twisting pitch, the effective area within the diffusion length of the carrier is limited, even though the illumination area becomes larger. The above issues make it challenging for the continuous fabrication of long fiber-shaped solar cells. The case is similar to interlaced fiber electrode textile structure. The future development of new materials, especially transparent conductive electrodes, is expected for continuous production of fiber and textile solar cells.

Another issue for wearable fiber solar cells lies in the encapsulation strategy. Effective encapsulation can benefit the long-term stability of the obtained devices. In addition, industrial standards should be used for stability evaluation. The encapsulation should fulfill the long-term requirements for power conversion efficiencies, breathability, and mechanical flexibility. Standard testing parameters and methods should be further established for the sustainable development of fiber and textile solar cells for wearable applications.

11.5 Potential Applications of Fiber and Textile Solar Cells

The future of wearable health monitoring systems has a huge demand for wearable power systems (Figure 11.3), which can efficiently collect environmental information and human physiological parameters in a real-time and wearable approach [18, 19]. The collected data can be further transmitted remotely and received by hospitals and doctors for real-time monitoring of human health [20, 21]. Generally, the powers required for these wearable sensors are very low, making fiber and textile solar cells ideal candidates to meet the requirements. Due to the intermittent nature of solar cells, integration with wearable supercapacitors and batteries for stable power supply has been developed, and lightweight, flexible, and efficient integration strategies have been actively pursued [22].

Another emerging application of wearable solar cells lies in the implantable powering system, which can be potentially used for powering the implanted electronic devices in human bodies (Figure 11.4), considering the concept of implanting solar cells under the skin has been proven [9, 23]. This kind of application might be of interest while might require further investigation. The bio-compatibility and bio-degradability of these transient electronic devices should be optimized, and the use of toxic elements and components in conventional solar cells should be replaced. On the other hand, the skin will shield most of the incident light. A promising strategy is the construction of epidermal solar cells, which can permit light harvesting and comfortability during use.

Flexible and wearable solar cells have been rapidly developed in recent years, and we can expect them to accelerate further in the near future. We believe that the fiber

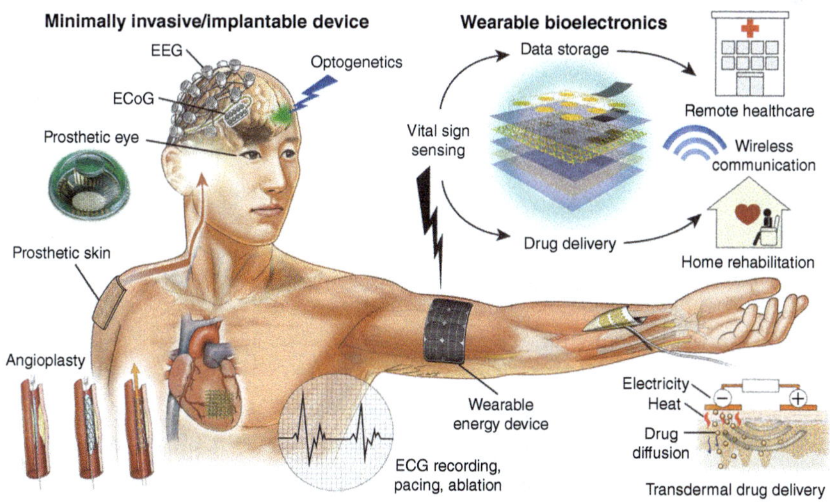

Figure 11.4 Wearable sensor systems requiring wearable energy supply systems for continues health monitoring. Source: Reproduced with permission Choi et al. [18] © 2016. John Wiley & Sons.

and textile solar cells can be the catalyst to drive wearable electronics to the world of the Internet of Things. We look forward to the practical applications of fiber and textile solar cells, and expect them to significantly change the style we live, work, learn, and entertain in the near future.

References

1 Schubert, M.B. and Werner, J.H. (2006). Flexible solar cells for clothing. *Mater. Today* 9 (6): 42–50.

2 Li, Q., Balilonda, A., Ali, A. et al. (2020). Flexible solar yarns with 15.7% power conversion efficiency, based on electrospun perovskite composite nanofibers. *Sol. RRL* 4 (9): 2000269.

3 Zhai, W., Zhu, Z., Sun, X. et al. (2022). Fiber solar cells from high performances towards real applications. *Adv. Fiber Mater.* 4 (6): 1293–1303. https://doi.org/10.1007/s42765-022-00184-1.

4 Hagfeldt, A., Boschloo, G., Sun, L. et al. (2010). Dye-sensitized solar cells. *Chem. Rev.* 110 (11): 6595–6663.

5 Li, G., Zhu, R., and Yang, Y. (2012). Polymer solar cells. *Nat. Photonics* 6 (3): 153–161.

6 Kim, J.Y., Lee, J.-W., Jung, H.S. et al. (2020). High-efficiency perovskite solar cells. *Chem. Rev.* 120 (15): 7867–7918.

7 Hashemi, S.A., Ramakrishna, S., and Aberle, A.G. (2020). Recent progress in flexible–wearable solar cells for self-powered electronic devices. *Energy Environ. Sci.* 13 (3): 685–743.

8 Um, H.-D., Hwang, I., Choi, D. et al. (2021). Flexible crystalline-silicon photovoltaics: light management with surface structures. *Acc. Mater. Res.* 2 (9): 701–713.

9 Song, K., Han, J.H., Yang, H.C. et al. (2017). Generation of electrical power under human skin by subdermal solar cell arrays for implantable bioelectronic devices. *Biosens. Bioelectron.* 92: 364–371.

10 O'Connor, T.F., Zaretski, A.V., Savagatrup, S. et al. (2016). Wearable organic solar cells with high cyclic bending stability: materials selection criteria. *Sol. Energy Mater. Sol. Cells* 144: 438–444.

11 Park, S., Heo, S.W., Lee, W. et al. (2018). Self-powered ultra-flexible electronics via nano-grating-patterned organic photovoltaics. *Nature* 561 (7724): 516–521.

12 Kaltenbrunner, M., Sekitani, T., Reeder, J. et al. (2013). An ultra-lightweight design for imperceptible plastic electronics. *Nature* 499 (7459): 458–463.

13 Seong, S., Liu, Y., and Gong, X. (2022). Mechanical study of perovskite solar cells: opportunities and challenges for wearable power source. *Opt. Mater. Express* 12 (2): 772–787.

14 Varma, S.J., Sambath Kumar, K., Seal, S. et al. (2018). Fiber-type solar cells, nanogenerators, batteries, and supercapacitors for wearable applications. *Adv. Sci.* 5 (9): 1800340.

15 Qiu, L., He, S., Yang, J. et al. (2016). Fiber-shaped perovskite solar cells with high power conversion efficiency. *Small* 12 (18): 2419–2424.

16 Cheng, X., Fang, X., Chen, P. et al. (2015). Designing one-dimensional supercapacitors in a strip shape for high performance energy storage fabrics. *J. Mater. Chem. A* 3 (38): 19304–19309.

17 Song, W., Wang, H., Liu, G. et al. (2016). Improving the photovoltaic performance and flexibility of fiber-shaped dye-sensitized solar cells with atomic layer deposition. *Nano Energy* 19: 1–7.

18 Choi, S., Lee, H., Ghaffari, R. et al. (2016). Recent advances in flexible and stretchable bio-electronic devices integrated with nanomaterials. *Adv. Mater.* 28 (22): 4203–4218.

19 Ling, Y., An, T., Yap, L.W. et al. (2020). Disruptive, soft, wearable sensors. *Adv. Mater.* 32 (18): e1904664.

20 Wang, C., Wang, C., Huang, Z. et al. (2018). Materials and structures toward soft electronics. *Adv. Mater.* 30 (50): e1801368.

21 Oh, J.Y. and Bao, Z. (2019). Second skin enabled by advanced electronics. *Adv. Sci.* 6 (11): 1900186.

22 Fan, X., Liu, B., Ding, J. et al. (2020). Flexible and wearable power sources for next-generation wearable electronics. *Batteries Supercaps* 3 (12): 1262–1274.

23 Gong, S. and Cheng, W. (2017). Toward soft skin-like wearable and implantable energy devices. *Adv. Energy Mater.* 7 (23): 1700648.

Index

Wearable Solar Cells: Mechanisms, Materials, and Devices, First Edition.
Hao Sun, Zhibin Yang, Longbin Qiu, and Huisheng Peng.